Hun

evolve

Learning System
Evolve Learning Resources for Students and Lecturers
See the instructions on the inside cover for
access to the Web site
Think outside the book ... evolve

Commissioning Editor: Rita Demetriou-Swanwick
Development Editor: Veronika Watkins
Project Manager: Sruthi Viswam
Designer/Design Direction: Stewart Larking
Illustration Manager: Merlyn Harvey
Illustrator: Amanda Williams and Chartwell

Human Movement

An Introductory Text

Edited by
Tony Everett MEd, BA, PG Dip Biomechanics, Grad Dip Phys, Cert HE, FHEA
Deputy Director, Department of Physiotherapy, School of Healthcare Studies, Cardiff University, Cardiff

and

Clare Kell MSc, PGCLTHE, Grad Dip Phys, FHEA
Senior Lecturer, Human Resources Directorate, Cardiff University, Cardiff

SIXTH EDITION

CHURCHILL LIVINGSTONE

EDINBURGH LONDON NEW YORK OXFORD PHILADELPHIA ST LOUIS SYDNEY TORONTO 2010

CHURCHILL
LIVINGSTONE
ELSEVIER

First published 1981
Second edition 1987
Third edition 1997
Fourth edition 2001
Fifth edition 2005
Sixth edition 2011

ISBN 978-0-7020-4477-9
Formerly 978-0-7020-3134-2

British Library Cataloguing in Publication Data
A catalogue record for this book is available from the British Library

Library of Congress Cataloging in Publication Data
A catalogue record for this book is available from the Library of Congress

Notice
Knowledge and best practice in this field are constantly changing. As new research and experience broaden our understanding, changes in research methods, professional practices, or medical treatment may become necessary.

Practitioners and researchers must always rely on their own experience and knowledge in evaluating and using any information, methods, compounds, or experiments described herein. In using such information or methods they should be mindful of their own safety and the safety of others, including parties for whom they have a professional responsibility.

With respect to any drug or pharmaceutical products identified, readers are advised to check the most current information provided (i) on procedures featured or (ii) by the manufacturer of each product to be administered, to verify the recommended dose or formula, the method and duration of administration, and contraindications. It is the responsibility of practitioners, relying on their own experience and knowledge of their patients, to make diagnoses, to determine dosages and the best treatment for each individual patient, and to take all appropriate safety precautions.

To the fullest extent of the law, neither the Publisher nor the authors, contributors, or editors, assume any liability for any injury and/or damage to persons or property as a matter of products liability, negligence or otherwise, or from any use or operation of any methods, products, instructions, or ideas contained in the material herein.

The publisher

The Publisher's policy is to use **paper manufactured from sustainable forests**

Printed in China

Dedication

We are grateful to our students and colleagues within the School of Healthcare Studies, Cardiff University for their patience and enthusiasm as we explored possible new directions for this, the sixth edition of *Human Movement*. These new ideas however could not have been considered without the strong foundation and readership base created for this text by its earlier editions. In this regard we are sincerely grateful for the work and commitment to *Human Movement* provided by Professor Marion Trew. From this secure foundation we have been able to extend our ideas to hopefully capture the interest and imagination of a breadth of readers curious about the wonders of human movement.

Contents

Contributors

Philippa Coales MSc, PGD Health Ergon, PGCE, MCSP, MMACP
Lecturer, Department of Physiotherapy,
School of Healthcare Studies, Cardiff University,
Cardiff, UK

Susan Corr PhD, MPhil, DipCOT, CMS, PGDipMedEd
Reader in Occupational Science, Division of Occupational Therapy, School of Health,
University of Northampton, Northampton, UK

Tony Everett MEd, BA, PG Dip Biomechanics, Grad Dip Phys, Cert HE, FHEA
Deputy Director, Department of Physiotherapy,
School of Healthcare Studies, Cardiff University,
Cardiff, UK

Bernhard Haas MCSP, MSc, BA (Hons)
Deputy Head, School of Health Professions,
Faculty of Health,
University of Plymouth,
Plymouth, UK

Lyn Horrocks MSc, MCSP, FHEA
Lecturer, Department of Physiotherapy,
School of Healthcare Studies,
Cardiff University, Cardiff, UK

Clare Kell MSc, PGCLTHE, Grad Dip Phys, FHEA
Senior Lecturer
Human Resources Directorate,
Cardiff University, Cardiff, UK

Roshni Khatri BSc (OT), Adv Dip (OT) Neurosciences, MSc (OT)
Senior Lecturer and Programme Leader Part-time Course (OT), Occupational Therapy, University of Northampton, Northampton, UK

Nicola Phillips MSc, MCSP, PhD
Senior Lecturer and Director, Postgraduate Healthcare Studies, Cardiff University,
Cardiff, UK

Sally Scott-Roberts MEd, DipCOT, FHEA
Senior Lecturer in Inclusive Education, School of Education, University of Wales Newport,
Newport, UK

Tim Sharp MSc, BSc (Hons)
Lecturer, Physiotherapy Department,
School of Healthcare Studies,
Cardiff University,
Cardiff, UK

Mike Smith BSc (Hons) (Physio), MSc, FHEA
Lecturer, Physiotherapy Department,
Cardiff University, Cardiff, UK

Valerie Sparkes PhD, MPhty, BSc (Hons), MCSP
Lecturer, Department of Physiotherapy,
School of Healthcare Studies, Cardiff University,
Cardiff, UK

Marion Trew BA, Med Dip TP, MCSP
Former Head, School of Health Professions,
University of Brighton, Eastbourne, UK

Robert W.M. van Deursen PhD
Director of Physiotherapy,
School of Healthcare Studies,
Cardiff University,
Cardiff, UK

Preface

Human movement is fascinating but complex. We recognize that there are many excellent textbooks and other learning resources to help readers get to grips with the theoretical components of human movements and do not use this book in the same way. This new edition of *Human Movement* has taken steps to offer readers an accessible source that draws together the essence of the underpinning theory in order to explore the practical application of the theory to everyday human movement. To this end, the sixth edition is structured about four central themes: anatomy, physiology, mechanical factors and the environment, and psychosocial factors and the environment. Topic chapters are clustered into related themes to help readers navigate through each and consider related issues together.

The major new development for this edition is the use of a family-based case study. The family is introduced consistently throughout the book, with chapter authors specifically selecting elements of the family's daily routine to help readers see the application of their reading to practice. Each chapter also offers readers thought-provoking activities and, in the Evolve package, the opportunity to test out their learning through further case study–based scenarios. The book concludes with a substantive family-based scenario that effectively draws together and applies the key messages from the chapters.

We hope that readers will be left sharing our marvel at the complexity of human movement: the ease with which many of us do it every day, and the multiplicity and interdependence of factors that can make the normal more challenging. We hope also that the text and related scenarios will help readers notice things in their environment, try to look behind the movements they see and engage meaningfully with the complexity of human movement. If we succeed in our aims, we will have opened *Human Movement* to a broad readership, which can only benefit the ongoing research and therapeutic developments in this area.

Tony Everett and Clare Kell
Cardiff 2009

Chapter 1

Introduction

Tony Everett and Clare Kell

CHAPTER CONTENTS

Human movement is complex, fascinating but complex. The multifactorial nature of the many manifestations of movement demands that its study, and consequent explanation, be framed in a way that diminishes the complexity while not losing any of the fascination. While there are several approaches that could be taken in studying and describing movement, this book promotes a method of movement analysis that deconstructs a movement into its component elements before rebuilding it into the complex behaviour we recognize. Each chapter will follow this approach by introducing the basic movement analysis before integrating and applying your learning to a more complex, functional activity.

Before you begin to analyse and describe movement, it would be useful to ensure that you are happy with what is meant by some of the core terms you will encounter in the text. Central to our discussions is *movement*. In general terms, human movement can be described as being related to mobility, function, occupation, leisure and/or communication, as outlined in Table 1.1. Movement is the end product of an intentional or unintentional consequence of joint motion and/or muscle activity and may take many forms or manifestations.

ACTIVITY 1.1

Using the categories in Table 1.1, look at the people in your current environment and try to categorize the activities that they are performing. Are there some activities that are more easily recognized? Are there a group of activities that are performed more frequently in your sample? Why might this be and what does it suggest about your sample?

Table 1.1 The categories of human movement

CATEGORY	DESCRIPTION
Mobility	Getting from A to B in all its forms
Function	The activities of daily living that are essential for our survival (e.g. dressing, washing and eating)
Occupation	An activity or group of activities that engage a person in everyday life, have personal meaning and provide structure to time (e.g. self-care, caring for others, productivity or sitting in lectures!)
Leisure	Not only sport and hobbies but also just reading or watching television requires movement
Communication	How often do we use our hands to talk and the rest of our body as a means of non-verbal communication?

During your observations, you might have noticed people using movement to get from one place to another. This activity is normally given the term *mobility*. The diversity of human needs elicits a broad range of mobility forms, for example walking, wheelchair use, swinging, crawling and even swimming. The term mobility may, however, be used in other ways. It can mean the ability or capacity to move and also the ease or freedom of that movement. (When studying subsequent chapters, you should ensure that you are clear in which way the chapter authors are using the term.)

Indeed, the difficulty of the activity (the ability to be mobile) will vary with respect to the environment within which it is occurring, i.e. the external factors (e.g. the space available for the movement) and/or the internal factors (e.g. muscle strength). Exploring human movement will require you to be alert to and curious about all these factors. To begin to explore these factors, try the following activity.

ACTIVITY 1.2

Revisit the activities that you observed in Activity 1.1. Make a list of all the internal and external factors that are affecting the execution and quality of these activities. Think of some of the functional activities that members of your extended family perform throughout the day. How do the age of these people and the environments they are in affect the internal and external factors influencing their activities? Add these additional factors to the list you have already made.

You have probably come up with many factors from Activity 1.2. There are no right or wrong answers to this task. One of the benefits of performing this activity is that you have begun to put your observations in order. This is an important skill to learn and will be needed again when the discussion turns to analysing human movement in Chapter 14. Table 1.2 presents our interpretation of the possible internal and external

Table 1.2 The influences on human movement

	DESCRIPTION
Internal influences	
The musculoskeletal system	An intact musculoskeletal system is essential for efficient movement. Freely moving joints that are supported by non-contractile structures. Muscles need to have the strength to produce the movement and the flexibility to allow it.
The peripheral nervous system	The peripheral nerves need to transmit the impulses that stimulate the muscles to contract. They also provide the feedback pathways for all the sensory information that helps refine movements.
The central nervous system	Movements need to be started, stopped and controlled. The brain performs these functions as well as storing, recognizing and refining movement patterns. The central nervous system also integrates the effects of the level of arousal, mood and personal interpretation on movement.
The cardiovascular system	All the systems need to be supplied with the energy and fuel to produce the work of movement. The heart pumps the energy-laden blood through the vessels to wherever it is needed. On its way, it picks up the oxygen fuel from the lungs. The waste products of respiration are delivered to the cleansing organs by this system.
External influences	
The physical environment	The world around us affects the way in which we move. Physical effects, such as the ground reaction force, are essential in initiating, changing and stopping movement. Other forces, such as gravity and other resistive forces, modify the way we move in both a negative and a positive way.
The social environment	The society we live in has a major effect on how we move. It affects the reason and meaning given to our movements. Our role within society and the context in which the movement occurs will have a profound effect.

factors affecting human movement. These six influences can then be broadly grouped into the themes about which this book is framed. Table 1.3 outlines these themes and their related chapter headings.

Table 1.3 The themes of *Human Movement*

THEME	CONTENT	RELATED CHAPTERS
Anatomy	Muscle strength Joint mobility	2, 3, 10, 11, 12, 13 and 14
Physiology	Motor control Posture and balance Motor learning	4, 5, 6 and 13
Psychosocial factors and the environment	Psychosocial Environment	7, 8, 13, 14 and 15
Mechanical factors and the environment	Biomechanics and the influence of the external environment	9, 13, 14 and 15

The book will describe each of the themes in Table 1.3 separately. It will soon become apparent, however, that the themes are inter-related and have an influence on each other. As we have seen in Activity 1.2, movement is very much influenced by our physical surroundings and the environment in which we perform the functional activities, occupation or leisure pursuits. Additionally, the culture and society we live in will not only have an influence on the physical aspects of our movement but also how we perceive and measure that movement. Finally, we must consider the effects of our muscle strength, joint mobility, neuromuscular coordination, posture, balance and cardiovascular capacity. The applied themes outlined in Table 1.4 aim to help you capture and record this complexity in a form that could be shared with others and draws the internal and external influences together to discuss functional human movement.

Table 1.4 Applied themes of human movement

THEME	CONTENT	RELATED CHAPTER
Functional implications	Upper limb	10
	Lower limb	11
	Spine	12
	Across the ages: how movement changes as a consequence of age	13
Describing and recording	Analysing movement	14
	Measuring movement	14
	Using measurement scales	15

WORKING THROUGH THIS BOOK

As you work through this book, you will meet a family performing movement in many different environments. This will help to make the study of the themes more relevant by relating them to your experience of the activities of daily living. Each chapter will include different case studies, which may be either a scenario involving the whole family or an activity of a single member with his or her contemporaries. Within each chapter, you will be invited to perform a number of activities. Please try to perform these activities, as they will help you understand the context in which the theme is being delivered.

MEET THE FAMILY

Represented in this extended family are three generations. Agnes is the oldest member; she is 83 years old. John and Liz share the care of their three children. Both parents work part time and are both in their mid forties. The three siblings are Chris, who is 19; Dan, who is 15; and Jenny, who is 9. All are in full-time education, with the older sibling, Chris, being home from university. They live in a house.

CASE STUDY 1.1

The family are just finishing their evening meal. Luckily, they were all able to eat together. Before they all move off to participate in their separate evening activities, they clear the table. The older sibling, Chris, is off to play badminton. The middle sibling, Daniel, is going to do his homework. John is going to clean the kitchen, while Liz, his wife, is dressed in formal clothes in preparation for a visit to the theatre. Agnes, the oldest member of the family, and the youngest sibling, Jenny, are going to watch the television.

How might the case study be used in this book? Let's take Chris as an example. To be able to play badminton, Chris is going to perform a lot of quite complex activity. There needs to be full-range movement of the shoulder joint (see Ch. 3) to reach the overhead shots, and considerable muscle strength (see Ch. 2) is necessary to perform a smash. Throughout the game, it would be evident that the small drop shots just over the net would require fine motor control (see Ch. 4), and that the whole game requires constant change of posture on a base of a stable posture (see Ch. 5). If the badminton example was a regular activity, the movement patterns would be different to those of a one-off game because of motor learning (see Ch. 6). In addition, the surface of the court and the lighting within the sports centre will alter the way the game is played, and the standard and/or motivation for the game will influence the intensity of the play: a competitive match would be quite different from a fun half-hour session with friends (see Chs 7 and 8).

Framed about such a practical example, each chapter will then go on to explore its content and use activity boxes such as Activity 1.3 to help you relate your new learning to your past experiences.

ACTIVITY 1.3

Look again at the badminton game example and the internal and external factors listed in Table 1.2. Can you map the scenario to the factors? Which ones have we not explored? Suggest an example that might arise from the badminton example.

Finally, look at Tables 1.3 and 1.4. If you wanted to find out more about a specific factor, would you know which chapter(s) to explore?

Chapter 13 could study Agnes and Jenny and explore the movements necessary to enable them to move from the table to watch television. While at first glance this activity may seem straightforward and banal, the chapter authors will help you consider the complexity and individuality of each person's movement. Agnes, for example, will probably have to take more care in movement, as balance reactions are known to be slower in the elderly. Jenny, however, might actually run to the room. The posture and support they choose to adopt to watch the television may also differ and be related to both their anatomy and their physical environment. Agnes will probably sit in a more upright armchair that will compensate for possible lack of joint range and any decrease in muscle strength to lower her down. Jenny, however, may squat cross-legged on the floor. This example demonstrates the complexity of the many integrated themes and will provide an opportunity for you to begin to study the analysis and recording of human movement (see Chs 14 and 15).

From these two examples, you may now be getting a picture of the complexity of movement even before the psychosocial aspects of John doing the washing up are considered! So as you read through this book, try to remember that although the themes may be self-contained they are a part of the bigger picture. Master the components but integrate them into the movement that you are all familiar with, but most of all enjoy the stimulating and fascinating study that is human movement.

Chapter 2

Skeletal muscle, muscle work, strength, power and endurance

Tim Sharp and Tony Everett

CHAPTER CONTENTS

LEARNING OUTCOMES

At the end of the chapter, you should be able to:
1. describe the structure and function of skeletal muscle
2. discuss the physiological process that occurs during a muscle contraction
3. discuss the different types of muscle activity
4. differentiate between muscle force and strength
5. discuss muscle work
6. differentiate between muscle strength and power
7. discuss muscle endurance
8. discuss the principles of measuring muscle strength and endurance
9. describe the methods of increasing muscle strength and endurance.

INTRODUCTION

In order for any member of our family to perform even the simplest of functional tasks, it requires a whole host of abilities – from motivation, in seeing the need to perform the task, to motor control to enable a successful execution while still maintaining balance and postural control within the context of environment. The anatomical basis of movement, knowledge of what allows us to do what we need to do, is essential. For even the simplest of tasks, movement is an essential component. This may be moving the shoulder to perform a smash in badminton, or movement of the rib cage and diaphragm when sitting still watching TV to ventilate the lungs providing the body with oxygen.

To understand movement, a basic knowledge of how we are put together is important. The skeleton is basically a solid subframe made of bone. This gives us our shape, and more importantly for movement, allows for the attachment of muscles. There are many individual bones in the body; how they are joined together may or may not allow for movement to occur. The place that they are joined is called a joint, and it is the shape of the bones and the type of joint that ultimately dictate if and what movement occurs at that joint. Running over the joints, attaching via tendons from one bone to the other, are the muscles. These muscles can be seen as motors that shorten and pull the two attachments together, causing the joint to move. More detail about joints can be found in Chapter 3; however, this chapter will deal with how muscle works, the different types of work, and how changes in muscle work can be brought about. First, it is important to have a basic understanding of the structure and function of skeletal muscle.

The primary function of muscle is to produce a force. The journey to discover how muscle tissue is able to contract to produce massive forces at high speeds is a fascinating one, especially when you consider how small each contractile part is. The picture becomes even more astounding when one considers the control and coordination that need to occur not only within a single muscle but also the coordinated contractions of other muscles to produce a precise movement.

ACTIVITY 2.1

To start you thinking about muscle work, think about the badminton smash, or even better watch someone performing a smash in badminton.

- Think about what is going on around the shoulder just in terms of muscle work.
- Which muscles accelerate the arm forwards, bringing the arm closer to the shuttle?
- Which muscles are slowing the arm down at the end of the shot?
- How quickly does the smash occur?

The force generated by a muscle is used to either produce or control movement, and to fulfil this function it must have components that are capable of contracting. As well as having contractile elements, muscles contain non-contractile elements; these form the structural proteins, various sheaths and the tendon. While these are not involved in producing forces, they play a large role in transmitting forces and passive tension in muscle. Although these structures are mainly made up of collagen, which is on the whole non-elastic, they do contain some elastin, which allows them to stretch (Lieber, 2002).

To understand how skeletal muscle works requires knowledge of its structure from the gross anatomy to the microanatomy. Conventionally, muscle structure is generally presented in a hierarchy from the gross structure to the microstructure; however, for the purposes of this text we shall start at the molecular level and build up to the gross anatomy of the muscle.

PROTEIN FILAMENTS

At a molecular level, muscle is composed of various protein filaments arranged in a precise fashion. This gives muscle a distinctively ordered array. There are two main types of filament: thick filaments, composed of the protein myosin, and thin filaments, which contain the three proteins actin, troponin and tropomyosin (Fig. 2.1). As can be seen in Figure 2.1b, myosin resembles a golf club, with a long filament handle and a head; unlike a golf club, the head is attached to the filament via a flexible neck. The thick filament consists of many

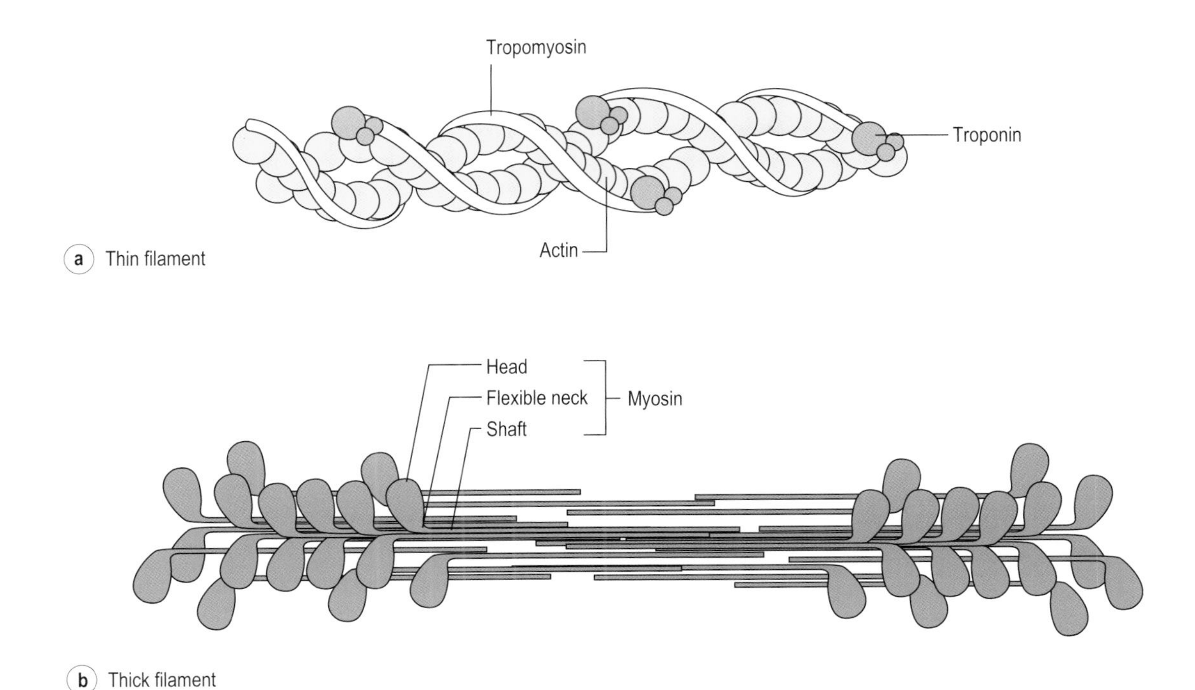

Figure 2.1 A thin filament (a), showing the double helix of the actin with the tropomyosin and troponin. A thick filament (b), showing the myosin molecules arranged in an antiparallel fashion. The resemblance of the myosin molecules to a golf club can clearly be seen.

myosin molecules packed together in a regimented fashion so that the heads project away from the shaft. They are also arranged in an antiparallel fashion, in which some lie pointing one way and others will point in the opposite direction. The heads end up at the ends of the filament, with the shafts in the centre. This makes it look like a double-headed toilet brush (see Fig. 2.1b). The three proteins of the thin filament are arranged with the actin as a core to the filament, the tropomyosin twisting around the actin, and the globular protein troponin sitting on the tropomyosin every so often (Fig. 2.1a).

So far, we have two separate filaments, the thick and thin filaments, each of which possess specific characteristics. It is these specific characteristics, along with the regimented organization of the thick and thin filaments, that give muscle its properties. Before looking at the organization, we need to know how these unique characteristics allow the thick and thin filaments to interact.

The actin of the thin filament possesses binding sites for the myosin heads, but these are covered by the tropomyosin. Calcium ions have the ability to bind to the troponin, changing its shape, which in turn pulls the tropomyosin out of the way of the binding site, allowing the myosin head to attach to the actin. Once attached to the actin, the myosin neck flexes, pulling the thin filament along. As can be seen in Figure 2.2, this flexing has the effect of sliding the thick and thin filaments over each other. Once the myosin neck flexes, it is released from the actin using energy from ATP; the neck is restored to its original position and is free to bind with the actin again and repeat the process. In effect, the myosin walks along the actin filament. This is a very basic description of the sliding filament theory as first described by Huxley and Simmons (1971), and further reading is recommended (see Hunter, 2000; Astrand et al. 2003).

The thick filaments are grouped together in bundles with structural protein binding the filaments together at the centre of the filament. These structural proteins, as well as holding the filaments together, space them apart. This space is important, as it is where the thin filaments slide. The structural proteins look like a line when viewed from

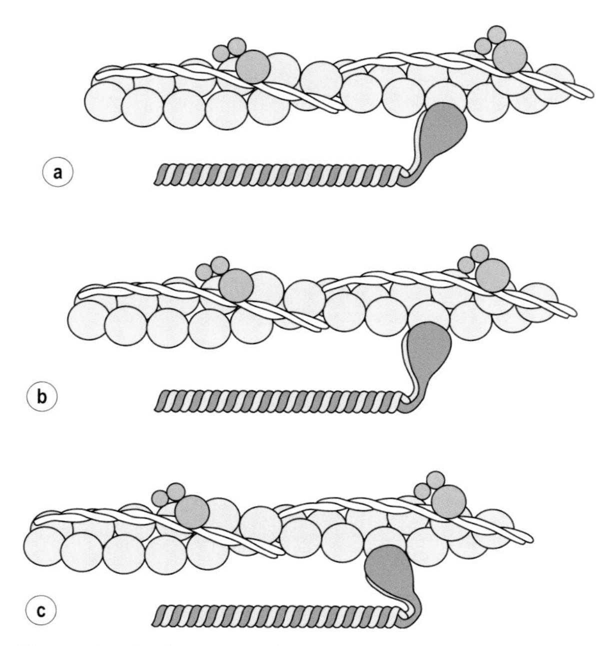

Figure 2.2 During a muscle contraction, the cross-bridge of the myosin molecule attaches to a binding site on the actin filament (a) and the myosin head rotates (b and c) and pulls the actin filament to the left.

the side but are more like a disc, with the thick filaments running through it and sticking out either side (see Fig. 2.3). This disc is known as the M line. The thin filaments are arranged in a similar way; the same type of disc can be found about halfway down the thin filaments, and this is known as the Z line or Z disc (Lieber, 2002).

Imagine these bundles to be stickle bricks; the thick ones can be represented by blue bricks, and the bundle of thin filaments by yellow bricks. Imagine taking the bricks and standing them on end on a table. Now line them up, alternating blue then yellow to form a long column of blue and yellow stripes. The stickle bricks fit together much like the thick and thin filaments; the filaments overlap from one brick to another. The difference in a muscle compared to stickle bricks is that the filaments have the ability to slide over each other, allowing the column to change length. This arrangement of alternating bundles of thick then thin filaments stacked in a long line produces the regimented organization to the histological appearance of muscle tissue and, more importantly, allows the column to change length.

THE SARCOMERE

The sarcomere is the basic functional unit of muscle and is defined by adjacent Z lines. Starting at one Z line travelling along the filaments, we first travel along a light area of thin filaments. This light area of solely thin filaments is known as the I band. Next to this is a darker area where we come across thick filaments as well as thin filaments, and this is the area of overlap. Then there is again a slightly lighter area known as the H zone, where there are just thick filaments, before we reach the M line and then go through the reverse of this to get back to the Z line. The light and dark areas of the sarcomere are clearly visible in Figure 2.4. As well as I bands, H zones, M lines and Z lines, there is also an A band, which is basically the zone that contains the thick filaments (including the H zone and the overlap area with the thin filaments). These bands change size during contraction, indicating that the

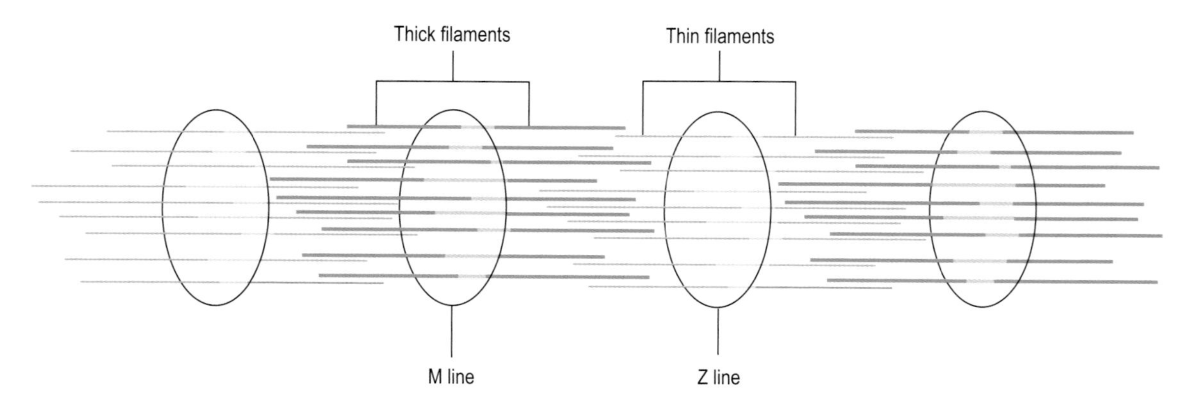

Figure 2.3 Alternating bundles of thin and thick filaments in series. Note how the thick and thin filaments overlap. A sliding of the filaments over one another would bring the M line and the Z line closer together.

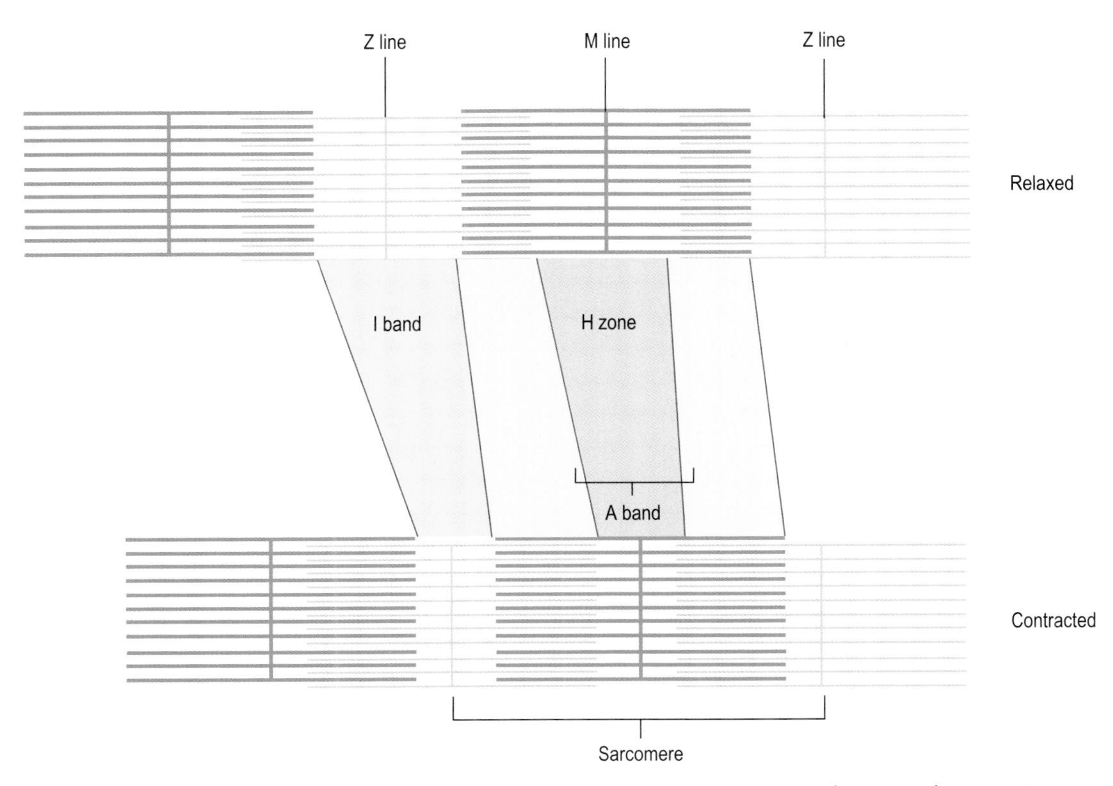

Figure 2.4 The divisions within a sarcomere and the effect on the various zones during a (concentric) contraction.

filaments are sliding over one another, making the sarcomere shorter (Astrand et al. 2003).

The column that we had described earlier of the alternating bundles of thick and thin filaments is repeated many times, making a long, thin, cylindrical column. This is called the myofibril and, as can be seen in Figure 2.5, it has a striated appearance. A muscle fibre is made up of lots of myofibrils lying parallel to each other, as can be seen in Figure 2.5. The sarcomeres are all lined up, so the entire muscle fibre has a striated appearance. The muscle fibre is, in essence, a single muscle cell and contains all the normal things found in a cell, such as a nucleus, mitochondria and a cell membrane (sarcolemma) (Astrand et al. 2003). The function and abundance of these structures, as well as others, is very important in the production of a contraction; however, it is not the place of this text to discuss this, as there are many other texts dedicated to this topic.

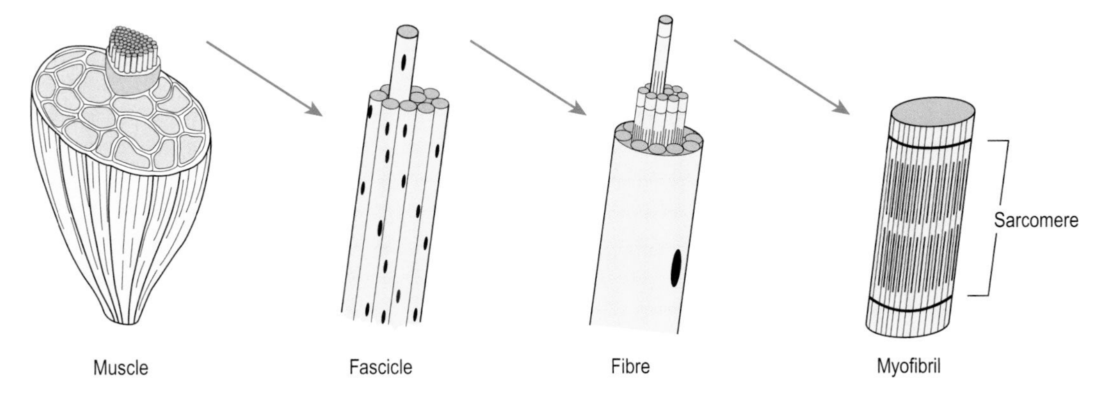

Figure 2.5 Hierarchical structure of a muscle.

Muscle fibres are surrounded by the sarcolemma and a connective tissue sheath called the endomysium. Anatomically, the muscle fibres are grouped into fascicles, which again have a connective tissue sheath (called a perimysium); the fascicles are all bundled together to form the muscle, which again is surrounded by a sheath (called the epimysium). The connective tissue sheaths that surround the various divisions of the muscle are all linked together and are important in the transmission of passive forces through a muscle, such as when a muscle is stretched. Figure 2.5 gives an appreciation of this structure.

Before we move on to muscle work, there are two important concepts about the arrangement of the sarcomeres in a muscle that are worth mentioning at this point. First, remembering back to the blue and yellow stickle brick construction of a myofibril, the sarcomeres are said to be in series. As each sarcomere has an ability to change length by a certain degree, the more sarcomeres you have in series the greater the degree and speed a myofibril can shorten (see Fig. 2.6). Under certain circumstances, the number of sarcomeres in series can change. This would change the range of movement that the muscle can work through and the speed of contraction. As we will see later, there are other aspects that can influence the speed of contraction.

Second, thinking back even further to the myosin–actin cross-bridged interaction, you may remember that ultimately this is the motor that drives the shortening of a muscle. Each sarcomere has the ability to generate a specific amount of tension. To increase force, you need more sarcomeres pulling side by side and not in a line. The easiest way to imagine this is to think of pulling a car or heavy object. If you see someone pushing a car and you want to help, you tend to push the car and not the person pushing the car. All of your force is thus transmitted into the car directly and not wasted squashing the other person. Basically, the greater the number of sarcomeres in parallel the greater the force generation capacity of the muscle. This is not quite as simple, as the wider the muscle the stronger it is, but it is close. This makes sense, as the stronger you get the bigger your muscles become; however, as we will see later, muscle force and strength are two different concepts. The examples presented above are to illustrate how the functional unit (sarcomere) arrangement can influence the muscle's force production capacity. However, as muscle force production and strength are conceptually different, there are many factors that can affect strength, and this shall be presented in the next section.

We know why muscles have to be developed in terms of strength and endurance when participating in sport at a high level. It is self-evident why a shot putter needs strong arms that can deliver a powerful thrust to propel an 8-kg shot 20 m, or a marathon runner needs muscle endurance to run over 26 miles. Although these athletes sometimes appear superhuman, they actually function in the same way as you or I, and the processes they need to build up their muscular systems are essentially the same as those for someone who has become weak and out of condition through trauma, disease or any other form of disuse.

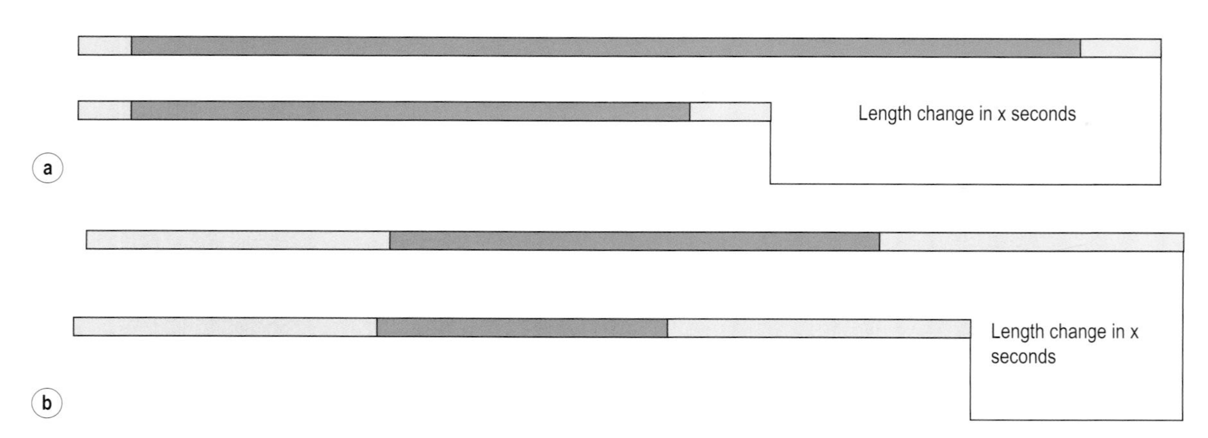

Figure 2.6 The effect of the number of sarcomeres in series on muscle range of movement. The muscle with more sarcomeres in series (a) is able to shorten through a greater range and at a greater velocity compared with the muscle in (b).

ACTIVITY 2.2

Think of some of the major diseases or illnesses that affect people. Make a list of the major ways in which these diseases manifest themselves. Are there any patterns emerging? What happens to this list when you add trauma?

Did you notice when undertaking Activity 2.2 that movement was affected in many, if not all, of the diseases and traumas in your list? This decrease in movement occurs even when the disease or injury does not directly affect the muscles themselves. If a person does not move or moves incorrectly for any length of time, then the muscular and cardiovascular systems will deteriorate. This process is technically known as *deconditioning*. So, just like the top-class athlete, deconditioned patients will need to improve their condition by increasing the strength, power and endurance of the muscles relevant to their functional needs.

In the following section, mechanisms by which muscle work brings about increases in strength and endurance will be discussed. Basic definitions will be given, as well as the physiological process and physical procedures required to produce these changes in condition.

THE DIFFERENCES BETWEEN MUSCLE FORCE AND MUSCLE STRENGTH

Muscle strength and *muscle force* are sometimes erroneously used as interchangeable terms. They are, however, two quite distinct concepts. As demonstrated in Chapter 9, a force is an entity that is generated by an action – a push or a pull, for example – or imparted as a kick. The object to which the force is imparted may move or deform. Therefore a force can be defined as *an influence that changes the state of rest or motion of a body or object*.

These forces are a product of the mass of the object producing them multiplied by its acceleration ($F = m \times a$). The unit of measurement is the newton (N).

Forces can be thought of as being either external or internal. External forces are those that happen outside the body, usually acting on the body, such as gravity, friction or other people. Internal forces are those mainly produced by the muscles. It is the muscle's ability to produce force that is a measure of the muscle strength. A fuller definition would be *the ability of a muscle or group of muscles to produce tension and a resulting force in one maximal effort, either dynamically or statically, in relation to the demands placed on it*.

The production of the internal force was explained in the previous section. The sliding filament theory explains how the actin and myosin protein fibrils slide over one another and form a series of cross-bridges that rotate and pull the sarcomeres closer together, thus shortening the muscle and producing the force (Hunter, 2000).

MUSCLE WORK

A muscle has to contract to produce a force. The force it produces can be either very small, so that the resultant action is correspondingly fine (e.g. picking up a feather), or quite large, with a result of deforming or moving a large object with a large movement (as in throwing a cricket ball across a field). The shortening of the muscles (bringing the proximal and distal attachments closer together) produces the force. If this movement, i.e. the change in length of the muscle, can be measured, then this can be multiplied by the force generated to give the work done by the muscle (work = force × distance). The unit of measurement of the work done is the joule (J).

From the definition, it can be seen that the muscle will be doing work only if there is a change in length. If the muscle is getting shorter as it is performing its task, it is said to be performing a *concentric contraction*, and the work done is therefore *concentric work*. If, on the other hand, the muscle is getting longer as it is performing its task, it is said to be doing *eccentric work*, as the contraction is eccentric. The former will be positive work and the latter negative. Paradoxically, if a muscle is producing an isometric contraction then mechanically no work is done, as there is no movement. If you hold a weight in an outstretched upper limb, you do feel as if you are doing some work! This is probably explained by the fact that there will be very small concentric and eccentric contractions of different groups of muscle fibres, giving an overall effect of no movement and thus technically no work being done.

The different amounts and types of muscle work, which result from the muscle's ability to alter the forces produced and the direction of its action, are important when we consider the activities that the human has to perform.

ACTIVITY 2.3

Think about these two activities.

1. Sitting in a bar lifting a full pint of fluid, taking a drink and lowering it back down to the table.
2. While walking down the street, you see the bus standing at the bus stop, so you have to run for the bus.

If you analyse the muscle activity in the first example in Activity 2.3, it can be seen that biceps brachii appears to be doing most of the work throughout the movement. As you take the drink up to your mouth, then the work being done by the muscle is concentric. As you lower the drink, the biceps again performs most of the work, but this time the muscle work is eccentric. Energy is also used when the muscle has to contract isometrically to hold the drink to your mouth. As you drink more, the load that you have to move becomes less, so although the type of work that the biceps produces does not change, the amount will decrease.

In the second example, the muscles of the lower limb are working to produce walking. As you study gait in Chapter 11, you will see that there is a systematic and regular change between concentric and eccentric muscle work to produce the gait cycle. If you suddenly have to change the amount of muscle work in order to run for the bus, then this is done by producing more force within the muscles.

These two examples show that there is a constant interplay between the type and quantity of muscle work that produces propulsion and control so that humans can perform the infinite variety of functional tasks that are needed throughout the day.

MUSCLE STRENGTH

The two activities in Activity 2.3 show the amount of force that needs to be generated by a muscle or muscle group used to carry out the functional activities of everyday life. The same muscles are used to walk and run, and the biceps are capable of lifting a heavier weight than a pint. The shot putter, on the other hand, needs to use all the force that could be generated in their muscles to give impetus to the shot. Therefore it can be concluded that a muscle has the ability to change the force generation for different activities. In other words, each muscle has a range of strength available to it. How much strength has to be used to perform a task is dependent on many influences.

RECRUITMENT

As we saw earlier, muscle tissue is grouped anatomically into fibres and fascicles and muscles. It can also be grouped into functional units. These units are called motor units. As illustrated in Chapter 4, muscles need a signal in order to contract. This signal is delivered by the alpha motor neurons originating from the spinal cord. A single alpha motor neuron may innervate from between 10 and 1000 muscle fibres, depending on the muscle. A single motor neuron and the muscle fibres that it innervates make up a motor unit, illustrated in Figure 2.7. The force produced by a single muscle fibre is very small and quite insignificant. It makes sense to group fibres together so that a single stimulation of a motor neuron will cause a simultaneous contraction of all the muscle fibres in that motor unit (the all or nothing principle), producing a useful amount of force. The motor unit has muscle fibres spread throughout the muscle, so they are not necessarily adjacent to each, which is why it is a functional collective rather than an anatomical one. The amount of force produced by a single motor unit depends on the number of muscle

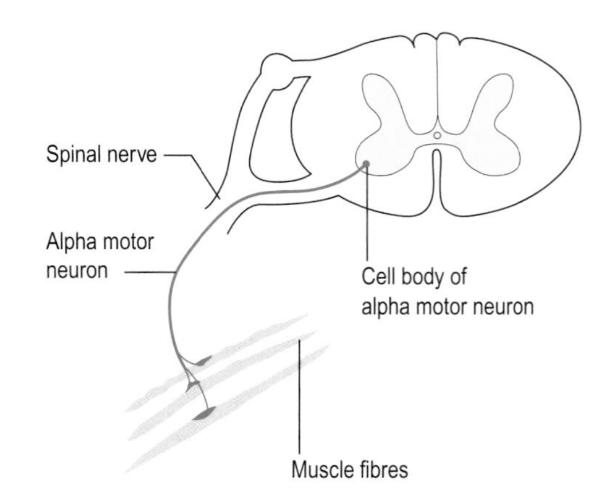

Figure 2.7 A single motor unit is composed of the motor neuron and the muscle fibres that it innervates. The cell body of the motor neuron is in the anterior horn cell in the spinal cord, and its axon is one of the motor nerves in a mixed peripheral nerve.

fibres it contains. There is a large variance in the number, because some muscles are associated with delicate and fine movements and some are quite gross. When precise and fine movements are needed, such as in the hand or eye, the motor units contain a small number of muscle fibres; when a relatively high degree of force is required and the movement can be quite gross, such as in the lower limb, there are a high number of fibres in a single motor unit.

MUSCLE FIBRE TYPES

Before considering how muscle force can be precisely controlled, it is necessary to be aware of one more factor that can influence the amount of force that a motor unit can produce. This relates to the different muscle fibre types. It is accepted that in humans there are three types of fibre, and they are classified by either their force characteristics or the isoforms of the contractile proteins (specifically myosin). Luckily, the two classifications are different only in their name. For the purposes of this text, the muscle will be classified according to the protein isoforms: type I, type IIa and type IIb. Table 2.1 lays out the various characteristics of these fibre types. It can be seen that for a given stimulation, type I muscle fibres reach the peak force at a slow speed and produce a low force but are fatigue-resistant. Type IIa fibres reach their peak force quicker, produce a greater force per stimulation and are fatigue-resistant. Type IIb fibres produce the highest amount of force in the shortest period of time but have a higher degree of fatigability. Type I fibres are therefore suited to activities that require a low amount of force for a long duration, whereas type IIb fibres are suited to quick, forceful contractions. A single motor unit will contain only one fibre type, and the alpha motor neuron will have functional characteristics that mirror those of the muscle fibre. So the motor unit that contains slow fibres will be small in diameter, making conduction velocity relatively low; this type of motor unit is known as a slow (S) motor unit. The motor units that contain type IIa fibres are known as fatigue-resistant (FR) motor units, and those that have type IIb fibres have the largest axons and fastest conduction velocities and are known as fast fatigable (FF) motor units. Each muscle contains a mix of the three types of motor unit in varying proportions according to the functional role of the muscle. Postural muscles (e.g. the soleus) that work to maintain posture will typically have a majority of S motor units, whereas the gastrocnemius has a high proportion of FR and FF motor units.

GRADATION OF MUSCLE FORCE

Past experience is a vital component. Lifting a full glass of liquid is a common activity, so the brain has a good idea of how much strength is needed and therefore how much effort to use to lift the glass. On the other hand, if we lift a closed box expecting it to be heavy, the result may be that the box is thrown up in the air because it is, in fact, empty. The brain had been expecting a heavy load and the muscles had contracted accordingly.

Table 2.1 Examples of differences between fibre types[a]

PROPERTY	TYPE I	TYPE IIA	TYPE IIB
Muscle fibre type	Slow oxidative (SO)	Fast oxidative glycolytic (FOG)	Fast glycolytic (FG)
Motor unit type	Slow (S)	Fast fatigue-resistant (FR)	Fast fatigable (FF)
Motor unit size	Small	Medium	Large
Twitch tension	Low	Moderate	High
Mechanical speed	Slow	Fast	Fast
Fatigability	Low	Low	High
Mitochondrial enzyme activity	High	Medium	Low
Glycogenolytic enzyme activity	Low	Medium	High
Myoglobin content	High	Medium	Low
Capillary density	High	Medium	Low

[a]Note that different terminology exists for types of muscle fibre and motor unit.

The above scenarios are dependent on the number of motor units used during the muscle contraction. The contraction of the motor unit is termed *recruitment*, and it is this recruitment that enables us to use our muscles to produce fine movements such as writing, larger movements such as picking up a glass, or the whole of our strength in putting the shot. So the task itself or the load to be overcome is the main component in determining the strength exerted by the muscle. However, as seen earlier, the motor unit has an all or nothing contraction. Therefore there must be a way to grade the muscle force according to the demands. Indeed, the body has two ways that vary the amount of force: motor unit recruitment and the frequency of stimulation.

Motor units within a muscle are recruited in a set order according to the force requirements. For activities requiring a low amount of force, the motor units that produce the lowest force are recruited first (the S motor units). This has two effects: first, the fibre type is fatigue-resistant and can therefore go on for a long period of time; second, subsequent recruitment of S motor units results in a smaller step increase in force (see Fig. 2.8), allowing for a finer gradation of muscle force production. As the force production increases, the fast FR motor units are recruited next. As the force reaches near maximal, the FF motor units are recruited. However, this cannot be sustained for very long, as they are susceptible to fatigue, and a sustained contraction will be unable to maintain the maximal force output. Note that in Figure 2.8 the steps associated with the FF units are much larger. This recruitment of motor units follows the Henneman principle; this is also known as the size principle, as the neurons that are easiest to stimulate are the smaller ones associated with the S motor units and are therefore recruited first. The larger the neuron, the harder it is stimulate, and therefore more effort is required to produce a contraction of the larger motor neurons of the FR and FF motor units.

The second point is the frequency of stimulation; for any given motor unit, a single stimulation will produce a given force output. We have already seen that if force needs be, increased recruitment of more motor units will achieve this; however, the motor unit can also produce more force (up to certain point) if the frequency of stimulation increases. These options give the body more ways of controlling the force output of a muscle. These two are both at work during sustained low force contraction; motor units are switched on and off according to demand and changes in stimulation frequency also occur, both of which can finely tune the force output. This type of activity is typical of everyday activities such as standing or walking.

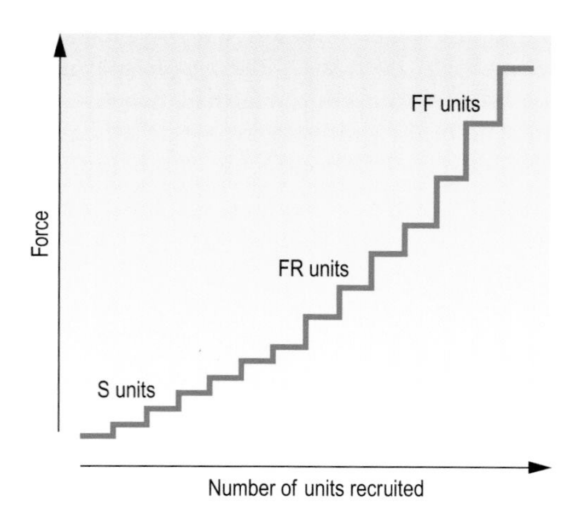

Figure 2.8 The regulation of force by recruitment of motor units. At low forces, only the small, slow (S) units are recruited. As force increases, the larger fatigue-resistant (FR) and then the largest fast, fatigable (FF) units are recruited. The S units have relatively few muscle fibres and the FF units have the most. Therefore the increment in force as a new unit is recreated also varies.

LENGTH–TENSION RELATIONSHIP

As well as recruitment, other physical properties of the muscle are important in strength generation. Among these is the length–tension relationship of the contractile unit. It has been shown that the production of force in the muscle is proportional to the number of cross-bridges that occur between the actin and myosin fibrils. If few myosin heads are in contact – as when the actin and myosin fibrils are stretched apart, for example – then force production will be decreased. If the fibrils are too contracted, the tapered ends of the myosin filaments push against the Z bands, and again the force that can be generated is decreased (see Fig. 2.9). These length–tension characteristics are enhanced by the intracellular titin filaments that run through the length of the myosin protein filament between the Z lines. These titin filaments have an intrinsic elastic property that can alter the propensity of the muscle to contract (Lakomy, 1999). It is tempting to think that during movement when the

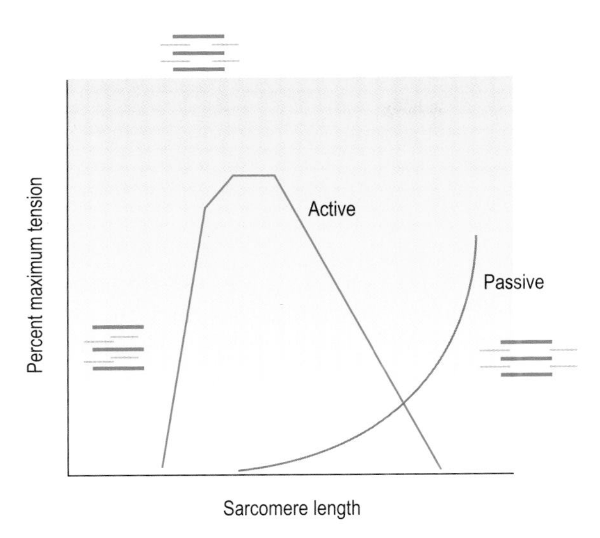

Figure 2.9 The sarcomere length–tension curve during sequential isometric contractions of a muscle fibre, with diagrammatic representation of the thick and thin filament overlap. The curved line represents the passive tension of the muscle.

muscle shortens, the muscle will behave exactly like the graph in Figure 2.9. In other words, when a muscle is in a lengthened position (outer range) there are very few cross-bridges between the myosin and actin, as the overlap is small, and the muscle is unable to generate much tension. Similarly, if the muscle is in a shortened position, there is too much overlap, and the myosin filaments cannot move any further down the thin filaments as the Z disc is in the way, again limiting the amount of tension. The length–tension curve as represented in Figure 2.9 is valid for isometric contractions only (when no movement occurs); it should be viewed as a series of points rather than a curve.

The full range of movement through which a muscle works is usually divided into thirds: inner, middle and outer. Inner range is when the muscle is in a shortened position, outer is when it is in a lengthened position, and middle range is between the two (Fig. 2.10). So for an isometric contraction, the outer range is where there is very little overlap between the thick and thin filaments, the middle range is where there is optimal overlap, and the inner range is where the thick filaments run into the Z disc. It has been known for some time that the muscle is weaker in inner and outer range and strongest in middle range. Part of the reason is the length–tension relationship; however, the biomechanics in relation to the anatomy plays a large role. The angle of pull of the muscle on the bone changes through the range such that the muscle is usually strongest in middle range. For more details on this, see Chapter 9. To be able to appreciate the tension in a muscle during a movement, the distinction between active and passive tension and velocity needs to be discussed.

ACTIVE AND PASSIVE TENSION

As can be seen in Figure 2.9, there is a curved line that represents the passive tension of a muscle. When a muscle is passively lengthened, the muscle resists the lengthening, without stimulation. This resistance is thought to be a result of the protein mentioned earlier: Titin. When a muscle is taken from mid range to outer range, the passive tension increases slowly initially, and as the end of range approaches the tension increases dramatically. Therefore as it approaches outer range, the tension of the muscle will increase irrespective of the

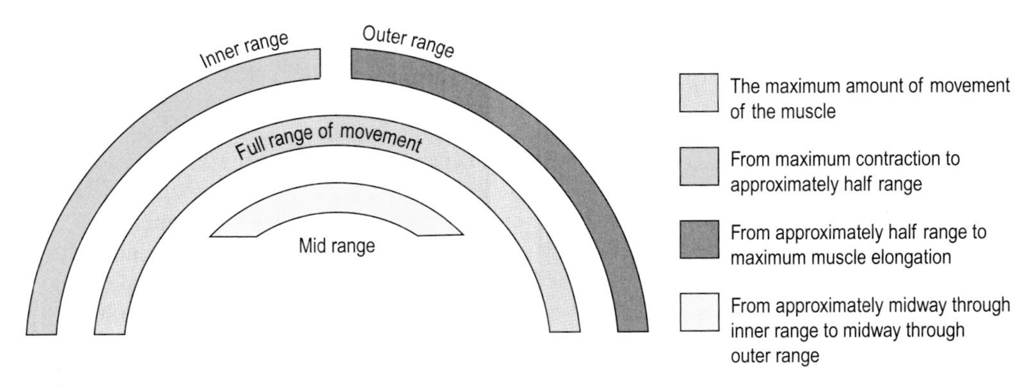

Figure 2.10 Ranges of movement.

activity of the muscle. As the tension is related to a protein that forms the basic skeleton of the sarcomere, actin, this could be viewed as a mechanism to stop the sarcomere from being overstretched and keep some degree of overlap between the thick and thin filaments.

FORCE–VELOCITY RELATIONSHIP

As most muscle contractions are involved in movement, it is important to understand how a muscle behaves during movement. As mentioned earlier, the length–tension relationship is not helpful here, as it is valid only when there is no length change. If a muscle is allowed to change length during a contraction, it can do so at a variety of speeds. The force–velocity relationship is another aspect that affects muscle strength. Different forces can be generated at different speeds (Fig. 2.11). If the force generated by the muscle is equal to the load, then no movement will occur (velocity = 0, an isometric contraction). If the muscle force is greater than the load, then the muscle will shorten (concentric contraction). The greater the muscle force generated in a concentric contraction against a load, the quicker the muscle can contract. In other words, the greater the velocity of contraction the smaller the tension in the muscle. It can be seen from the force–velocity graph that concentric

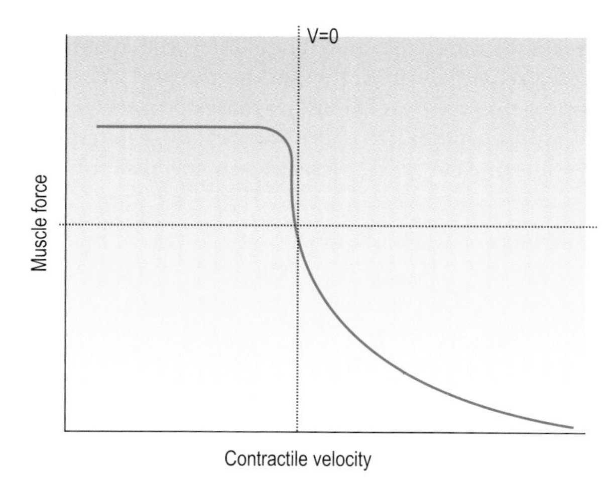

Figure 2.11 Force–velocity relationship graph. Isometric contraction is represented at $V = 0$. To the right of this is concentric contraction. As the velocity increases, the force reduces. To the left of $V = 0$ are eccentric contractions as the velocity of lengthening increases.

contractions do not develop as much tension as isometric contractions. If the load is greater than the muscle force, then the muscle will lengthen (eccentric contraction). This type of muscle contraction generates a greater amount of force than isometric or concentric contractions. Unlike concentric contraction, in eccentric contraction as the velocity of lengthening increases the tension increases, up to a point. This is quite poorly understood, but this relationship is also applicable to the energetics at a molecular level. Therefore this increase in tension is probably because the muscle uses very little ATP to break the bonds holding the cross-bridges.

ANGLE OF PULL

Other anatomical and biomechanical aspects of muscle function will affect the strength generation. The angle of pull of a muscle at the time of its action will affect its strength. The angle of pull of a muscle is defined as *the angle between the segmental axis and the line of pull of the muscle.*

An angle of pull that is nearer to 90° means that more of the resolved muscle force would rotate the segment (vertical component). If the angle of pull was greater or less than 90°, then the distractive or compressive force (horizontal component) would increase. This is explained in Chapter 9. Luckily for us, the angle of pull never reaches 90°. If it did, then all the muscle force would be used to move the segment and there would be no horizontal force to stabilize the joint. This would mean that the joint would actually be damaged along with other musculoskeletal structures!

STABILITY AND SEQUENCING

For the muscle to work efficiently, it should work from a stable base. The strength of the muscle will then be used for the intended task and not diverted to ensure that a stable base is maintained. This is very important during early rehabilitation, as the muscles are usually very weak. This is quite commonly seen when performing exercises on a gym ball. People often perform weight-lifting exercises on a gym ball for various reasons; however, the weight that they can lift on the gym ball (the unstable surface) is much less than if they adopted the same position on a stable surface. During many functional movements, many muscles have a role to provide this stable base, for example some of the abdominal muscles. These muscles are helped to achieve this stable base and work more efficiently by working in predefined sequences. These sequences are learned as movement is refined during the maturation process or we learn the pattern as a new skill. Chapter 6 explains these processes.

ANATOMY

The gross structure of the muscle seems to be well adapted to provide the appropriate range, direction and force of contraction. The muscles that produce the precision movements appear to have fine muscle fibres, whereas the gluteus maximus, for example, has coarse muscle fibres. How the fibres are structured within the muscle is also important. They are usually parallel, oblique or spiral relative to the direction of pull of the muscle. In muscles, the angle of pennation (the angle at which the fibres join the central spine of connective tissue) of the muscle itself will affect its ability to produce force. The greater the angle of pennation, then the more sarcomeres there are in parallel. This will lead to an increase in strength but a decrease in shortening velocity (Hunter, 2000). Examples of the different forms of muscle fibre arrangement can be seen in Figure 2.12.

The length–tension relationship, the angle of pull, and the sequencing and patterning of muscle action are all most effective in producing greater strength when the muscle is in its middle range. Of course, these biomechanical factors result in efficient strong muscle work only if the neuromuscular and muscular systems are intact. The physiological systems, such as the circulatory system, must also be functioning optimally to initiate, maintain and terminate muscle action. If any of these systems malfunction, as in many pathologies, then muscle strength and subsequent efficient movement are decreased.

AGE AND GENDER

Age and gender also have an effect on absolute muscle strength. It will be seen in Chapter 13 that the changes caused by ageing are complex and are a combination of the physiological process, disease and lifestyle. The general consequence, however, is that muscle strength decreases with age.

PSYCHOLOGICAL FACTORS

It is also important to remember that as well as the physical and physiological aspects discussed above, there can also be a psychological element to muscle strength. The shot putter will win the gold medal only if all the psychological elements are right, if they have produced enough adrenaline (epinephrine) and believe that they are able to do it. Sometimes the apparent limit of physical ability to produce muscle strength is overcome by psychological influences. There is the tale of the mother who lifted a car under which her child was trapped. This was a feat that was obviously beyond her perceived physical ability.

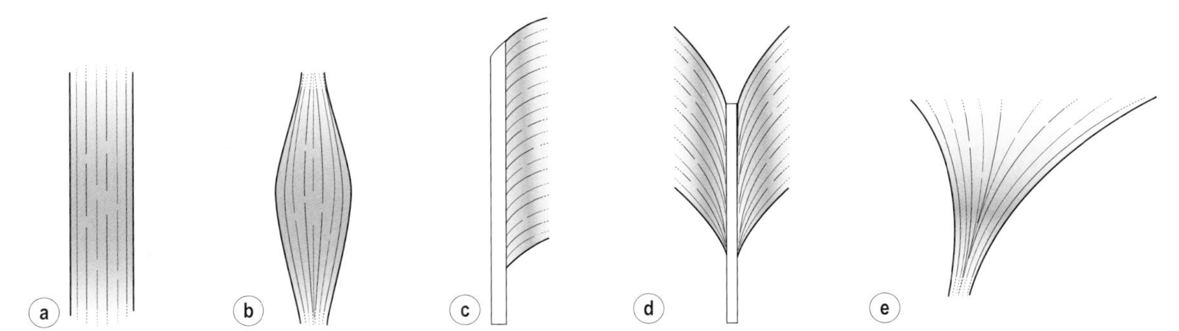

Figure 2.12 Some of the common arrangements of muscles. Strap (a) and fusiform (b) muscles have a tendon at either end and relatively long muscle fibres that run between the tendons. The muscle fibres are arranged in parallel to or are very similar to the angle of pull of the tendon. Unipennate (c) and bipennate (d) muscles have a central tendon into which short muscle fibres are inserted. The fibre direction is different to the angle of pull of the tendon. Triangular muscles (e) display mixed characteristics.

MUSCLE POWER

In Chapter 9, power is defined as the rate at which work is being done. The unit of work is the watt (W). The rate at which a muscle works is termed *muscle power* and can be calculated by using the following formulae:

$$\text{power} = \text{force (of contraction)} \times \text{velocity (of contraction)},$$

in which velocity is distance moved/unit time, i.e. distance of contraction, and

$$\text{power} = \text{net joint moment} \times \text{joint angular velocity}.$$

Positive power is generated in a concentric contraction, and negative power is generated in an eccentric contraction. Earlier, it was explained that the velocity of muscle contraction is largely determined by the composition of that muscle, i.e. the fibre type(s) it contains. The type II (phasic fast twitch) fibres that generate large amounts of tension in a short time are geared towards anaerobic metabolic activity and tend to fatigue quickly. The power produced by these muscles will therefore be anaerobic and will produce high-intensity activity over short periods of time. Type I (tonic slow twitch) fibres generate a low level of muscle tension but can sustain contraction for a long time. These fibres are geared towards aerobic metabolism and are slow to fatigue. The power they produce is therefore aerobic power.

ACTIVITY 2.4

For two examples of sporting extremes (e.g. a high jump and a marathon) and two examples of everyday activities (e.g. writing and opening a tight jam jar), analyse the differences between each upper and lower limb activity in terms of muscle strength, force, work and power.

MUSCLE ENDURANCE

If activities that are carried out throughout the day are analysed, there are very few occasions when maximum strength has to be used. Most activities that are performed will require the muscle action to be repetitive, as many activities will be carried out a number of times; just think of walking as an example. The muscle's ability to keep on working over long periods of time is a measure of its endurance and is defined as *the ability of a muscle to maintain isometric contraction or continue dynamic contractions.*

The muscular endurance can be thought of as the ability of a muscle not to tire; in other words, to resist fatigue.

In the preceding text, strength was described in biomechanical and physical terms. Endurance, however, needs to be described with a more physiological perspective. As was described for aerobic power, endurance is dependent on the type of respiration performed by the working muscle, i.e. the aerobic energy systems. For the aerobic system to work, there must be an adequate energy supply to the working muscle (Daniels, 2001). This energy may be in the form of creatine phosphate and glycogen stored in the muscles themselves or the liver, which last for a short while (through anaerobic respiration). This is then replaced by energy that is brought to the area from the metabolism of foodstuffs derived from digestion (aerobic respiration). Because this foodstuff is being burned to produce energy aerobically, a key ingredient for the process is a constant supply of oxygen brought to the area. Thus oxygen transport from the lungs to the working muscle is a vital part of the muscle's respiratory function. During continuous muscle respiration, waste products such as carbon dioxide, potassium, acetylcholine and lactic acid are being produced. It is just as important for the functioning of this system that this waste is removed from the area. If the waste is not removed, then its build-up will cause the muscle to fatigue. The longer the muscle can resist fatigue, the greater is its aerobic capacity (Daniels, 2001).

FATIGUE

Everyone has felt tired at times when there seems to be a lack of energy at the end of a task or at the end of a long day at college. This must, however, be distinguished from true fatigue, which means that there is an inability to carry on with the task.

Fatigue can take two forms: *general body fatigue* and *local muscle fatigue*. The former involves the depletion of energy stores and/or the build-up of waste products in many regions and systems of the body. This is characterized by physical and mental exhaustion and is an extremely serious situation. Local muscle fatigue is caused by the depletion of energy and/or the build-up of waste products within the functioning muscle, a decrease in the availability of oxygen, and any disturbances in the contractile mechanisms such as inhibition by the central nervous system and a decrease in conduction at the neuromuscular junction. It is defined as *a diminished response to repeated stimuli characterized by a decrease in amplitude of the motor fibre units or the ability to sustain a force.* This is a much more common event than general fatigue and provides the biochemical and neurological stimuli necessary for the development of the aerobic capacity of the muscle. Local muscle fatigue is recognized by the inability to generate force and carry out the full range of movement of the task being undertaken, a decrease in the speed of contraction, shaking or fasciculation within the particular muscle, pain and a loss of coordination.

A full explanation of exercise physiology can be gained from many standard texts, but from the brief description above and the preceding text it can be seen that for the muscle to carry out its functions properly there have to be intact neuromuscular, musculoskeletal and cardiorespiratory systems. Failure of any or all of these systems will lead to the muscle being unable to adequately perform its function (Wilmore and Costill, 2004).

ACTIVITY 2.5

List the situations that could affect the functioning of a muscle. For each of the items on your list, think of which component of muscle function could be impaired. Try also to analyse why this impairment is taking place.

MEASURING STRENGTH AND ENDURANCE

To enable us to determine if a muscle is working to its maximum capacity for both strength and endurance, it is important that both these components can be measured. For each of the problems you have listed in Activity 2.5, you will need to measure the strength or endurance of the muscle or muscle groups used. This is important for several reasons: first, to determine if there is any deviation from the normal; second, to assess whether the condition is getting worse, better or remaining the same; third, to give a base measurement to the muscle condition; fourth, to assess whether any treatment you implement is having a positive or negative effect on the muscle; and last, to give an insight into the mechanism underpinning any loss of strength.

MEASURING STRENGTH

While studying the shot putter of the earlier example, it could be seen that the size of the muscle was important in producing strength. Those people with large muscles are usually stronger than those with smaller muscles. Therefore it can be assumed that the larger the muscle, the greater the force that the muscle can generate. This would seem pretty obvious, as the larger the muscle, the more actin and myosin would be present and therefore the greater the propensity for the contractile unit to form cross-bridges.

Physiological cross-sectional area

It has been described by many people (e.g. Jones and Round, 1990) that the physiological cross-sectional area (PCSA) of the muscle is proportional to its strength. This, however, is not a convenient measure, as only the anatomical cross-sectional area (ACSA) can be ascertained when measuring the girth of the limb using a tape measure. The true PCSA can be found only through dissection or modern scanning procedures. However, a loose relationship between size and strength does exist.

The results of measuring the girth of a limb to discover the ACSA must be read with caution, as the whole of the limb may be asymmetrical when compared with the other side. This would mean that there is difficulty in distinguishing which component is responsible for the anomaly.

Atrophy and hypertrophy

When the muscle has decreased in size through injury, disease or disuse, it will usually be weaker than normal. This decrease in size is termed *atrophy*. When we build our muscles up in size, we usually find that the muscle is stronger. This increase in size is termed *hypertrophy* and is caused by an increase in the amount of protein in the muscle (Harris and Dudley, 2000). Measurement of the PCSA is possible with the use of techniques such as magnetic resonance imaging, but these are not really practicable.

Indirect measures of strength by measuring force can be obtained using handheld dynamometers, but these are limited by the strength of the operator. Another valuable tool in measuring force production in the form of muscle torque (turning effect) is the isokinetic dynamometer, although this machine is often too expensive for general therapeutic departments. Chapter 14 gives a description of this and other devices used for measuring strength. If the muscle is capable of performing its full range of movement, it is convenient to use the Medical Research Council Scale (the Oxford Grading Scale) to measure muscle strength. This is classified on a scale of 0–5 (Hollis and Fletcher-Cook, 1999):

- 0, no movement at all
- 1, a flicker of movement
- 2, full range of movement with the effect of gravity eliminated
- 3, full range of movement against gravity
- 4, full range of movement against gravity and resistance
- 5, full range of movement with maximal resistance.

Although useful, the Oxford Grading Scale has many limitations, and many variations of this scale have been developed (see Ch. 14). A more useful measure of strength in a wider context may be the 1 repetition maximum (1 RM), defined as *the maximum amount of weight a muscle can lift once only*. The 10 RM is a derivation of the 1 RM and is *the maximum weight a muscle can lift 10 times*. Both these quantities are found by trial and error. A more subjective measure that is frequently employed is isometric testing. This is carried out by manually resisting the muscle contraction in various positions within its range and not allowing any movement. The resistance that has to be used to prevent movement is compared with that of the unaffected side. This method is obviously applicable only if the problem is unilateral and the subject normally symmetrical.

MEASURING ENDURANCE

As discussed earlier, there is a greater involvement of physiological processes in muscle endurance. Therefore testing endurance is going to involve physiological testing. Measurements of expired air, heart rate, respiratory rate, blood gases and waste products after maximal and submaximal exercise on treadmills are the procedures usually used (McArdle et al. 1996). These measurements would be fairly difficult in an ordinary department. Although it is difficult to measure muscle endurance directly, measuring the time to fatigue (or number of repetitions to fatigue) and the time to recovery will give an adequate estimate of the endurance of the muscle and how it is progressing and regressing. The isokinetic dynamometer described in Chapter 14 also has a facility to measure, indirectly, muscle endurance.

WHEN THINGS GO WRONG

There are times when muscle strength and endurance cannot be maintained. The most common causes of decrease in strength and endurance are injury, disease and disuse. Injury can be to the muscle itself, its nervous or vascular supply, or the skeletal support. Direct trauma that disrupts the contractile unit, such as a muscle or ligament tear, will mechanically affect the ability of the muscle to either produce or transmit force. Disruption of nerve supply will mean that no recruitment is able to take place and hence no muscle contraction. Disruption of blood supply will mean that energy, in the form of foodstuff, and oxygen will not be delivered to the muscle, and waste products will not be taken away. This will mean a decrease in the respiratory capacity of the muscle and hence a decrease in its endurance. There are many diseases that affect the muscle or its neurological supply. Muscular dystrophy is an example of a muscle-wasting disease in which the muscle protein is affected. A neuropathy would disrupt muscle stimulation, leading to a decreased contraction.

The examples above would lead to an inability of the muscle to function to its optimal capability. The muscle may also not be used if another body part, proximal or distal to the muscle, has to be kept immobile (as in a joint problem or bone fracture). Pain, muscle spasm, habit and certain psychological factors will lead to weakness caused by disuse. If the muscle is not used, then the actin and myosin protein will be reabsorbed. This will diminish the ability to form cross-bridges and hence the force production and strength will decrease. This decrease in strength is mirrored by a decrease in size, which is termed *atrophy*. Therefore the case above will be termed *disuse atrophy*. Concurrent with the reabsorption of actin and myosin, there will also be a collapse and eventual reabsorption of the small blood and lymph vessels. This, together with the decrease and inefficient use of energy stores, will cause a decrease in muscle endurance.

INCREASING STRENGTH AND ENDURANCE

For a muscle to perform any functional activity, it must be able to generate enough force to overcome the resistance of the task. This could be one maximal effort or a series of submaximal efforts over a time period. For the former, the muscle strength is paramount, while the latter requires both strength and endurance.

MUSCLE STRENGTH

For a muscle to increase in strength, it has to work to fatigue with a load placed on it that exceeds its usual metabolic work rate. This is known as the *overload principle*.

PHYSIOLOGICAL PROCESSES

If the muscle is stimulated to work hard, then this information is relayed to the central nervous system, which in turn stimulates the ribosomes to replicate more actin and myosin protein. The myofibrils are therefore increased in length and width. There will be an increase in myocyte number and size and in the number of sarcomeres, which will increase the strength of the muscle. There will also be a change in the density of the mitochondria within the muscle tissue. More muscle glycogen, creatine phosphate and ATP substrate will be laid down. There will also be an increase in the concentration and activity of glycolytic enzymes, myokinase and creatine phosphokinase (the enzymes needed for growth) (Maughan and Gleeson, 2004). This process usually takes about 4 weeks.

INCREASED VASCULARIZATION

If there is enough biochemical stimulation through activity, then increased vascularization of the area will also occur (up to 50%), and thus the supply and utilization of oxygen and energy will also change so both strength and endurance will increase (Greenhaff and Hultman, 1999).

INCREASE IN SIZE

As the muscle increases in strength, it also increases in size, which is termed *hypertrophy*. However, it must be noted that initially, increases in strength are not accompanied by a corresponding increase in size. It is thought that recruitment of motor units becomes more efficient, and there is also an increase in the number of motor units that are recruited. There also appears to be an inhibition of the antagonist muscle groups, together with a more efficient integration of synergists. The training effects described in Chapter 14 also play their part. Some authors believe that hyperplasia, or the splitting of developed fibres, takes place, but there is debate concerning this (Conroy and Earle, 2000).

INCREASING STRENGTH

Muscle weakness is the inability to generate force. If a muscle is weak, then the best method to increase strength is exercise in the form of training. Training can be defined as *the facilitation of biological adaptations that improve the performance of specific tasks*.

There are three main categories of exercise that are used to increase muscle strength: assisted exercise, free active exercise and resisted exercise. Assisted exercise could be manual or mechanical and is used when the muscle is so weak that the segment cannot be moved through a sufficient range against the force of gravity. Manual assistance may be given by the therapist (therapist-assisted) or by the patient (autoassisted). The assistance given may be just eliminating the effect of the force of gravity. The assisting force must be in the direction of the required movement and must only assist the movement. Care must be taken that the movement does not become passive. Mechanical assistance could be assisting the movement by eliminating or decreasing friction by using a sliding board or sling suspension. Pulleys can also be used to facilitate the movement. Such assistance is useful for muscles that have been measured as grade 2 on the Oxford Grading Scale. If the muscle has been measured at grade 3 or above, then it can be strengthened by using free active exercise, defined as *exercise without the use of assistance or resistance except gravity* (some people make body weight the exception to this).

Fatigue, and therefore muscle strengthening, can be achieved with active exercise by changing the parameters of the exercise, such as repetition, speed, rhythm and leverage. The exercise can be progressed and regressed by altering the starting position.

Once a muscle is able to move a segment against gravity, then the more usual form of exercise used to strengthen muscles is resisted exercise. To be most effective, the resistance must be directed against the movement of the muscle and if possible at 90° to the axis of the segment.

The resistance may be given manually or mechanically. Manual resisted exercise can be thought of as *active resistance exercise in which the resistance force is applied by the physiotherapist (therapist-resisted) or the patient (autoresisted) to either a dynamic or a static muscular contraction.*

ACTIVITY 2.6

Work in pairs. One of you sits over a plinth with your thigh supported, acting as the patient. The other one, acting as the therapist, gives resistance to the leg, beginning at the knee and moving down towards the ankle in increments. Think about the force required to resist maximal effort and the effect this has on the patient. What trick movements do you notice when the activity becomes more difficult? How could these be corrected?

The position of resistance is an important consideration, as the further away from the axis of rotation the resistance is given on the segment that is moving, the less effort has to be made by the therapist. This can be explained by considering leverage and the effect the length of the lever arm has on the effort required. As will be seen in Chapter 9, the moment of force is defined as the force multiplied by the distance from the pivot. Therefore if the distance from the pivot is increased, then to maintain the same moment the force could decrease. The forward-sliding force that causes the anterior glide at the knee joint is increased the further down the leg the resistance is given. This may have adverse consequences for the patient, so this has to be taken into consideration when performing the technique.

Mechanical resistance is any resistance against which the body can exercise. The main categories are weighted resistance, free weights (barbells, dumb-bells, cuff weights, ankle weights, weighted boots or sandbags), multigyms, isokinetic dynamometer and springs made, for example, from materials such as elastic latex. Exercise cycles, body weight and hydrotherapy may all be used as forms of mechanical resistance.

Strength training

When a muscle has to undergo exercise to improve strength, it is usually because there has been a problem, and it is very unusual for this problem to manifest itself as only a decrease in muscle strength. Many other systems may be involved, so we need to approach rehabilitation holistically. Strength and endurance are closely linked and really should be treated together. Other factors – such as muscle elasticity, joint range, coordination and balance, and cardiovascular fitness – must also be addressed. Although this text does not cover the skills involved in undertaking a training programme, it would be useful to cover the main principles even if there is an artificial separation between strength and endurance.

Assessment

Muscle strength, joint range of movement and integrity, pain and functional ability all need to be assessed before any treatment is planned. This is a good time to work out the 1 RM and 10 RM of the muscle or the level of cardiovascular fitness or muscular endurance.

Principles of strength training

The fundamental principles of strength training are as follows.

Overload

As described earlier, the system must work at a level greater than normal function. For strength, this would mean moving a greater resistance and maintaining a contraction for a longer period. The stimulus for muscle growth is thought to be related to the tension generated, therefore it is important to consider aspects of how this tension varies with speed and length.

Specificity

It has been shown that training that mimics the activity for which the action is needed is more effective than if the conditions are different (Ackland and Bloomfield, 1995). Conversely, training for one factor, for example strength, would not necessarily improve another factor, such as endurance. There is not necessarily an overlap. This is also true for different fibre types and possibly speeds of activity. Therefore the training needs to be specific for the muscle action, range of movement, type of contraction, energy system and functional need.

Progressive (or principle of diminishing returns)

For any given training sessions, a certain degree of change will occur. If the training is simply repeated, then the degree of change will reduce with each session. Therefore the training sessions need to be assessed and progressed regularly to continue strength or endurance improvements (see *Progressive resistance exercises* below).

Reversibility

Generally, it can be said that the adaptations gained through training are lost rapidly once training stops (MacDougall et al. 1980). However, there is a difference between endurance and strength gains. Endurance capacity quickly reduces when activity levels are reduced through either bed rest or cessation of training. However, for strength, other aspects – such as age, genetics and nutritional, hormonal and environmental factors – can all play a role in regulating muscle mass. It has also been suggested that different fibre types respond to decreased use in different ways (Astrand et al. 2003). As a general rule, the maxim 'use it or lose it' applies to the principle of reversibility. If the muscle is not being used to its full capacity, strength gains will be reversed. However, as little as one exercise session per week is sufficient to maintain muscle strength.

Motivation and learning

This is an important aspect of any exercise programme and is discussed in Chapter 14.

Progressive resistance exercises

These are specific exercise regimens that ensure overload of the muscle is achieved and generally use weights in a strengthening programme. Progressive resistance exercise is defined as *load or resistance to the muscle as applied by some mechanical means and quantitatively and progressively increased over time.*

Below is a list of the variables that may be changed in addition to the load.

- Repetition: the number of a specific exercise, usually in sets.
- Sets: a number of repetitions.
- Frequency: the number of times the exercise session is performed.
- Duration: the length of time of each session.
- Speed: the speed at which the exercise is carried out.
- Muscle action: concentric, eccentric or isometric.
- Starting position: this is the position from which the exercise is performed.

Specific regimens

Many training regimens were developed during and just after the Second World War to treat the many injured soldiers. They were based on the fundamental principles given above, and some are shown in Table 2.2.

Although these programmes are probably not used these days in their original form, derivations are used, and the basic principles underlying them are certainly still valid.

Table 2.2 The DeLorme and Watkins programme (DeLorme and Watkins, 1948), the Oxford programme (Zinovieff, 1951) and the MacQueen programme (MacQueen, 1954, 1956)

NAME	REGIMEN	EFFECT	CONDITIONS
DeLorme & Watkins	10 lifts at ½ 10 RM 10 lifts at ¾ 10 RM 10 lifts at full 10 RM	Increases strength	× 3 each session, 4 or 5 per week, retest 10 RM weekly
Oxford	10 reps at 10 RM 10 reps at ¾ 10 RM 10 reps at ½ 10 RM or 10 reps at 10 RM then reduced by 5 kg for 10 sets	Increases strength	× 5 sessions per week
MacQueen (1)	4 × 10 reps at 10 RM	Hypertrophy	
MacQueen (2)	10 reps at 10 RM 8 reps at 8 RM 6 reps at 6 RM	Power	

RM, repetition maximum.

INCREASING ENDURANCE

Conditioning is the augmentation of the energy capacity of the muscle through an exercise programme. This is produced by exercise of sufficient *intensity*, *duration* and *frequency*. The methods used to increase endurance differ between the fit athlete and the patient who has become deconditioned, but the principles of treatment remain the same. The overriding consideration is the overload principle. As for strength training, the muscle must work above its usual metabolic function for adaptation to take place.

Intensity is the rate at which the exercise is carried out. Intensity is easier to quantify if we are trying to improve cardiovascular endurance. The maximum volume of oxygen uptake (VO_{2max}) is a function of the intensity, and because heart rate and VO_{2max} are linearly related, heart rate can be considered as a function of intensity. So the intensity of the exercise could be described through the heart rate. Local muscle endurance is usually measured as the length of time the muscle can function or the number of repetitions of the activity before fatigue occurs. The muscle's ability to recover post exercise is also used as a measure of its endurance. General conditioning is said to take place when the heart rate is between 60 and 90% of maximum. For local muscle endurance to increase, the number of repetitions needs to be high but against a low resistance.

The *duration* of an exercise is the length of time for which the exercise is carried out. The greater the intensity, the shorter will be the duration, and vice versa.

Frequency is the number of times that the exercise programme is carried out per week. Although this will vary from patient to patient depending on the assessment, usually a minimum of three times per week is necessary to ensure that the physiological adaptations take place (Watham and Roll, 2000).

ACTIVITY 2.7

Choose a muscle or group of muscles, the shoulder abductors for example, and work out a strengthening regimen as the muscle progresses from a grade 1 to a grade 5 on the Oxford Grading Scale. Set aims and objectives for each stage of the recovery and use assisted, free active and resisted exercise. Discuss how you would assess the condition and progress and regress the exercises as necessary.

ACTIVITY 2.8

For the lower limb, devise a circuit that would improve local muscle endurance for the major groups of muscles.

Endurance training

When training for endurance, the principles that were used for strengthening still apply. Exercise is still used, but the repetitions have to be greater to ensure that the overload principle is met. Therefore it must be ensured that the aerobic system is utilized, so low or even no resistance is used. These exercises can be progressed or regressed with some degree of objectivity by using numbers of repetitions, duration of exercise or a combination of both as markers. The rest or time interval between sets of exercises can also be used as a method of progression or regression. The rest interval can be shortened as the recovery from fatigue is speeded up.

For the earlier stages of rehabilitation, assisted and free active exercise can be used to increase endurance. As rehabilitation progresses, free active and resisted exercise are used.

Delivery of exercise

The programme used to deliver the exercises for increasing strength or endurance could be either individual (patient and therapist) or in a group situation (therapist and class). Each has its advantages and disadvantages. In the one to one situation, the therapist could give individual encouragement and ensure that the exercise is being carried out properly. The exercise could be progressed and regressed as soon as applicable, as the therapist would be constantly assessing capability. In the group situation, however, the patient will be motivated by competition with others and the variety of exercise could be greater.

Circuit training as a group exercise is particularly effective when increasing cardiorespiratory endurance, but it is also very effective for increasing local muscle endurance, as the delivery of oxygen and the ability of the muscle to use it are closely linked (Maughan and Gleeson, 2004). The circuit is made up of a series of well-defined activities with a predescribed rest period between each. The circuit is usually repeated a set number of times with a rest period between each (usually enough time for recovery) (Astrand and Rohdal, 1988).

CONCLUSION

Muscle strength, power and endurance, although described separately, can be seen to be closely integrated and should be thought of as equally important in affecting movement of the musculoskeletal system and ultimately locomotion of the human. We have seen through this chapter that the production of muscular force, movement initiation, and control and continuation of useful integrated and functional movement require an intact neuromusculoskeletal system. These systems have to be studied through anatomical, physiological, biomechanical and psychological perspectives so that if any are disrupted through trauma or disease we can, after thorough assessment, return them to an optimal condition to perform functional human movement.

References

Ackland, T.R., Bloomfield, J., 1995. Applied anatomy. In: Bloomfield, J., Fricker, J., Fitch, K. D. (Eds.), Science and Medicine in Sport. Blackwell Scientific Publications, Oxford.

Astrand, P., Rohdal, K., 1988. Textbook of work physiology. Physiological Basis of Exercise. McGraw-Hill, Singapore.

Astrand, P., Rohdal, K., Dahl, H.A., Stromme, S.B., 2003. Textbook of work physiology. Physiological basis of exercise, 4th ed. McGraw-Hill, Singapore.

Conroy, B.P., Earl, R.W., 2000. Bone, muscle and connective tissue adaptations to physical activity. In: Baechle, T.R., Earle, R.W. (Eds.), Essentials of strength training and conditioning. second ed. Human Kinetics, Champaign, Illinois.

Daniels, J., 2001. Aerobic capacity for endurance. In: Foran, B. (Ed.), High performance sports conditioning. Human Kinetics, Champaign, Illinois.

DeLorme, T.L., Watkins, A.L., 1948. Techniques of progressive resistance exercises. Arch. Phys. Med. 29, 263–273.

Greenhaff, P.L., Hultman, E., 1999. The biomechanical basis of exercise. In: Maughan, R.J. (Ed.), Basic and Applied Sciences for Sports Medicine. Butterworth Heinemann, Oxford.

Harris, R.T., Dudley, G., 2000. Neuromuscular adaptations to conditioning. In: Baechle, T.R., Earle, R.W. (Eds.), Essentials of Strength Training and Conditioning. Human Kinetics, Champaign, Illinois.

Hollis, M., Fletcher-Cook, P., 1999. Practical Exercise Therapy, 4th ed. Blackwell Science, Oxford.

Hunter, G.R., 2000. Muscle physiology. In: Baechle, T.R., Earle, R.W. (Eds.), Essentials of Strength Training and Conditioning. Human Kinetics, Champaign, Illinois.

Huxley, A.F., Simmons, R.M., 1971. Proposed mechanism of force generation in striated muscle. Nature 233, 533–558.

Jones, D.A., Round, J.M., 1990. Skeletal Muscle in Health and Disease. Manchester University Press, Manchester.

Lakomy, H.K.A., 1999. The biomechanics of human movement. In: Maughan, R.J. (Ed.), Basic and Applied Sciences for Sports Medicine. Butterworth Heinemann, Oxford.

Lieber, R.L., 2002. Skeletal Muscle Structure, Function and Plasticity. The Physiological Basis of Rehabilitation. Lippincott Williams & Wilkins, Philadelphia.

MacDougall, J.D., Elder, G.C., Moroz, J.R., Sutton, J.R., 1980. Effects of strength training and immobilization on human muscle fibres. Eur. J. Appl. Physiol. Occup. Physiol. 43 (1), 25–34.

MacQueen, I.J., 1954. Recent advances in the techniques of progressive resistance exercise. Br. Med. J. 2, 1193–1198.

MacQueen, I.J., 1956. The application of progressive resistance exercise. Physiotherapy 40, 83–89.

Maughan, R.J., Gleeson, M., 2004. The Biomechanical Basis of Sports Performance. Oxford University Press, Oxford.

McArdle, W.D., Katch, F.I., Katch, V.L., 1996. Exercise Physiology. Lea and Febiger, Philadelphia.

Watham, D., Roll, F., 2000. Training methods and modes. In: Baechle, T.R., Earle, R.W. (Eds.), Essentials of strength training and conditioning. Human Kinetics, Champaign, Illinois.

Wilmore, J.H., Costill, D.L., 2004. Physiology of Sport and Exercise, 3rd ed. Human Kinetics, Champaign, Illinois.

Zinovieff, A.N., 1951. Heavy resistance exercise. Br. J. Phys. Med. (June), 129–133.

Chapter 3

Joint mobility

Tony Everett

CHAPTER CONTENTS

LEARNING OUTCOMES

At the end of this chapter, you should be able to:
1. describe the structure and function of joints
2. discuss ranges of joint movement
3. classify joint movement
4. describe how movement is produced at joints
5. discuss factors that influence normal facilitation and restriction of joint range
6. discuss the causes of abnormal restriction of joint range
7. discuss the rationale for the use of movement to increase joint mobility.

INTRODUCTION

Biomechanically, the body can be considered as composed of segments divided between the axial and appendicular skeleton. There are many models that describe this segmental arrangement; the one that is used in this book divides the body into eight segments. The head and neck and the trunk make up the two segments of the axial skeleton. The other six segments form the appendicular skeleton and are equally divided between the upper and lower limbs; the upper limb consists of arm, forearm, and wrist and hand, while the lower limb consists of the thigh, leg, and foot and ankle. All movements, including locomotion, involve the motion of these bony segments, be they in the appendicular or axial skeleton.

Junctions between these segments are provided by the joints (juncture, articulations or arthroses), which are themselves classified into three groups: fibrous or fixed (synarthroses), cartilaginous (amphiarthroses) or synovial (diarthroses), the last being the only freely moveable joints (Strandring, 2005). It is at these joints that the motion actually takes place. Movements of the segments are produced by forces, mainly the internal forces provided by muscles but also the force of gravity, which is modified by the internal muscle forces acting in opposition.

This combination of the bones that form the core of the segments and the muscles that produce the force that provides movement is described as the musculoskeletal system. It is vital that this musculoskeletal system is intact for functional movement to occur. The role of the muscles producing the movement was covered in Chapter 2 and specific muscles will be covered in Chapters 10–12 of this text, but the vital components of this system, the synovial joints, will be described below.

The major characteristics of a synovial joint include the surface of opposing bones being in contact, but not in continuity, and covered in hyaline cartilage. These bony ends are joined together via ligaments, and the whole joint complex, which may or may not include the ligaments, is surrounded by an extensive synovium-lined fibrous joint capsule. The viscous synovial fluid secreted by this synovial membrane not only provides the articular cartilage with nutrition but acts with it to decrease the coefficient of friction within the joint to a level that is low enough to reduce the possibility of joint surface destruction. Intracapsular structures are usually covered by synovium. Intra-articular discs or menisci may be found within the synovial joint, helping congruity and acting as shock absorbers. Labra and fat pads may also be found within the joints, having the function of increasing joint surface area (and possibly stability) and shock absorption, respectively.

There are a large number of different types of synovial joint, classified according to their shape – for example plane, saddle, hinge, pivot, ball and socket, condylar and ellipsoid (Palastanga et al. 2006) – but movement at all of these joints can be considered as either physiological or accessory. A physiological movement is the movement that the joint performs under voluntary control of the muscles or is performed passively by an external force but still within the available range of the joint. Maitland (1986) defines accessory movements as those movements of the joints that a person cannot perform actively but that can be performed on that person by an external force. They are an integral part of the physiological movement that cannot be isolated and performed actively by muscular effort.

Although there may seem to be a large number of directions in which joints may move, a system of description has been devised to make the visual analysis of movement more simple. From the anatomical position (standing upright with the upper limbs at the sides and palms and head facing forwards), movements can be described as occurring in three planes and around three axes (Fig. 3.1).

The frontal plane splits the body into front and back halves, the sagittal plane splits the body into right and left halves, and the transverse plane splits the body into top and bottom halves. Movements within these planes take place around three axes. The axes can be described as being perpendicular to the plane of movement. Therefore there are two horizontal axes and one vertical axis. The sagittal axis is at 90° to the frontal plane and therefore allows movements within that frontal plane. These movements consist of abduction, adduction, deviation and lateral flexion. The frontal axis allows movements of the segments within the sagittal plane and consists of the movements of flexion and extension. Both frontal and sagittal are horizontal axes. The vertical axis is at 90° to the horizontal plane, and movements around this axis give rotatory motion.

The actual movements performed can be described in terms of the degrees of freedom the joint allows. A uniaxial joint will possess only one degree of freedom, i.e. rotation about only one axis. An example of this is flexion and extension at the elbow. A biaxial joint has two degrees of freedom; for example, the radiocarpal joint has flexion and extension at the wrist about one axis and ulnar and radial deviation about the other axis. The movement available at a multiaxial joint can be described as having three degrees of freedom. This type of movement can be described at the shoulder, where flexion, extension, abduction, adduction and internal and external rotation all take place.

ACTIVITY 3.1

Movements of the body segments are described in terms of their planes and axes. For each of the major joints of the body, describe the planes in which the segments move, followed by the axes that they move around.

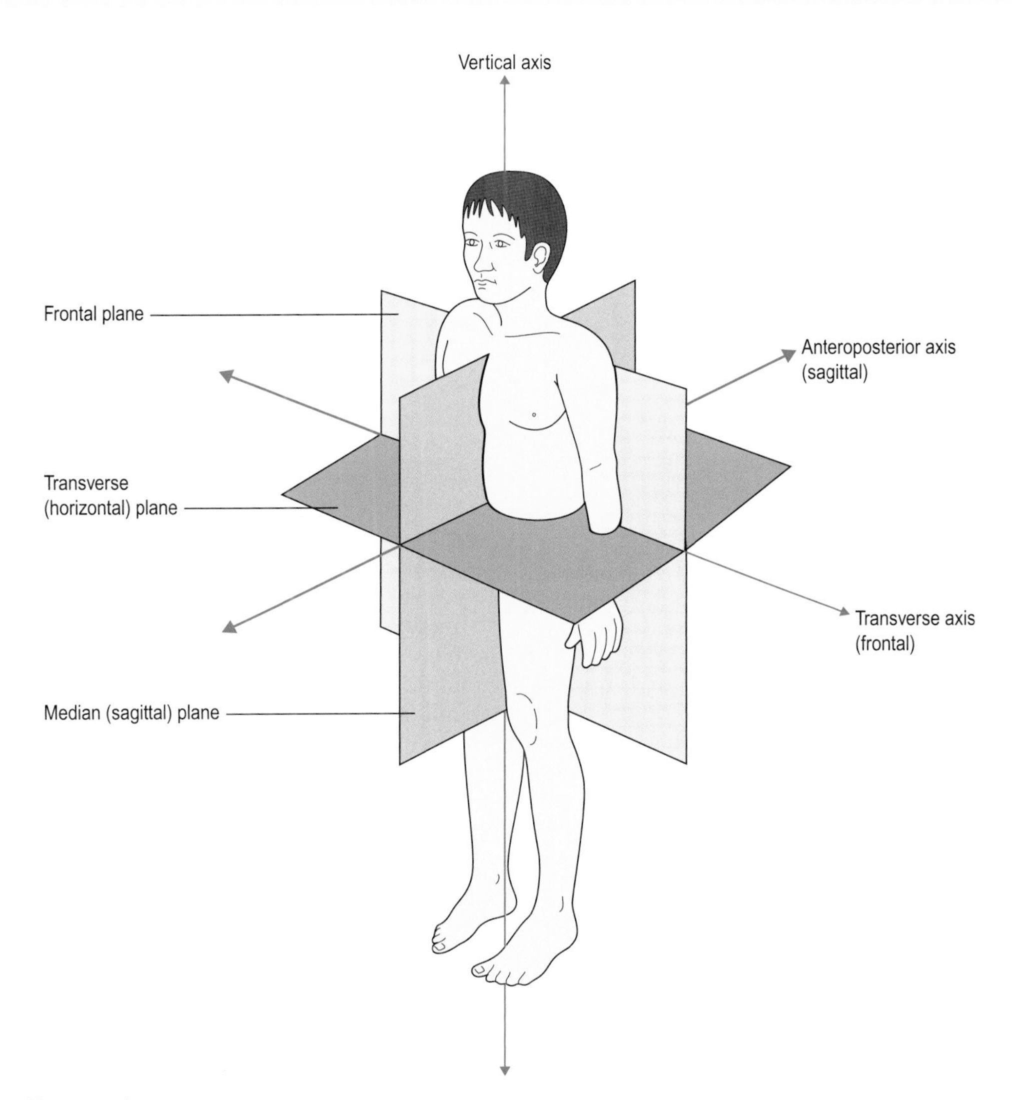

Figure 3.1 Planes and axes.

The above system of movement analysis is truly applicable only if the movement takes place within the anatomical position and does not cross planes and axes. This, however, is not really the case. The shape of most joints is rather complex, and they have axes that permit movement in more than one direction. Most movements are also functional in nature and therefore do not take place from the anatomical position. This makes the analysis system rather artificial.

An alternative system for describing movement is the Cartesian coordinate system, which allows any permutation of movement to be described in the three planes. It has, as shown in Figure 3.2, three coordinates in the directions of:

1. anteroposterior (the x coordinate)
2. mediolateral (the z coordinate)
3. superior–inferior (the y coordinate).

RANGE OF MOVEMENT

With the combination of uniaxial, biaxial and multiaxial joints, the body may adopt a multitude of functional positions. When these positions or movements are analysed, the components are

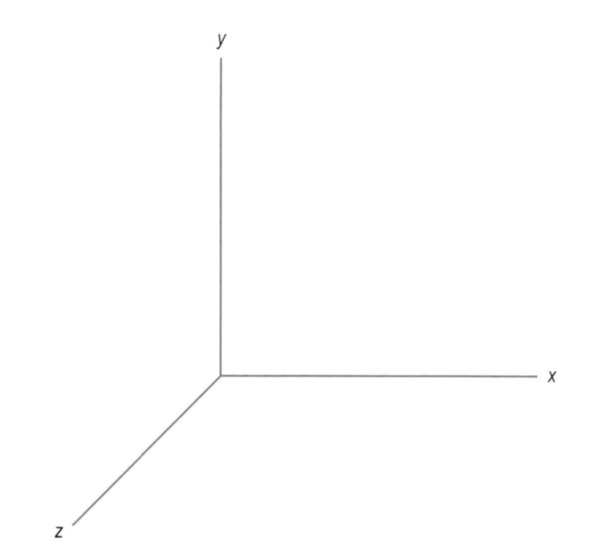

Figure 3.2 The Cartesian coordinate system.

broken down to each individual joint with the range of movement (ROM) of that joint being described. This movement may not be the maximum movement that the joint is capable of achieving, i.e. its full ROM, but only a functional component of it.

Movements of the joints are dependent on many factors, and the description of these factors is largely dependent on the discipline by which the movement is being studied. One such discipline is arthrokinematics. This is the intimate mechanics of the joints and is dependent, for its description, largely on the shape of the joint surfaces. Most of the synovial joints are complex in their formation, having more than one axis within the joint. These joints, being ovoid in different axes of the same joint, have the ability to bring about these different movements. This means that, although joints have roughly reciprocally shaped surfaces, the maximum congruity of the articular surfaces occurs at specific positions within the ROM and does not necessarily equate with the end of range of the physiological movement. This position of maximum congruity is called the close-packed position and is the position of greatest joint stability. At this close-packed position, not only is there most joint surface contact but the ligaments are often taut. The loose-packed position, conversely, is that in which the apposition of the joint surface is the least; muscle, ligaments and capsule are usually lax, and the joint is in its least stable position (Hall, 1995).

Physiological movements always contain different combinations of physiological or accessory movements. These may be a combination of physiological movements, such as side flexion of the cervical spine, which, if examined closely, will be seen to involve both side flexion and rotation in combination. By studying the arthrokinematics of the joint, it can be shown that there is a combination of accessory-type movements occurring during the physiological joint movement. These accessory movements are considered to be of three types: spin, roll and glide. A roll refers to one surface rolling over another, like a ball rolling over a surface. An example of roll is seen when the femoral condyle rolls over a fixed tibial plateau during knee extension. Gliding is a pure translatory movement, one fixed point sliding over the other joint surface. A glide usually takes place in an anteroposterior or mediolateral direction, and this type of movement is again seen when the femur slides forwards on a fixed tibia at the knee joint. Spin is like a top spinning, a pure rotatory motion. These movements are illustrated in Figure 3.3. Accessory movements enable the ROM to be increased at the joint and also maximize the congruency of the joint surfaces to improve stability (Norkin and Levangie, 2005). Descriptions of these movements and their combinations for specific joints can be found in many anatomical texts.

So it can be seen that movement at joints is not the straightforward unidimensional process that it may at first appear. Movement is caused by a force acting on the bony segment, which in turn produces the movement at the joint. This force may be an internal force, the concentric or eccentric work of the muscle, or an external force, the force of gravity for example.

When the joint is moved by the force of muscle contraction, either concentrically or eccentrically, the range of joint movement may be described in terms of the excursion of the muscle. This excursion consists of the full range of the muscle, i.e. the inner, middle and outer range, each being roughly a third of the full ROM. The inner range is where the muscle is at its shortest, the middle range is the middle third of the muscle excursion, and the outer range is where the muscle is at its longest. This is graphically illustrated in Figure 3.4.

The actual range through which the joint moves, either actively or passively, is measured in degrees of a circle. ROM is usually measured by goniometry. This gives an accepted objective measurement

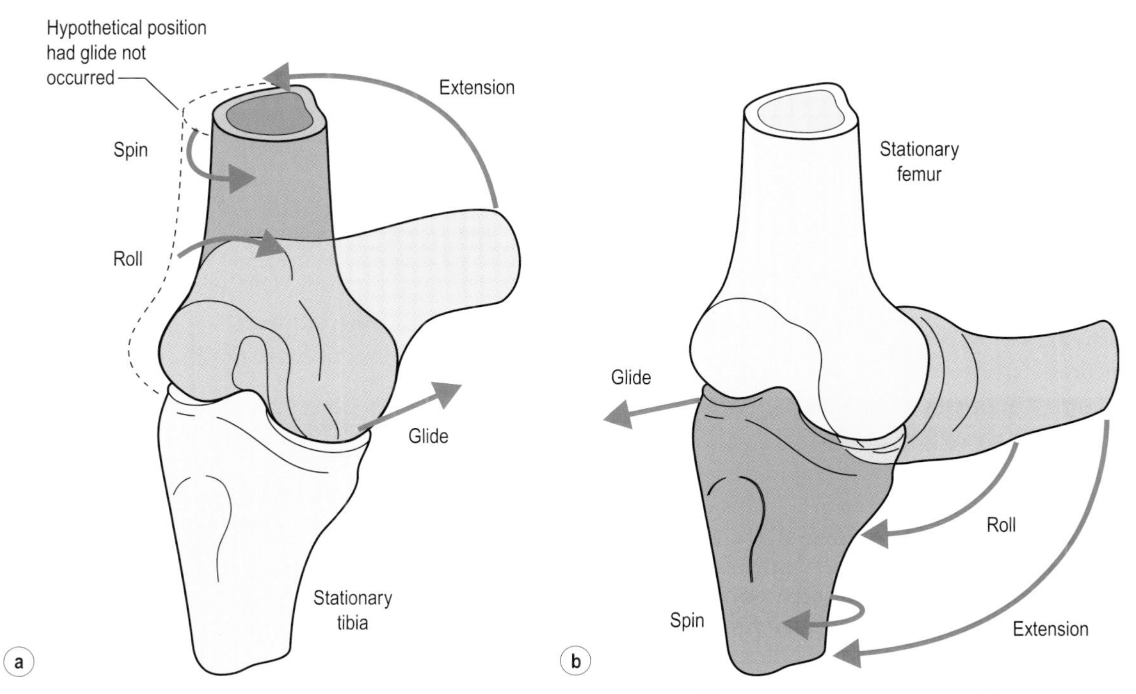

Figure 3.3 Diagrammatic representation of roll, spin and glide at the knee: (a) with stationary tibia, and (b) with stationary femur.

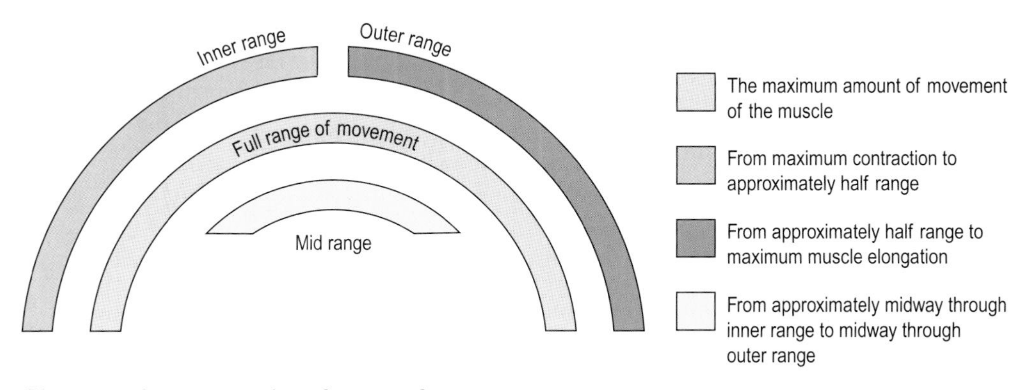

Figure 3.4 Diagrammatic representation of ranges of movement.

that may be used when analysing joint motion as a part of movement analysis or used as an objective marker when assessing patients; see Chapter 14 for a description of goniometry.

The following activity will be useful to reinforce your understanding of the above text. It will deepen your understanding if you could do this task twice, once now and then after reading the visual analysis section of Chapter 14.

ACTIVITY 3.2

Work in small groups and choose some simple activities. One person performs the activity while the others observe the movement very closely. Try to estimate the ROM of each joint and identify the muscles that are working to produce the movement. Describe the ROM of the muscle and the type of muscle contraction throughout the activity.

FACILITATION AND RESTRICTION OR LIMITATION OF MOVEMENT

Anatomical, physiological and arthrokinaesiological factors combine to give normal facilitation to, and restriction of, movement within a joint.

NORMAL FACILITATION

Range of movement is facilitated by the following:

- bony shape
- hyaline cartilage
- capsule supporting synovial membrane
- additional structures
- elastic ligaments
- intact neuromuscular and musculoskeletal system
- muscle strength.

The reciprocal convex and concave shapes of the joint surfaces – combined with the accessory movements of roll, spin and glide – provide a greater surface area over which the two bone ends can move. Hyaline cartilage, present on the articulating surfaces of the bone ends, has the dual function of providing a smooth surface over which the bone ends can glide and affording the joint protection from wear and tear. Both the functions of the hyaline cartilage are enhanced by the presence of synovial fluid within the joint. The fluid layer over the two joint surfaces will reduce the amount of friction between the bone ends when movements occur and act as a form of shock absorber to reduce the trauma of constant impact, particularly in weight-bearing joints. Synovial fluid produced by the synovial membrane also provides the joint with some of its nutrition. Most of the capsule surrounding joints is lax, thus permitting a large ROM by the joint.

Many joints have additional structures within them; examples of these are the menisci in the knee or the glenoidal labrum within the shoulder. Both these structures, and others like them, can act as a mechanism for increasing joint surfaces, increasing congruity or affecting stability within the joint. Occasionally, the ligaments surrounding the joints contain yellow elastic fibres in addition to the usual collagen, enabling the ligament to allow the joint an increase in range by providing more flexibility. An example of this is the ligamentum flavum, which connects adjacent laminae of the spine. It permits separation of these laminae in flexion and ensures that the end range is not reached abruptly. It also assists in returning the spine to the erect position after flexion has occurred (Standring, 2005).

NORMAL LIMITATION OR RESTRICTION OF AN INTACT SYSTEM

Normal limitation of ROM is brought about by:

- articular surface contact
- limit of ligament extensibility
- limit of tendon and muscle extensibility
- apposition of soft tissue.

The two main factors that normally limit joint mobility are the shape of the joints and the type of structures that run over them. Articular surface contact could mean that the joint is in the close-packed position, as in the elbow, where the joint is actually prevented from extending beyond approximately 180° by the olecranon of the ulna impinging on to the humerus. Flexion of the elbow, on the other hand, is limited by the bulk of the biceps brachii pressing against the forearm, this being an example of soft tissue apposition.

Most ligaments and all tendons are primarily composed of white fibrous collagen. One of the properties of collagen is that it is fairly inelastic, and stretching achieved by deformation requires strong forces. Therefore within normal activities, if the ligaments or tendons are at their maximum length, no more movement is possible at that joint. This is illustrated by McMahon et al. (1998), who point out that the inferior band of the inferior glenohumeral ligament is the primary restraint to anterior stability post shoulder dislocation. When considering normal movement, Branch et al. (1995) point out that the anterior and posterior components of the glenohumeral capsuloligamentous complex limit the external and internal rotation of the glenohumeral joint, respectively.

It must be remembered, as O'Brien et al. (1995) describe, that differing positions of joints enable different structures to limit movement. This point is taken a stage further by Warner et al. (1999), who show that glenohumeral compression through muscle contraction provides stability against inferior translation of the humeral head, and this effect is more important than intercapsular pressure or ligament tension. This agrees with Wuelker et al. (1998), who found that the rotator cuff force significantly contributes to stabilization of the glenohumeral joint during arm motion.

The structure of muscle, on the other hand, offers the opportunity of more stretch, as it crosses the joint and thus affords greater mobility, but it still has a limit of extensibility. The limitation that muscle offers to joint mobility is seen particularly when the muscle stretches over two joints, such as the hamstrings stretching over the hip and knee. If the hip is flexed, then the amount of knee extension is limited, as the muscle is already near its maximum possible length. If the hip is extended, then the hamstrings are no longer near the limit of their potential length and will therefore allow for a greater range of knee extension.

It is important that the person analysing the movement, particularly if they are a therapist, becomes aware of the normal limitations and can recognize these by both visual analysis and recognizing how the joint feels at the end of passive range. This is referred to as the *end feel* of the joint and will be different according to the circumstance of the particular joint. For example, the feel of a bony end block, as in the elbow extension, is quite different from that of soft tissue apposition, as in full flexion of the elbow. There is also the springy end feel of normal tendon and ligamentous and other joint structures. It is only by recognizing the normal end feel of joints that the therapist will become skilled at recognizing pathological joint changes.

CASE STUDY 3.1

The whole family are sitting down having their evening meal comfortably around the table. There appears to be no apparent stress to their bodies, and if we analyse the lower limbs of the family we may be able to explain this. The sitting position allows the knees to be in mid range. The thigh is supported, so the muscle work to maintain this position is minimal. The hips and ankles are also at 90° and are well within their physiological ROM, and once again the segments are supported so the muscle work to maintain this position is minimal.

As soon as the meal is over, there is a burst of activity. Chris rushes off to play badminton, and Agnes and Jenny make their way into the lounge to watch the television. Jenny slides off the chair and squats down to pick up the spoon she dropped earlier. Her quadriceps needs to work eccentrically to allow her to lower herself to the floor without collapsing. Her knees will probably be flexed to their full range, stopped only by the soft tissue apposition of her calf against her hamstring muscles. She is able to bounce a little at this position, as her knee joint is not in the close-packed position. Her ankle, on the other hand, has also reached its full range, but there is no bounce available here, as the block to this movement is bony and the ankle seems firmly stopped; in fact, it is in its close-packed position. In fact, Jenny is quite eager to get to the television, as her favourite programme is on, so she does actually bounce from the floor and shoots up and is off to the television. As the foot is locked in full plantar flexion, it acts as a rigid base on which the rest of the lower limb can act. In a full squat, the quadriceps muscles are in their outer range, which makes initiation of the contraction and work through this range quite difficult. With the momentum caused by the little bounce at the end of range and the elastic energy built up during the squat, Jenny is able to almost leap into the air. During this movement, her knees will probably go through their full range. Her knees will be stopped by the contiguation of the joint surfaces, going into the close-packed position. Her ankle will probably not go through full range, but there will be great interplay between the eccentric and concentric muscle work. She does not need to regain her balance before she shoots off at pace for the lounge.

Agnes, on the other hand, has to take much more care when getting up from the table and moving off. She may not be able to achieve full-range extension, as there is the possibility of a degenerative disease within the joint, roughening the joint surfaces at the periphery and making close-packed difficult. The accessory movements, particularly roll and glide, may be difficult, and therefore Agnes will have to compensate for this lack of range, probably by changing the posture of the hips.

John and Liz, on the other hand, clear the table and wash the dishes. John loads the dishwasher; initially, he bends his trunk to place the dishes in the bottom shelf. John finds this difficult as, although his hips are not going through their full range, he feels a tightness and pulling at the back of his thigh and his hips appear to be stuck. John's problem is that he is not bending his knees as he bends forwards, and it is the lack of elasticity in his hamstring that is the problem, as it needs to stretch over a fully extended knee and also over the hip. If John bent his knees, then the hamstring would have a shorter distance to travel and allow greater movement at the hip, making his task easier.

ACTIVITY 3.3

Work with a partner and move the major joints of the body passively to their physiological limits. Identify (by end feel if possible) what is preventing further movement, and remember what the end range feels like. Move to a different partner and perform the same passive movements. See if you can detect any differences and similarities. Discuss your findings with your colleagues.

ABNORMAL LIMITATION

Although both normal facilitation and limitation are important factors to consider when discussing joint range, joint range becomes an issue when there is an abnormal limitation in that range. Abnormal limitation of joint range is usually brought about by either injury to or disease of its structure, surface or surrounding soft tissue, i.e. the muscles producing the movement or their functioning.

The above factors can be summarized as:

- destruction of bone and cartilage
- bone fracture
- foreign body in joint
- tearing or displacement of intracapsular structures
- adhesions or scar tissue
- muscle atrophy or hypertrophy
- muscle tear, rupture or denervation
- pain
- psychological factors
- oedema
- neurological impairment.

Destruction of bone and cartilage

Any disease that destroys the articular cartilage, such as osteoarthritis or rheumatoid arthritis, will impair the functioning of the joint and thus the movement will be limited. This may be for two reasons. Either the destroyed surface will physically prevent the movement, or the pain produced when the two exposed surfaces grind together may lead to a reduction in range or a deterioration in the quality of the movement. Either singly or in combination, these factors may actually prevent movement altogether.

Fracture

A fracture near to or within the joint will also prevent movement via mechanical obstruction or pain. The same applies to a foreign body within the joint complex.

Tearing or displacement of intracapsular structures

Field et al. (1997) demonstrate that recurrent anterior unidirectional shoulder instability is most commonly associated with an avulsion of the glenoidal attachment of the labroligamentous complex (Bankart lesion). This would limit the ROM available. Tearing of the meniscus of the knee is a common example of this.

Soft tissue lesions

If there has been an injury to the soft tissue surrounding or within the joint, then repair to that tissue usually takes place by the formation of fibrous or scar tissue, which does not have the same extensibility as the tissue it is replacing. Fibrous adhesions may also form; these would bind structures together, and hence movement would be restricted.

Injury or immobilization

If immobilization of soft tissue occurs, there are biomechanical, biochemical and physiological changes that take place within 1 week. These changes are magnified in the presence of trauma or oedema (Cyr and Ross, 1998). These structural changes are a result of stress deprivation, which causes the matrix of the tissue to be remodelled to its new resting length while being held immobile (Hardy and Woodall, 1998). The net result of this will be a decrease in the ROM. If muscle tissue is held in a shortened position, there appears to be absorption of the sarcomeres, causing a change in length. This shortening in length is termed *adaptive shortening* and will limit joint movement.

Muscular changes and pathology

The joint itself may be intact, but if the muscles that produce the movement have a dysfunction then the net result is a decrease in ROM. If the muscle is atrophied to a large degree, then it would not create sufficient force to move the joint through its full range. Conversely, if there was a large amount of muscle hypertrophy, ROM would also be decreased because of the increased amount of soft tissue apposition.

Neurological impairment

The muscle itself may be intact, but its neural control may be impaired. This impairment could range from total denervation, causing flaccidity of the muscle, to lack of higher centre control, which may cause spasticity. Local spinal reflexes may also have the effect of limiting movement by causing the muscle to be in spasm.

Pain

The body's response to pain is usually to keep the part still and avoid movement. This may be only short term, but if the pain becomes chronic then adaptive shortening may occur. The pain may disappear, but the pattern that the brain adopted as a result of the memory of pain that occurred on movement may continue. Other psychological problems, such as depression or lack of motivation or self-confidence, may also be responsible for the subject not moving. This may also be transient and cause no physical limitation of movement, but if the condition persists then adaptive shortening may occur.

Hypertrophy

When a muscle is overdeveloped, as in a bodybuilder or sportsperson, for example, soft tissue apposition may cause a decrease in ROM.

Hypermobility

It is important to remember that what has so far been discussed describes a decrease in movement (hypomobility), but the opposite may also occur. This is termed *hypermobility* when the ROM exceeds that of the expected physiological range. This could be caused by pathological change either at the joint or elsewhere within the musculoskeletal or neuromuscular systems. It may, however, be a natural phenomenon caused mainly by laxity of ligaments or a congenital joint deformity, and it can also result from a deliberate attempt to stretch the joint range well beyond that which is functionally acceptable, as in a gymnast or ballet dancer, for example. As Lewit (1993) states, this may be an advantage to these sportspeople, but with increased mobility there may be a decrease in stability, with the disadvantage of possible problems in the future. There is also the possibility of subluxation of the joint occurring during movement, which could result in neurological damage. If the joint is hypermobile, there is also the possibility that the joint will articulate on bone that is not designed for this function, and therefore there is a great risk of degenerative changes occurring at the joint surfaces.

EFFECTS OF DECREASED RANGE OF MOVEMENT

The effects of a decreased range of joint movement will obviously depend on how much the range is decreased and the importance that joint plays in functional activity. Limited flexion at the knee, for example, will have a major effect on essential functional activities such as toileting and walking. Limitation of ROM at one of the metatarsals, however, may have little obvious major effect on any functional activity, even gait. The human body is very adaptable, and decrease in the range of one joint is sometimes compensated by hypermobility at another joint.

TREATMENT

Before treatment can be given for any decrease in ROM, it is obvious from the above that the cause of the decrease will have to be known. It has been shown that the cause may be in the joint structure (surface or intracapsular), the structures surrounding or running over the joint (ligaments or tendons), or the neuromuscular system that produces the movements. So, to establish the pathological changes that have occurred, it is vital that the therapist performs a full and detailed assessment. Once the pathology is known, the therapist can choose a method of treatment whose physiological effects alter the pathological changes that have occurred to limit joint movement.

Once the assessment has been made, it is important to know the physiological effects of the possible treatment options and match them with the effects they will have on the pathological changes. This is the rationale of the treatment.

Limitation of movement, from whatever cause, impairs function of the joint and the muscles producing the movement. Measures that increase the ROM must also include methods that strengthen the muscles in their new, lengthened position. The degree of ROM gained must be able to be controlled, and the stability of the joint maintained, or further injury may result.

It is important that details of the anatomy and arthrokinaesiology are understood as well as the pathology of the joint, as these have an effect on the rationale of treatment choice when there is a pathological reduction in joint range.

Many studies have looked at actual joint ranges of physiological ROM and presented tables of values (Norkin and White, 2003). It is accepted, however, that each ROM is specific to each individual, and discrepancies may even exist when comparing both sides of an individual. Two of the factors that have an effect on the ROM are age and gender. Younger children appear to have a greater amount of hip flexion, abduction and lateral rotation, and ankle dorsiflexion than an adult. Elbow movements are also greater than those of an adult, while there is less hip extension, knee extension and ankle plantar flexion. Older age groups seem to have a generalized appendicular and axial joint decrease. Connective tissue may become stiffer, but lack of activity through full range may be a major cause (Robergs and Keteyian, 2003). Gender appears to have different effects on different joints depending on what movement that joint is performing (Norkin and White, 2003).

There are many physiotherapy modalities to increase the ROM. The most obvious is the use of movement itself. The main classifications of the therapeutic movements are described below.

CASE STUDY 3.2

Chris is a keen badminton player and is at the sports centre engaging in a tough match with his friend from university. In the past, Chris has had an injury to the soft tissue of his right shoulder. The injury is well healed now, but Chris is left with a slightly restricted range of shoulder flexion. Chris is right-hand dominant. This limitation of movement makes it a little difficult for Chris to reach behind his body when returning an overhead shot. Chris's opponent is a regular playing partner and is aware of this problem so increases the amount of overhead shots Chris is forced to return. There are two ways in which Chris can compensate for this slight decrease in range. First, as Chris is reaching his full flexion he can rotate his spine to the right. This will reposition Chris's upper limb so that it is taken further back. Second, Chris could run faster to the back of the court so that the shuttle would not be so high and he could hit it slightly in front of him. Neither of these actions is ideal; the former is energy inefficient and would eventually put considerable strain on Chris's spine. The lateral requires a more fitness, and the stroke Chris would play would not be as good as if he hit it above his head.
Careful analysis of Chris's play would lead you to testing and treating the decrease in movement at the shoulder. This may solve the problem of Chris's poor form of late!

ACTIVITY 3.4

In groups, look at the major joints of the body and assess visually the differences in the ROM between each person. Can you discover what is limiting the movement for each joint? Can you find anyone with hypermobility? Are there any differences between men and women in ROM or hypermobility?

TYPES OF MOVEMENT OF JOINTS

Movement is one of the main methods that therapists use to increase joint range. The therapist will use the different ways in which the joint moves as a basis for these different methods. There are two main types of movement:

1. passive movement
2. active movement.

Passive movement is defined as those movements produced entirely by an external force, i.e. with no voluntary muscle work. These can be subdivided into:

- relaxed passive movements
- stretching
- accessory movements
- manipulations.

Active movements are those movements within the unrestricted range of a joint produced by an active contraction of the muscles crossing the joint. These can be subdivided into:

- active assisted exercise
- free active exercise.

PASSIVE MOVEMENT

Relaxed passive movements

These are movements performed within the unrestricted range by an external force and involve no muscle work of the particular joint, or joints, at which the movement takes place. These movements can be performed in three ways.

1. Manual relaxed passive movements are movements performed by another person, usually the physiotherapist, within the unrestricted range.
2. Autorelaxed passive movements are performed within the unrestricted range by the patients themselves, i.e. with their unaffected limbs.
3. Mechanical relaxed passive movements are performed by a machine but still occur within the unrestricted range.

Manual relaxed passive movements

Relaxed passive movement has been a core skill of the physiotherapist for many years and is still widely used today. As with many of the traditional skills, little evidence of its clinical effectiveness has been published (Basmajian and Wolf, 1990), but the following is the accepted rationale for its use.

Indications

These movements are indicated when the patient is unable to perform an active full-range movement. The reasons for this inability may include unconsciousness, weak or denervated muscle, spinal injury, pain, neurological disease or enforced rest.

Effects

- Maintain ROM
- Prevent contractures
- Maintain integrity of soft tissue and muscle elasticity
- Increase venous circulation
- Increase synovial fluid production and therefore joint cartilage nutrition
- Increase kinaesthetic awareness
- Maintain functional movement patterns
- Reduce pain

Maintaining ROM and preventing contractures. If muscle is not moved through its full range, then it will adapt to the demands being placed on it. The actin and myosin protein filaments (the contractile element) will be reabsorbed and thus the area for cross-bridge formation will be decreased. This will cause muscle weakness and the inability to perform the movement. The muscles will adopt the new position and will be shortened, i.e. they will have adaptive shortening. The non-contractile elements within the muscle, the connective tissue, will add to this effect by altering the collagen turnover rate, which is the balance of collagen production and destruction (Basmajian and Wolf, 1990). If more collagen is produced, as in immobilization, it increases the stiffness of the muscle and decreases its propensity to stretch. If no movement takes place, then the muscle will adapt to the new position and thus contractures will occur.

Other soft tissues, such as ligaments and tendons, will also be similarly affected. As they have a greater proportion of collagen, the increase and change in consistency will also lead to stiffness and eventually to contractures.

Maintaining integrity of soft tissue and muscle elasticity. By placing stresses on these tissues, the collagen turnover rate is normalized and the elasticity of the tissues is maintained. As Cyr and Ross (1998) conclude, early controlled motion is vital to prevent the negative effects of immobilization and maintain normal viscoelasticity and homeostasis of connective tissue. Passive movements will not, however, increase the strength of the muscle, as this requires the greater physiological demand of active and resisted work.

Increase venous circulation. If a limb, particularly the lower limb, is not moved, venous congestion may occur. This is because the muscle pump does not work to aid venous return. Pooling occurs and is increased through dilation of the vessels caused by the physical pressure of the blood on the veins and the possible lack of sympathetic tone. This decrease in flow can lead to an increased risk of deep vein thrombosis. Passive movements will act as a prophylaxis to prevent stagnation. This is achieved by physically compressing the veins, and one-way flow is achieved via the valves within the veins themselves. Lymph is also encouraged to move. Compression of the tissues increases the hydrostatic pressure and thus encourages tissue perfusion and fluid reabsorption. This may be useful in reducing oedema.

Joint cartilage nutrition. If the synovial fluid is swept over the articular cartilage, it will provide nutrition and help prevent deterioration of the surface. Production and absorption of the synovial fluid by the synovial membrane are stimulated by movement of the joint. This is quite an important effect and is lost if the joint is immobile for any length of time. If the immobility is caused by injury, then movement is of greater importance, as one of the consequences of injury is inflammation and repair by fibrosis, which in itself will increase the risk of adhesions forming within the joint. Therefore, with passive movement, this risk will be decreased.

Kinaesthetic awareness. To perform coordinated, energy-efficient and safe movement, it is important for the central nervous system to receive information about the position and movement of the joints and soft tissue. This information is supplied by sensory nerve endings in the many structures in and surrounding the joints. This kinaesthetic awareness may be lost if there is a long period of immobilization. The kinaesthetic pathways may be maintained when performing passive movements by stimulating the nerve endings within the joint complex.

Maintain functional patterns. The brain is said to recognize gross movement patterns, so if these patterns cannot occur then the memory of the pattern may be lost. Passive movement in these patterns will decrease this risk.

Reduce pain. Rhythmical movements are said to reduce pain by causing a relaxation effect within the muscles (Gardiner, 1981). This may partly be achieved by removing waste products and chemical irritants from the area through increased circulation. Stimulation of the joint mechanoreceptors may also subserve the sensations from the pain nerve endings and thus decrease their effect (see the section on accessory movements, below, for further explanation).

Contraindications

- Immediately post injury, as this may increase the inflammatory process.
- Early fractures, when movement may cause disruption of the fracture site.
- When pain may be beyond the patient's tolerance.
- Muscle or ligament incomplete tears when further damage may occur.
- When the circulation may be compromised.

Principles of application

Passive movements may be performed either in the anatomical planes or in functional patterns. The choice and type of movement will depend on the findings of the assessment and the aims of the treatment. The same basic principles of application need to be considered whichever movement is chosen, as follows (after Hollis, 1989).

- The segment should be comfortable, supported and localized to the specific joints.
- The patient should be comfortable, warm and supported.
- Handholds should support the segment and protect the joint.
- The motion should be smooth and rhythmical.
- Speed and duration should be appropriate for the desired effects.
- Range should be the maximum available without stretching or causing pain.
- Segments should be positioned so that muscles that stretch over two or more joints are not restricting joint range.

Autorelaxed passive movements

Although the rationale is the same, the method of application must be modified. Patients who have to perform their own passive movements are usually those with a long-term problem. People with spinal injuries, for example, must retain their joint range and muscle length if they are to perform the functions necessary for daily living (Bromley, 2006).

Mechanical relaxed passive movements

Unlike manual or autopassive movements, which, by their nature, have to be carried out intermittently, mechanical passive movements may be carried out continuously. Mechanical devices for producing continuous passive movement were first used by Salter in 1970 (McCarthy et al. 1993). Although their designs and protocols of use may differ, they all have essentially the same function.

The rationale is the same as for any relaxed passive movement, but the benefits of continuous movement are particularly evident following surgery (Kisner and Colby, 2007). Basso and Knapp (1987) found that continuous passive movement decreased joint effusions and wound oedema while increasing ROM and decreasing pain in post-operative knee patients.

Stretching

Stretching differs from relaxed passive movement in that it takes the movement beyond the available range. This available range may be limited because of disease or injury. Stretching may also take the joint beyond the normal physiological range.

Whereas relaxed passive movements are designed to maintain length of soft tissue and hence joint range, stretching should result in a change in length of the soft tissue structures crossing over the joint, with the consequent increase in joint range. Passive stretching is not the only method of increasing joint range via the soft tissues. This can also be attained via active stretching, which will be discussed later.

Stretching of biological material

Most biological materials are viscoelastic. This means that they exhibit both viscous and elastic properties. Viscosity is the property of a fluid that is a measure of the resistance to flow. Elasticity is the property of a solid. Therefore, viscoelastic materials possess both solid and liquid properties, which means that stress is not the only function of strain. There is also a strain rate. This differs from most of the purely elastic materials discussed in Chapter 9, in which there is no time dependency, i.e. how quickly the stress was applied to the material. This time dependency is described in the equation below

$$E = de/dt$$

where E is the strain rate and t is the time.

As will be discussed in Chapter 9, a useful way of describing the relationship between stress and strain is to plot a graph of the stress versus strain and discuss the resulting curves.

Loading and unloading paths

For elastic materials, strain energy is stored within the substance as potential energy. When the load is released, it is this energy that returns the material to its original stress. This is represented graphically in Figure 3.5.

For viscoelastic materials, some of the strain energy is stored as potential and some is dissipated as heat. Therefore once the applied load is removed, there is not enough stored energy to regain the normal configuration. This is shown in Figure 3.6.

Within the enclosed area shown in the graph is the hysteresis loop, which represents the energy dissipated as heat when the material is stretched and the stress released, allowing a return to the non-stressed condition. Therefore continual loading and unloading will produce heat. The amount of hysteresis (heat produced) is dependent on the strain rate. This loss of energy does not allow the material to return to its original state, hence permanent deformation has taken place.

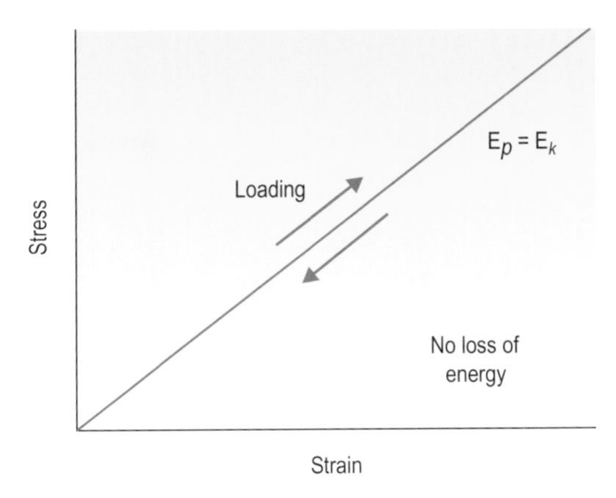

Figure 3.5 Loading and unloading paths (where $E_k = E_p$).

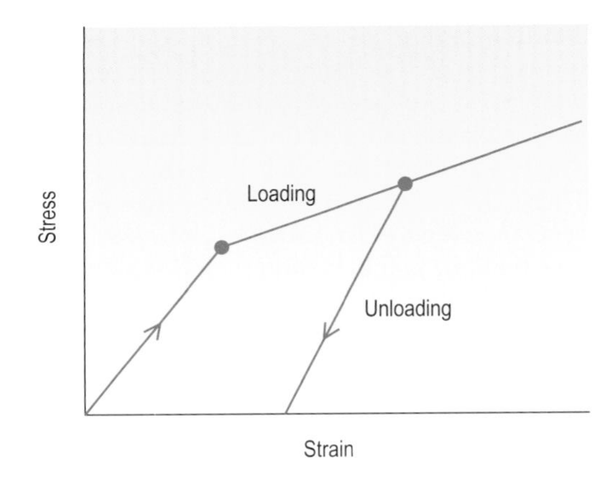

Figure 3.6 Elastic materials exhibiting hysteresis.

When considering biological tissue, it is important to know the micro- and macrostructure of the tissue concerned. Not many biological tissues are pure, as most are a composite of materials. This may mean that they have fibres in an aqueous matrix (which therefore makes them viscoelastic), or that they have a combination of different fibres, or both. Orientation of the composite fibres is also of importance. If the fibres are arranged in parallel and the stress is applied in the direction of the fibres, then the material will be strong and possibly stiff (a measure of the resistance to stress). If there is irregular orientation of the fibres, then there will be less strength but what strength there is will be multidirectional.

The stress–strain curve for a 'general' biological tissue can be seen in Figure 3.7. Although not presenting the graph of any particular tissue, it shows the characteristics that are common to most biological tissues.

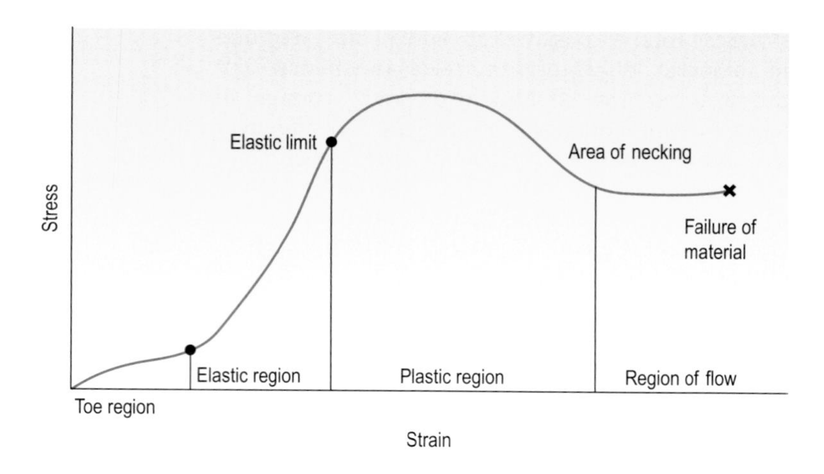

Figure 3.7 Typical stress–strain curve for biological material.

The toe region of the graph represents the straightening of the wavy collagen. It represents no change in the structure of the tissue under stress. The elastic range is the area under the graph that obeys Hooke's Law in that the tissue will return to its original length once the tension is released (between elastic region and elastic limit). The elastic limit is that point beyond which the tissue does not return to its original length when the stress is removed. Plastic range refers to that area in which the tissue will undergo permanent deformation and will not return to its original position. The strength of the tissue at this point is referred to as its yield strength, whereas the ultimate strength is the greatest load the tissue can sustain before strain occurs without further stress. There is a point, which is usually greater than the ultimate strength, at which necking occurs. Necking is when considerable weakening occurs and strain continues to increase even if the stress or loading is greatly reduced or even removed. At the point of failure of the tissue, it has reached its breaking strength, i.e. the load at the time the tissue fails and rupture occurs.

Another property of tissue that is important when considering stretching is its stiffness, which is a measure of the resistance offered by the tissue to deformation. The stiffness of a tissue is often rate- and speed-dependent. The ductility of the tissue is its capacity to absorb plastic deformation before failure occurs. If the tissue has an increase in strain with a constant stress, then this increase in length is referred to as creep. This phenomenon is often used in serial splinting, when the tissue is held in a cast under constant load and over time the tissue undergoes further lengthening. The resilience of material is its ability to recover quickly from its deformation, whereas damping refers to the slow return to shape (Soderberg, 1997).

Bone

Bone is non-homogenous, therefore it will vary in its response to stress. It is a composite of compact and cancellous material, and during loading compact bone is seen to be stiff, with a high ultimate strength and a large modulus of elasticity (see section on *Deformation of materials* in Ch. 9). It can resist rapidly applied loads better than loads that are applied slowly. Cancellous bone, on the other hand, is more compliant, hence it has greater shock-absorbing capacity. These properties are shown in Figure 3.8, in which the different amounts of stress needed to produce the same amount of strain and

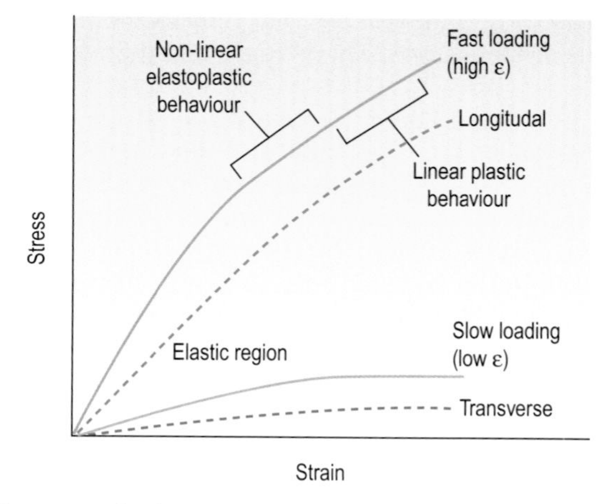

Figure 3.8 Strain rate-dependent stress–strain curves for cortical bone under longitudinal stress (continuous line) and direction-dependent stress (dotted line).

ultimate failure are shown for loads in the longitudinal and transverse directions. The graphs of fast and slow loading also illustrate the time-dependent response of the bone to stress (Fig. 3.8).

Soft tissue

Most soft tissue found in the musculoskeletal system is a composite material of mainly collagen, elastin and the aqueous ground substance. The collagen is composed of crimped fibrils that are aggregated into fibres, and its prime function is to withstand axial tension.

On stretching, the crimps straighten out and the collagen then stores the potential energy that returns the fibril to the original position. As the collagen is surrounded by fluid (gel-like ground substance), it also possesses fluid properties of creep and hysteresis.

Elastin is highly elastic even at high-stress strains, i.e. it possesses a low modulus of elasticity, whereas that for collagen is high. The importance of these properties is illustrated in Figure 3.9, which shows the stress–strain curves for the ligamentum flavum (70% elastin) and the anterior cruciate ligament (90% collagen).

Tendons

Fibrous connective tissue is mainly composed of collagen and ground substance, functioning primarily as a passive transmitter of the force produced by muscle contraction. Compared with muscle, the tendon is stiffer, has higher tensile strength and can endure larger stresses. The tendon can support a large stress with only a small strain and thus makes muscle contraction more efficient, as not much of the muscle action is wasted on movement of the tendon. This facilitates greater apposition of the bones. The properties of a tendon are dependent on the type and proportion of its fibres, as shown in Figure 3.10. The resultant strain of a tendon is also time-dependent and the tissue exhibits a hysteresis loop, showing that deformation of the tendon will be permanent. This can be seen from Figure 3.11. Tendons have a low shear modulus, so they can act as pulleys to redirect high forces.

Ligaments

Most ligaments contain a greater proportion of elastin than tendons and therefore have higher flexibility and lower strength and stiffness than

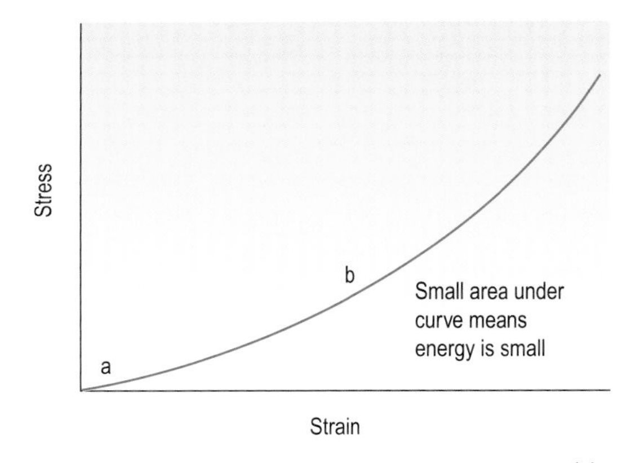

Figure 3.10 Stress–strain curve for a typical tendon: (a) low strain (elastic fibres dominate and crimping straightens) and (b) stiffer (viscoelastic matrix takes over).

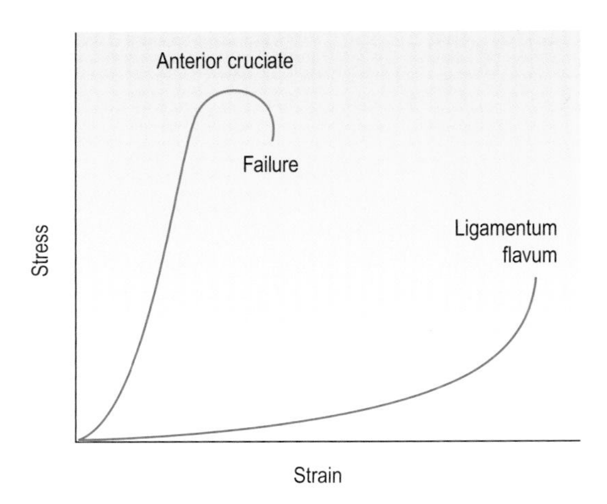

Figure 3.9 Stress–strain curves for anterior cruciate ligament and ligamentum flavum.

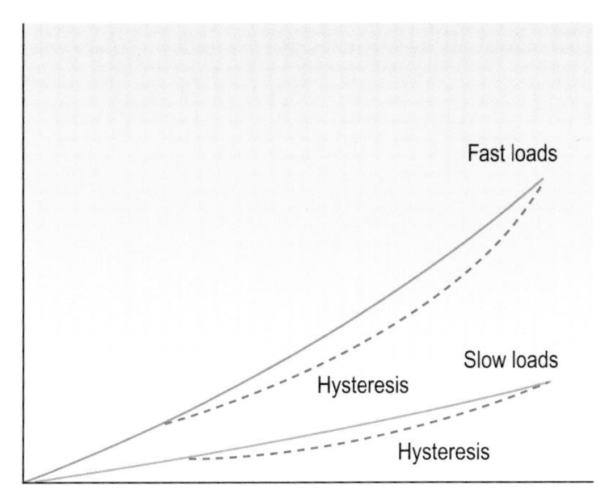

Figure 3.11 Stress–strain curve showing rate dependency and hysteresis for the tendon.

tendons. Sometimes, however, ligaments have greater strength if resistance is applied quickly.

Muscle

Muscles vary in their reaction to stress, depending on which muscle is being studied and the age of that muscle (see Ch. 13). The structure of the muscle with its attendant tendon also has a role in the strain that occurs when stress is applied to the muscle–tendon–bone complex (Soderberg, 1997).

Muscle, however, is an active tissue, and the force produced within the muscle is caused by its contraction. Active tension is developed in the muscle as a result of cross-bridge formation, and passive tension is developed as a result of the stress placed on the connective tissue elements of the muscle when it is stretched past its resting length. Active tension is decreased, however, as the cross-bridges are pulled apart. Therefore as muscle lengthens, passive tension increases.

Cartilage

Cartilage has a high fluid content, so it has greater viscoelastic potential and becomes very variable in its properties.

Therapeutic stretching

Therapeutic stretching of soft tissue is a passive movement and is an important physiotherapeutic skill that can be carried out for a variety of reasons. The usual indications for performing stretching are to regain ROM or to facilitate an increase in the available ROM.

Changes in collagen that affect the stress–strain response

Immobilization

During immobilization, there will be a decrease in the collagen turnover rate with a consequent weak bonding between the new, non-stressed fibres. There will also be adhesion formation, with the consequent greater cross-linking between disorganized fibres, and ground substance will not retain its viscous properties. Decrease in normal activity will result in a decrease in the size and amount of collagen fibres, leading to weakened tissues and possibly an increase in elastin, with the consequent increase in compliance.

Effects of age

One of the natural consequences of the ageing process is the decrease in tensile strength and elastic modulus of all soft tissues. This may lead to a decrease in the rate of adaptation, with the increased chance of overuse injury, fatigue and trauma (Cribb and Scott, 1995).

Effects of drugs

Corticosteroids cause a long-term decrease in tensile strength.

All the above effects must be taken into consideration when stretching is performed.

As has been described, stretching of collagen is caused mainly by plastic deformation, as the material is fairly inelastic. Connective tissue will reorganize itself in response to a sustained stretch provided the stress is not too great or applied for too long (Basmajian and Wolf, 1990). Therefore any stretch that does occur will be fairly permanent. It can be seen, then, that care must be taken not to reach the necking phase, as this would compromise the integrity of the tissue being stretched and affect its function. This could lead to a decrease in stability and a greater risk of further injury to the joint.

To summarize, the effects of stretching are to:

- increase joint ROM
- increase soft tissue length
- relieve muscle spasm
- increase tissue compliance in preparation for an athletic event (Vujnovich and Dawson, 1994).

Performing a stretch

Passive stretching can, like passive movements, be performed as manual, auto- or mechanical stretching. In manual (or auto-) stretching, the therapist (or patient) produces the sustained stretch at the end of range with the patient relaxed. It is generally accepted that the stretch should be held for at least 15 s and repeated several times. When stretching a muscle tendon complex, it is important that the sarcomeres have an opportunity to give before the tendon can be stretched. Sarcomere give occurs when the force applied to the muscle is sufficient to separate the actin and myosin filaments so that the non-elastic components of the complex can be stretched. Mechanical stretching involves a low stress being applied by a machine over a prolonged period.

The above discussion refers to static stretching, which is defined as a slow sustained stretch at the end of range. This should be held, as stated, for up to 15 s, but it is possible that a patient will be unable to tolerate a stretch held for this long, and therefore the duration may have to be built up from a few seconds. This type of passive stretching is in contrast to ballistic stretching, which is defined as small end of range bounces. Despite having been used for many years, the ballistic method of stretching has been shown to increase the risk of injury and should not be performed (Vujnovich and Dawson, 1994).

Accessory movements

As described earlier, Maitland (1986) defines accessory movements as those movements of the joints that a person cannot perform actively but that can be performed on that person by an external force. Although these movements form an integral part of the normal physiological movement and occur throughout the ROM, they cannot be physically isolated by patients themselves. However, if the accessory movement was being performed by the physiotherapist, the patient would be able to stop the movement from taking place. These movements take the form of glides (medially, laterally or longitudinally), compressions, distractions or rotations.

Anatomists group accessory movements into two categories: type 1 are those that cannot be performed unless resistance is provided to the active movement, and type 2 are movements that can be produced only when the subject's muscles are relaxed (Strandring, 2005). The latter are the type used for therapeutic purposes. Accessory movements are used to increase the ROM at joints and also to decrease any pain that is present. These benefits are in addition to those of passive movements. The mechanism for decreasing pain depends, of course, on the cause of that pain. If the pain is caused by the decrease in ROM, then the rationale of treatment will be the same as that for increasing the range. If not, then the usual effect of small-amplitude movements, applied rhythmically to the joints, is to inhibit the afferent impulse traffic from articular receptors, thus blocking the pain (Grieve, 1988).

Twomey (1992) suggests that articular cartilage facilitates the ROM of joints. If there is joint immobility, the articular cartilage will degenerate more quickly, as it requires movement and loading to ensure adequate nutrition. This nutrition is facilitated by the synovial fluid, which is swept over the joint surface. As the stimulation of the synovial membrane declines, the amount of synovial fluid produced is decreased, becoming thicker with decreased osmolarity. Accessory movements assist in the maintenance of synovial production and, by the small oscillatory movements they make, produce the washing effect of the synovial fluid over the joint surfaces. Twomey (1992) also suggests that movement is important for the nutrition of all collagenous tissue as well as the prevention of adaptive shortening. Small-amplitude movements at the end of range will elongate connective tissue, ligaments, joint capsule and other periarticular fascia via the processes described in the previous section dealing with stretching (*Performing a stretch*). End range movements will also break down any intraarticular adhesions that have been formed (Grieve, 1988).

If joint stiffness is caused by fibrocartilaginous blocks and subsequent connective tissue shortening, Threlkeld (1992) claims that the relative positions of the joint surfaces can be altered to regain normal accessory movement and restore displaced material.

Manipulation

Joint manipulations differ from all other types of therapeutic movement, because the patient has no control over the procedure. These movements are potentially dangerous and should therefore be performed only by skilled professionals with much experience in the mobilization of joints. Manipulations are small-amplitude forceful movements that take the joint past the available physiological range. As Maitland (1986) states, manipulations are performed very quickly before the subject has time to prevent the movement from taking place. Lewit (1993) describes manipulation as a technique for treating end range blocking of joints. Displaced intra-articular material may be one reason for the restriction of movement. Joint misalignment may be another. This blocking or restriction of movement, he claims, has two effects: one is the restriction of the subject's functional movement; the other is the effect on the accessory movement or joint play. Restrictions caused by meniscal or other material within the joint are termed *loose bodies*.

Lack of movement of a joint may lead to adaptive shortening of the soft tissue structures surrounding it: capsule, ligaments, tendons or muscles, for example. This secondary consequence will in itself cause a restriction of movement and possibly pain when movement is attempted. Thus a vicious circle is set up. The initial immobility may be a consequence of pain. If the pain is caused by trauma to the joint, then the problem may be compounded by the presence of adhesions within and around the joint that are the natural consequence of the inflammatory process.

Shortened soft tissue structures may also be manipulated to physically break the structures.

Effects

The primary effect of manipulating the joint is to restore its mobility. A secondary effect may be to decrease pain, if the pain was a direct result of abnormal tension on structures caused by incorrect functioning of the joint.

These effects are brought about in a variety of ways. As the joint is manipulated, space is created between the surfaces. The movement of the surfaces, together with the greater space created, possibly causes any physical obstruction between the joint surfaces to be moved clear. Joint surfaces may then be realigned, causing the correct afferent information to be sent to the spinal cord. Fibrous adhesions resulting from the inflammatory exudate, the organization of the synovial fluid, or the adaptive shortening of any soft tissue structure will be physically torn. This takes the structure through the plastic phase and very rapidly past the breaking point. The intended consequence of this procedure is to free the joint to perform its full functional excursion.

Contraindications for the above technique are the same as those for relaxed passive movements.

ACTIVE MOVEMENT

Active movement can be thought of as movements of the joint within the unrestricted range that are produced by the muscles that pass over that joint. The movements may be active assisted or free active. Therapeutically, these movements are performed as exercise.

Active assisted exercise is exercise carried out when the prime movers of a joint are not strong enough to perform the full ROM. The forces that need to be overcome are friction, gravity and the effects of the mechanical disadvantage of lever length. Assistance may be given by any external force, but it is important that the external force provides assistance only so that the movement is simply augmented and does not become a passive movement. The external force has to be applied in the direction of the muscular action but not necessarily at the same point.

External assistance may be:

- manual assistance
- mechanical assistance
- autoassistance.

Manual assisted exercise is when the subject's muscular effort is assisted by the therapist. The therapist changes the assistance as the muscles progress through their ROM, compensating for such factors as angle of pull and length–tension relationships. The amount of assistance may also be changed as the muscle strength increases.

Mechanical assistance may be provided by a variety of apparatus. Isokinetic equipment has the facility for active assisted movement provided the trigger forces are set low enough. The most useful mechanical assistance is sling suspension, in which the assistance is given in two ways. First, the resisting force of friction is reduced by physically lifting the segment clear of its resting surface, which also helps to counteract the force of gravity by supporting the segment. Second, depending on the point of fixation of the sling suspension, gravity may be used to assist the movement, provided the desired movement occurs on the downward arc of the curve produced by the segment within the sling suspension.

Autoassisted exercise may be performed by the subject using the same principles as those of manual assistance. More common, however, is for the assistance to be a combined form of auto- and mechanical assistance. This is seen in the use of bicycle pedals for lower limb mobility or pulleys for upper limb mobility.

Free active exercise differs from assisted exercise in that the movement is carried out by the subject, with no assistance or resistance to the movement except that of the force of gravity. There are many ways in which free active movements can be performed.

- Rhythmical: this uses momentum to help perform the movement taking place in one plane but in opposite directions.
- Pendular: these are movements performed in an arc and are useful for improving mobility as, on the down-curve of the arc, the movement is assisted by gravity.

Single or patterned

Depending on the aims of the intended movement, the choice of single or patterned movements is made. As a general rule, single movements are used to demonstrate or restore actions, whereas patterned movements are used for functional activities. The use of biceps brachii to flex the elbow as a pure movement is an example of a single movement. Reaching out to pick up an object (food) and taking it to the mouth also involves flexion of the elbow, but this movement also uses other joint movements in a functional pattern (the feeding pattern). These movements may also be classed by their effect.

- Localized: designed to produce a local or specific effect, such as mobilizing a particular joint or strengthening a particular muscle.
- General: gives a widespread effect over many joints or muscles; running, for example (Gardiner, 1981).

Exercise affects all the systems of the body and is covered elsewhere in this and many other textbooks. What must be remembered, however, is that the effects produced cannot be isolated to one particular system or even one effect within that system. Exercise, for example, will maintain muscle length and joint range, but it may also alter the strength (aerobic and anaerobic capacity) of that muscle and have a more widespread effect on the cardiovascular system.

One of the major local effects of exercise is the increased rate of protein synthesis, thus producing more actin and myosin as a response and facilitating an increase in muscle length. Connective tissue also responds to increased exercise by becoming stronger in order to cope with the increase in function that is required of that muscle (Basmajian and Wolf, 1990).

If active exercise is performed regularly and through the available physiological range, it has all the effects of passive exercise that were previously explained, including maintaining joint range, increasing joint nutrition and decreasing pain. Active movement has the advantage of strengthening muscles to some extent, thus providing stability for the joint with its increased range. Another advantage of performing active movements is that, if they are performed in a rhythmical manner, they may promote relaxation of the muscles surrounding the joint. If this happens, then the joint range may well be increased, especially if the restriction was caused by muscle spasm.

CONCLUSION

Movement occurring at joints depends on a variety of anatomical and biomechanical factors that can facilitate and/or limit the ROM available. How movements are classified and described will depend on which discipline is being studied. For this text, movements are classified as either active or passive, with subdivisions of each. Restriction of movement, caused by pathological changes, can be successfully managed by the therapist using different forms of movement. This is achieved once the rationale of the chosen method is known and correctly applied to the fully assessed patient.

The preceding descriptions by no means represent an exhaustive survey of the therapeutic modalities available to improve joint range and muscle length. They are, however, representative of the basic principles of the techniques that are used based on normal joint movement.

References

Basmajian, J., Wolf, S., 1990. Therapeutic Exercise, 5th ed. Williams & Wilkins, Baltimore.

Basso, D., Knapp, L., 1987. Comparison of two continuous passive motion protocols for patients with total knee implants. Phys. Ther. 67, 360–363.

Branch, T.P., Lawton, R.L., Iobst, C.A., Hutton, W.C., 1995. The role of glenohumeral capsular ligaments in internal and external rotation of the humerus. Am. J. Sports Med. 23 (5), 632–637.

Bromley, I., 2006. Tetraplegia and Paraplegia: A Guide for Physiotherapists, 5th ed. Churchill Livingstone, Edinburgh.

Cribb, A.M., Scott, J.E., 1995. Tendon response to tensile stress: an ultrastructural investigation of collagen: proteoglycan interactions in stressed tendon. J. Anat. 187 (Part 2), 423–428.

Cyr, L.M., Ross, R.G., 1998. How controlled stress affects healing tissues. J. Hand Ther. 11 (2), 125–130.

Field, L.D., Bokor, D.J., Savoie, F.H., 1997. Humeral and glenoid detachment of the anterior inferior glenohumeral ligament: a cause of anterior shoulder instability. J. Shoulder Elbow Surg. 6 (1), 6–10.

Gardiner, M.D., 1981. The Principles of Exercise Therapy, fourth ed. Bell and Hyman, London.

Grieve, G., 1988. Contraindications to spinal manipulations and allied treatment. Physiotherapy 75 (8), 445–453.

Hall, S.J., 1995. Basic Biomechanics, second ed. Mosby, St Louis.

Hardy, M., Woodall, W., 1998. Therapeutic effects of heat, cold and stretch on connective tissue. J. Hand Ther. 11 (2), 148–156.

Hollis, M., 1989. Practical Exercise Therapy, third ed. Blackwell Science, Oxford.

Kisner, C., Colby, L., 2007. Therapeutic Exercise, Foundations and Techniques, second ed. FA Davis, Philadelphia.

Lewit, K., 1993. Manipulative Therapy in Rehabilitation of the Locomotor System, second ed. Butterworth-Heinemann, Oxford.

Maitland, G., 1986. Vertebral Manipulation, fifth ed. Butterworths, London.

McCarthy, M., Yates, C., Anderson, M., Yates-McCarthy, J., 1993. The effects of immediate continuous passive movement on pain during the inflammatory phase of soft tissue healing following anterior cruciate ligament reconstruction. J. Sport Phys. Ther. 17 (2), 96–101.

McMahon, P.J., Tibone, J.E., Cawley, P.W., et al., 1998. The anterior band of the inferior glenohumeral ligament: biomechanical properties from tensile testing in the position of apprehension. J. Shoulder Elbow Surg. 7 (5), 467–471.

Norkin, C., Levangie, P., 2005. Joint structure and function: a comprehensive analysis, fourth ed. FA Davis, Philadelphia.

Norkin, C., White, J., 2003. Measurement of Joint Motion: A Guide to Goniometry, third ed. FA Davis, Philadelphia.

O'Brien, S.J., Schwarts, R.S., Warren, R.F., Torzilli, P.A., 1995. Capsular restraints to anterior–posterior motion of the abducted shoulder: a biomechanical study. J. Shoulder Elbow Surg. 4 (4), 298–308.

Palastanga, N., Soames, R., Palastanga, D., 2006. Anatomy and Human Movement Pocket Book. Churchill Livingstone, Elsevier, Edinburgh.

Robergs, R.A., Keteyian, S.J., 2003. Fundamentals of Exercise Physiology for Fitness Performance and Health, second ed. McGraw-Hill, Boston.

Soderberg, G.L., 1997. Kinesiology – Application to Pathological Motion, second ed. Williams & Wilkins, Baltimore.

Standring, S. (Ed.), 2005. Gray's Anatomy, thirtyninth ed. Elsevier, Edinburgh.

Threlkeld, J., 1992. The effects of manual therapy on connective tissue. Phys. Ther. 72 (12), 61–70.

Twomey, L., 1992. A rationale for the treatment of back pain and joint pain by manual therapy. Phys. Ther. 72 (12), 53–60.

Vujnovich, J., Dawson, N., 1994. The effect of therapeutic muscle stretch on neural processing. J. Sport Physical Ther. 20 (3), 145–153.

Warner, J.J., Deng, X.H., Warren, R.F., Torzilli, P.A., 1999. Static capsuloligamentous restraints to superior–inferior translation of the glenohumeral joint. Am. J. Sports Med. 20 (6), 675–685.

Wuelker, N., Korell, M., Thren, K., 1998. Dynamic glenohumeral joint stability. J. Shoulder Elbow Surg. 7 (1), 43–52.

Chapter 4

Motor control

Bernhard Haas

CHAPTER CONTENTS

LEARNING OUTCOMES

When you have completed this chapter, you should be able to:

1. outline the roles of the various nervous system centres in the control of movement
2. explain how movements are planned, generated and controlled
3. apply understanding to specific movements (e.g. balance and gait)
4. demonstrate knowledge of information transmission within the nervous system.

INTRODUCTION

The purpose of this chapter is to help you with understanding the human nervous system and how it participates in motor control. This chapter will give you an overview that will help you get the whole picture, and you should refer to specific texts to obtain detailed information. Traditionally, this has been done by reducing the function of this complex system to the properties of its individual elements: the neurons and control centres. Although it is essential to understand the language used to describe the system and to have a basic knowledge about which elements contribute to the complex affair of human movement, this reductionist view is not sufficient, and sometimes you will have to ignore the individual elements in order to better understand the system as a whole.

One of the traditional views of the nervous system has been that it is purely hierarchical, with a top down approach of control. According to this view, the cortical areas of the brain would exert a higher level of control and organize voluntary skilled movement. On the other end, the spinal cord would be fairly low down in the control stakes and mainly execute plans designed and refined above. A number of textbooks, such as Tortora and Derrickson (2008), Ganong (2003), Kiernan (2008) and Martini and Nath (2008), will provide you with further details. The section in this chapter on *The structures of the nervous system for controlling movement* will also give you an overview of the individual parts of the movement control system. In reality, there is, however, no real separation between voluntary movements and the background of postural control that maintains the body in an upright position with the aid of automatic reflexes and responses. See also Chapter 5 for more information on posture and balance. Therefore parallel systems of control, with integration of all levels rather than just a serial hierarchy, may be a more appropriate description. All levels of control, from the spinal cord up to the cerebral cortex, are necessary and integrated to provide the base of axial stability for more normal distal mobility and skilled or refined coordinated limb movements (Kandel et al. 2000). In addition, the environmental context and the movement task itself will influence how the nervous system organizes movement.

ACTIVITY 4.1

Chris reports that one of his friends at university has been involved in a car accident. He injured his leg during the accident, but his bones are now healed and he is back at university. Unfortunately, he still finds it difficult to move his foot because one of his nerves in his leg sustained an injury.

Work in small groups or on your own. Find a physiology textbook that has a diagram of peripheral nerves and find the common fibular nerve. Identify the function of this nerve and explain why Chris's friend may find it difficult to move his foot. He also cannot feel touch or pressure over the dorsum of his foot. Explain why.

INFORMATION TRANSMISSION

The vast numbers of neurons in the human nervous system need to communicate with each other, often very rapidly. Information within neurons and between neurons is carried by electrical and chemical signals. The rapid transmission of signals, which is vital for human movement, is a function of the action potential. This action potential is achieved by temporary changes of current flow in and out of cells, which then propagate a signal along the nerve axon. A necessary precondition for action potentials is the creation of a membrane potential, the resting potential (Tortora and Derrickson, 2008). Please note that no movement is possible if the action potential is completely interrupted, and that movement will be impaired if the signal propagation is abnormal. This may be the case if the myelin sheath that surrounds nerve axons is damaged, such as in multiple sclerosis. Figure 4.1 shows the concentrations of ions inside and outside the nerve cell during the resting potential. Figure 4.2 shows the changes in membrane potential during the action potential. Box 4.1 summarizes the key facts about the action potential.

Information transmission from one cell to another occurs at the synapse. The most important components of a synapse are the presynaptic membrane, the synaptic cleft and the postsynaptic membrane (Latash, 2008). An action potential arrives at the presynaptic membrane. This leads to the influx of Ca^{2+}, which in turn facilitates the fusion of neurotransmitter vesicles to the membrane for the release of the neurotransmitter into the synaptic cleft. Neurotransmitter molecules diffuse across the synaptic cleft and bind at specialist receptor sites in the postsynaptic neuron. This changes the potential in the postsynaptic neuron as ion channels are opened and the voltage across the cell membrane changes. Depending on the particular type of channel that is being activated, either depolarization or hyperpolarization may occur. This explains how an action potential in the presynaptic neuron can cause either excitation or inhibition of the postsynaptic membrane. Opening of Na^+ channels would lead to depolarization and therefore excitation, whereas opening of the Cl^- channels would hyperpolarize the postsynaptic neuron and lead to inhibition. Figure 4.3 summarizes the events that occur at a synapse.

ACTIVITY 4.2

Work in small groups or on your own. Find a physiology textbook that has a diagram of peripheral nerves and answer the following question: chemical synapses such as the one shown in Figure 4.3 transmit a signal in only one direction – why?

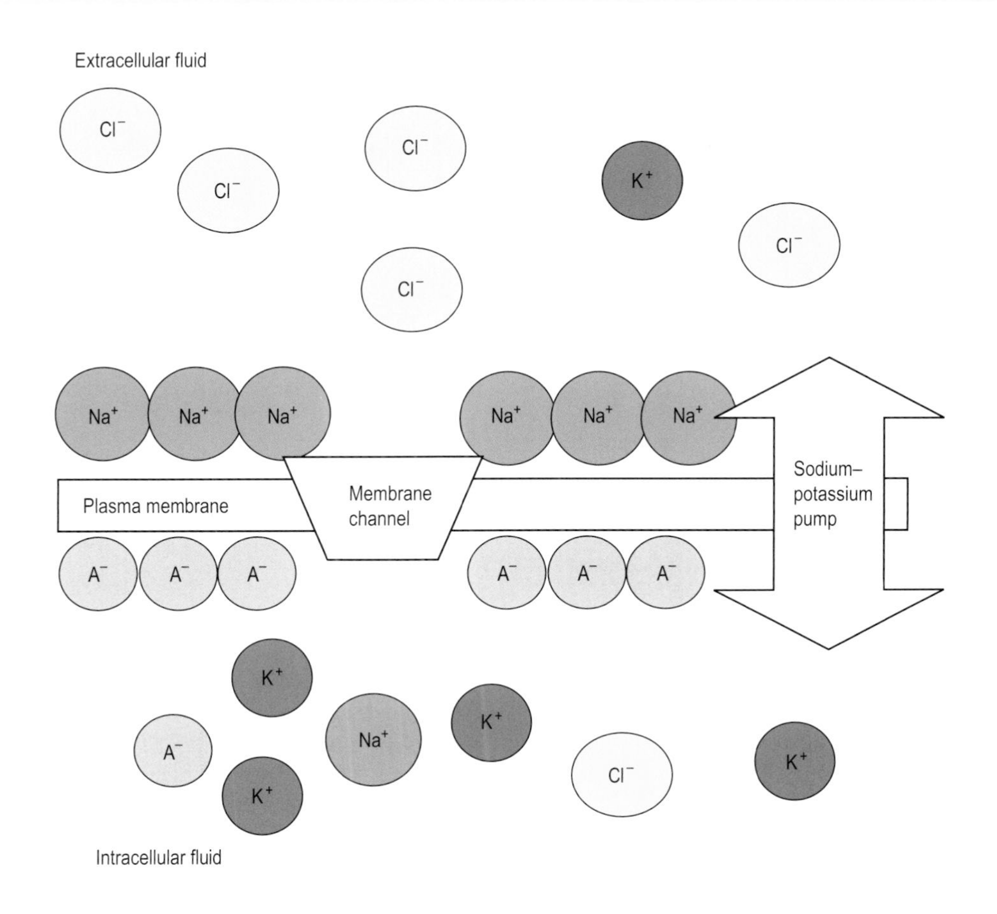

Ion	Concentration intracellular fluid	Concentration in extracellular fluid
K^+ – Potassium	150 mmol/l	5 mmol/l
Na^+ – Sodium	12 mmol/l	150 mmol/l
Cl^- – Chlorine	5 mmol/l	125 mmol/l
A^- – Organic anions	150 mmol/l	–

Figure 4.1 Distribution of ions across the cell membrane during the resting potential.

RECEPTORS

The central nervous system (CNS) needs to receive continuous feedback about movement. It receives this information in the form of the status of muscles, i.e. length, instantaneous tension, and rate of change of length and tension (Shumway-Cook and Woollacott, 2007). Muscle spindles detect the rate and changes in the length of a muscle, whereas Golgi tendon organs detect degree and rate of change of tension. Signals from these sensory receptors operate at an almost subconscious level, transmitting information into the spinal cord, cerebellum and cerebral cortex, where they assist in the control of muscle contraction.

The muscle spindle has both a static and a dynamic response. The primary and secondary endings respond to the length of the receptor, so impulses transmitted are proportional to the degree of stretch and continue to be transmitted as long as the receptor remains stretched. If the spindle receptors shorten, the firing rate decreases.

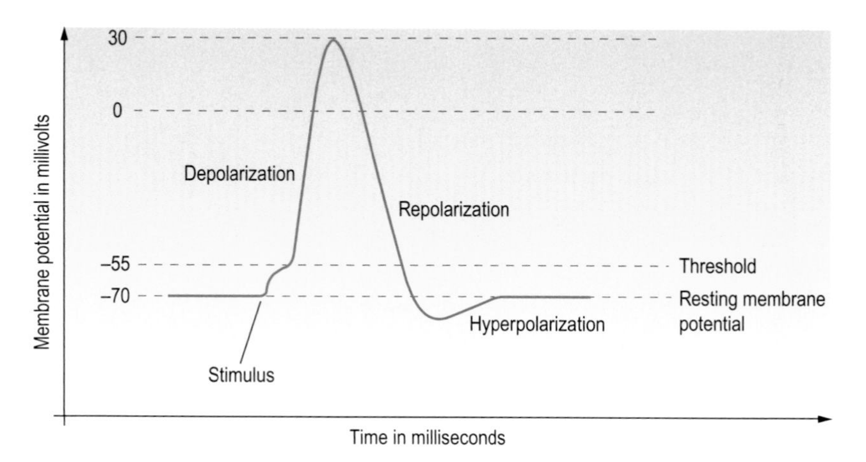

Figure 4.2 The action potential.

BOX 4.1	**Summary of important facts associated with the action potential**

- A necessary precondition for the creation of an action potential is the resting potential.
- The resting potential creates the excitability of the cell.
- The resting potential is the unequal distribution of ions across the cell membrane, with a negative charge of −70 mV in the cytosol (the intracellular fluid).
- The action potential emerges if a stimulus is large enough to take the membrane potential above the threshold (−55 mV).
- When the threshold level is reached, voltage-gated Na^+ channels open and Na^+ rushes into the cell, which produces the depolarization period.
- Voltage-gated K^+ channels open to allow K^+ to flow out of the cell and produce the repolarization period.
- Another action potential cannot be generated during the depolarization period and during most of the repolarization period.
- The action potential propagates along the axon segment by segment until it reaches the synaptic end bulb at the end of the axon.
- Propagation is more rapid in myelinated axons, where the signal leaps from node to node. Large-diameter axons also propagate signals faster than small-diameter axons.
- Axons of sensory neurons transmitting information about touch, pressure and movement, as well as the axons of motor neurons transmitting movement instructions to the skeletal muscles, are all large and myelinated.

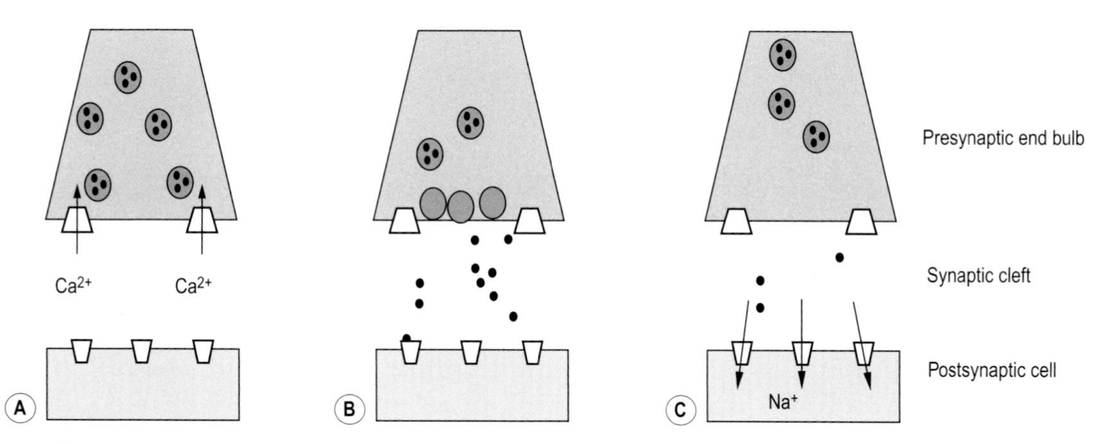

Figure 4.3 Synaptic events associated with the action potential.

Only the primary endings respond to sudden changes of length by increasing their firing rate, and then only while the length is actually increasing. Once the length stops increasing, the discharge returns to its original level, although the static response may still be active. If the spindle receptors shorten, then the firing rate decreases.

Control of the static and dynamic response is by the gamma motor neuron. Normally, the muscle spindle emits sensory nerve impulses continuously, with the rate increasing as the spindle is stretched (lengthened) or decreasing as the spindle shortens.

The spinal reflexes associated with the muscle spindle and Golgi tendon organ are the stretch reflex and the tendon reflex, respectively. Stimulation of the stretch reflex leads to a reflex contraction of the muscle that has been stretched, whereas the tendon reflex will lead to a reflex relaxation of the muscle if there is tension build-up. Both reflexes have a protective function.

The stretch reflex also has the ability to prevent some types of oscillation and jerkiness of body movements even if the input is jerky, i.e. a damping function (Hall, 2005).

When the motor cortex or other areas of the brain transmit signals to the alpha motor neurons, the gamma motor neurons are nearly always stimulated simultaneously, i.e. a coactivation of the alpha and gamma systems so that intra- and extrafusal muscle fibres (usually) contract at the same time. This stops the muscle spindle opposing the muscle contraction and maintains a proper damping and load responsiveness of the spindle regardless of change in muscle length. If the alpha and gamma systems are stimulated simultaneously and the intra- and extrafusal fibres contract equally, then the degree of stimulation of the muscle spindle will not change. If the extrafusal fibres contract less because they are working against a great load, the mismatch will cause a stretch on the spindle, and the resultant stretch reflex will provide extra excitation of the extrafusal fibres to overcome the load (Cohen, 1998).

The gamma efferent system is excited or controlled by areas in the brainstem, with impulses transmitted to that region from the cerebellum, basal ganglia and cerebral cortex.

The Golgi tendon organ, as a sensory receptor in the muscle tendon, detects relative muscle tension. Therefore it is able to provide the CNS with instantaneous information on the degree of tension of each small segment of each muscle. The Golgi tendon organ is stimulated by increased tension. When the increase in tension is too great, the tendon reflex response is evoked in the same muscle, and this response is entirely inhibitory. The brain dictates a set point of tension beyond which automatic inhibition of muscle contraction prevents additional tension. Alternatively, if the tension decrease is too low, then the Golgi tendon organ reacts to return the tension to a more normal level. This leads to a loss of inhibition, so allowing the A-alpha motor neuron to be more active and increase the muscle tension.

Box 4.2 lists key facts about receptors and reflexes.

BOX 4.2 Summary of important facts associated with receptors and reflexes

- Sensory feedback for movement control is mainly provided by receptors inside the muscle and between the muscle and tendon.
- The receptors are the muscle spindle and the Golgi tendon organ.
- The muscle spindle provides information about muscle length changes.
- The Golgi tendon organ provides information about tension changes.
- Both of these receptors are also closely linked to spinal reflexes.
- The stretch reflex relies on muscle spindle information and is triggered when a muscle is lengthened. It is designed to prevent overstretching of a muscle by causing a reflex contraction of the lengthened muscle.
- The tendon reflex relies on Golgi tendon information and is triggered when tension is building up at the interchange of muscle and tendon. It is designed to prevent tearing of a muscle by causing a reflex relaxation of the muscle.
- Stretch reflex and tension reflex therefore have opposite effects on a muscle.

MOTOR CONTROL

CONTROLLING 'SIMPLE' MOVEMENTS

Human movement is anything but simple. There is infinite variability, and any attempt to describe a complex system in simple terms is likely to tell you only part of the story. However, if you understand the 'simple', then you are more likely to grasp the more complex.

Most human voluntary movements require the design and planning of that movement by a control centre. This control centre will use previous experiences in the planning of movements. Once a movement plan has been designed, it will be supplied as a signal by the control centre in a feed forward manner to an execution centre responsible for activating muscles to produce the movement. Feed forward implies that the signal is independent of the output or any other variable (Latash, 2008). The feed forward signal uses knowledge of the dynamics of the musculoskeletal system and the environment it is in (Stroeve, 1999). Once the movement has started, receptors will be able to provide feedback about the movement. The controller may then be able to alter the signal according to this feedback. The addition of a comparator centre provides a mechanism to speed up the refinement of movement according to its feedback. Over a period of time, the feedback will in turn influence the feed forward signal designed by the control centre and motor learning will have taken place (Houk et al. 1997). It may be worth visiting Chapter 6 (*Motor Learning*) before you move on. Figure 4.4 shows such a simple movement control system using feed forward and feedback mechanisms.

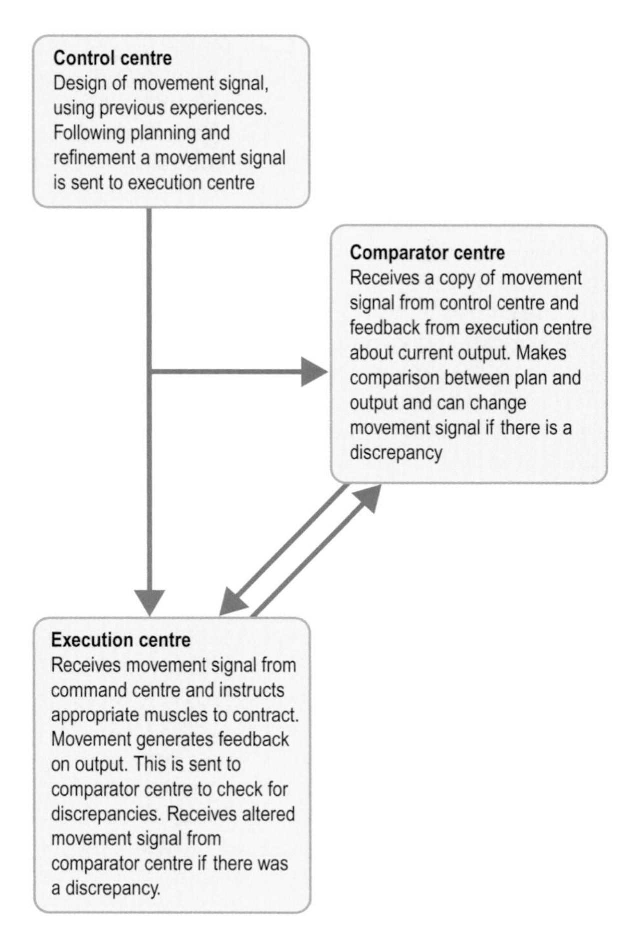

Figure 4.4 Feed forward and feedback system of a 'simple' voluntary movement.

POSTURAL CONTROL AND BALANCE

Postural control and balance involve controlling the body's position in space for stability and orientation (Shumway-Cook and Woollacott, 2007). The nervous system participates in postural control by designing command signals and by providing feedback through a number of receptors. The interpretation and integration of all the feedback signals would also be undertaken by the nervous system. The postural control requirements vary with the task. For example, sitting in a chair and watching television requires minimal stability control, whereas standing on one leg and watching television is a lot more demanding on the postural control system.

Therefore, we need to have a flexible control system that can adapt to these varying demands. Like the simple movement system above, postural control requires the production of movements or muscular contractions that help keep the body upright in space. Like the simple movement system above, postural control is also achieved by a combination of feed forward and feedback mechanisms. The control centre for posture will utilize previous experiences. These previous experiences will contribute to an internal representation of the body or body schema (Massion, 1994). This body schema provides reference points for body alignment, movement and orientation in space. The aim of the nervous system is then to maintain this body schema during changes in the environment or during movement. The feedback mechanisms for posture and balance involve more than just the receptors for movement in the muscles. In addition, there will be feedback about movements of the head through the vestibular system in the inner ear, visual feedback, and feedback about pressure changes through the support surfaces of the body (Kandel et al. 2000). The feed forward mechanisms will have to include signals that are able to anticipate disturbances to the postural control system that will arise as a consequence of movement (Aruin et al. 2001). Figure 4.5 shows the system of feed forward and feedback for the control of posture and balance. Please read Chapter 5 for more detailed exploration of these issues.

GAIT

Walking requires the cooperation of a large number of muscles and joints. Research on animals has shown that a neural network in the spinal cord is responsible for regulating the stepping motions during gait. There is controversy about whether such a network also exists in humans (Vilensky and O'Connor, 1997; Guadagnoli et al. 2000). The brainstem, together with the spinal cord, could provide such a network or central pattern generator to coordinate locomotion. The impulse for walking may come from higher cortical centres, but these central pattern generators could provide the motor pattern for walking. Figure 4.6 provides a proposed diagram of a central pattern generator for movement in the lamprey fish (Grillner et al. 1995). A similar neural network may also exist in humans. It may be worth also visiting Chapter 11.

Box 4.3 lists key facts about the nervous system and movement.

THE STRUCTURES OF THE NERVOUS SYSTEM FOR CONTROLLING MOVEMENT

This chapter has so far given you an overview of how the nervous system controls all types of movement and how the necessary signals for this control are generated and propagated. This has been the difficult part of the chapter, and once you have understood that, you should move on to the next part. This part will add some more detail about

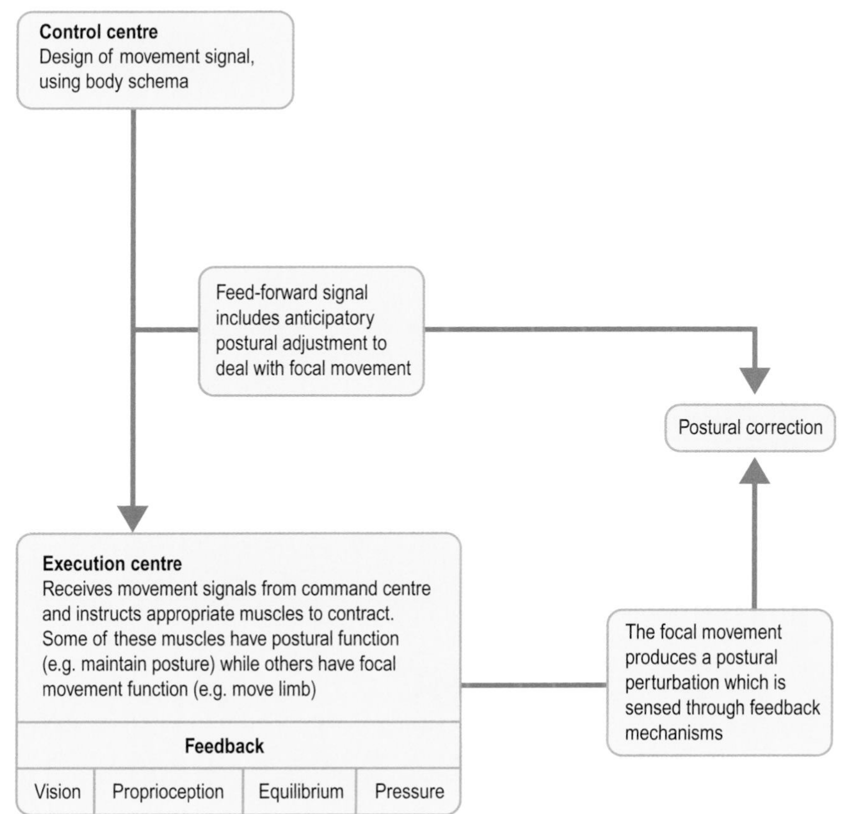

Figure 4.5 Feed forward and feedback system for postural control.

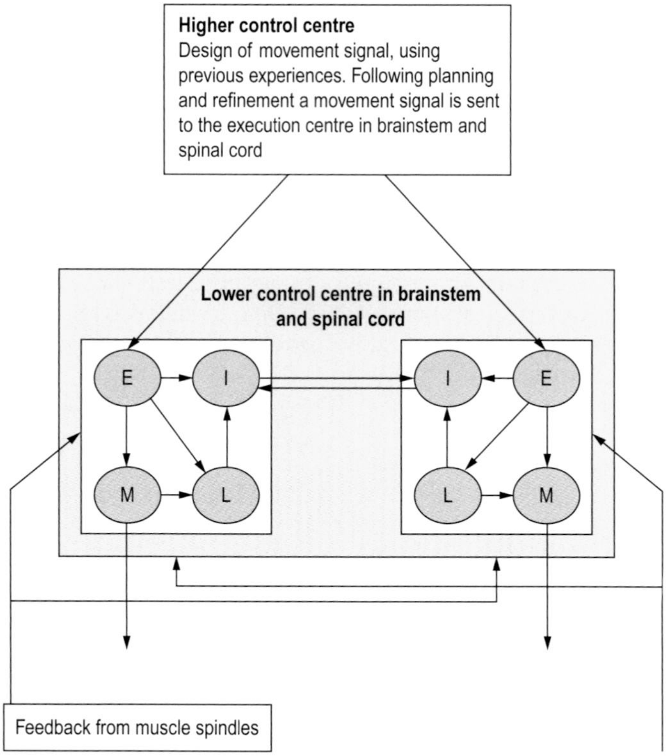

Higher control centre
Design of movement signal, using previous experiences. Following planning and refinement a movement signal is sent to the execution centre in brainstem and spinal cord

Lower control centre in brainstem and spinal cord

E I I E

M L L M

Feedback from muscle spindles

Signal bursts excite all interneurons (E) within the box. That means they excite the motor neuron (M) for muscle contraction, the inhibitory interneurons (I), which cross the midline to inhibit the activity in the contralateral side, and the lateral interneuron (L), which then in turn inhibits the interneuron (I). Feedback from muscle spindles provides feedback of movement which can provide further excitatory stimuli ipsilaterally or inhibitory signals contralaterally.

Figure 4.6 Hypothetical model of a central pattern generator for locomotion. *(After Grillner et al. 1995, with permission.)*

BOX 4.3 Key factors relating to the nervous system and movement

- The nervous system as a *whole* controls all types of movement.
- These movements can differ in complexity and characteristics. They can range from a relatively simple contraction of a muscle over one joint to multijoint and whole body movements such as those used during walking.
- Voluntary movements are designed utilizing previous experiences and use feed forward signals towards the muscles.
- In addition to the feed forward signals, there will also be feedback about movement and the body in relation to the environment.
- All voluntary movements, including posture, balance and gait, are based on these feed forward and feedback principles.
- The feedback generated through movement experience also provides for the possibility of motor learning.

the individual parts of the nervous system, which have been described only in very broad terms up until now. For example, you will find that the comparator centre described in Figure 4.4 is in reality called the cerebellum.

Cerebral cortex

The cerebral cortex is the main centre for the control of voluntary movement. It uses the information it receives from the cerebellum, basal ganglia and other centres in the CNS, as well as the feedback from the periphery, to bring movements under voluntary control.

The cerebral cortex, or more specifically the association areas of the cerebral cortex, provides the advanced intellectual functions of humans, having a memory store and recall abilities along with other higher cognitive functions. The cerebral cortex is, therefore, able to perceive, understand and integrate all the various sensations. This provides the transition from perception to action (Shumway-Cook and Woollacott, 2007). Its primary movement function is in the planning and execution of many complex motor activities, especially the highly skilled manipulative movements of the hand. This fact becomes clear when one considers the size of a cortical area for a particular part of the body.

The motor cortex occupies the posterior half of the frontal lobes. It is a broad area of the cerebral cortex concerned with integrating the sensations from the association areas with the control of movements and posture. It is closely related to other motor areas, including the primary motor area and the premotor or motor association area. The primary motor area contains very large pyramidal cells that send fibres directly to the spinal cord and anterior horn cells via the corticospinal pathways. In contrast, the premotor area has a few fibres connecting directly with the spinal cord, but it mainly sends signals into the primary motor cortex to elicit multiple groups of muscles, i.e. signals generated here cause more complex muscle actions usually involving groups of muscles that perform specific tasks, rather than individual muscles. This area connects to the cerebellum and basal ganglia, which both transmit signals back, via the thalamus, to the motor cortex. Projection fibres from the visual and auditory areas of the brain allow visual and auditory information to be integrated at cortical level to influence the activity of the primary motor area.

Each time the corticospinal pathway transmits information to the spinal cord, the same information is received by the basal ganglia, brainstem and cerebellum. Nerve signals from the motor cortex cause a muscle group to contract. The signal then returns from the activated region of the body to the same neurons that caused the contraction, providing a general positive feedback enhancement if the movement was successful and recording it for future use. The role of the cerebral cortex and its subdivisions is described in Figure 4.7.

Basal ganglia

The basal ganglia consist of five nuclei deep inside the brain (putamen, caudate nucleus, globus pallidus, subthalamic nucleus and substantia nigra). They serve as side loops to the cerebral cortex, because they receive their input from the cerebral cortex and project exclusively back to the cerebral cortex. The basal ganglia are involved in all types of movement but have a predominant role in the provision of internal cues for the smooth running of learned movements (Morris and Iansek, 1996).

It is believed that the basal ganglia play an essential role in the selective initiation of most activities of the body as well as the selective suppression of unwanted movements. A number of distinct loops have been described, and the interconnections of inhibitory or excitatory neurotransmitters explain the variety of symptoms that emerge in disorders of the basal ganglia. The direct pathway is responsible for the facilitation of movement, whereas the indirect pathway is more involved in the inhibition of unwanted movements (Rothwell, 1994). Figure 4.8 shows the direct loop through the basal ganglia.

Cerebellum

The cerebellum is vital for the control of very rapid muscular activities such as running, talking, typing, playing sport or playing a musical instrument. Loss of the cerebellum leads to incoordination of these movements such that the actions are still available but no longer rapid or coordinated. This is caused by the loss of the planning function.

The cerebellum makes comparisons between the movement plan and output and can change movement signal if there is a discrepancy.

Extensive input and output systems operate to and from the cerebellum. Input pathways to the cerebellum from the cerebral cortex, carrying both

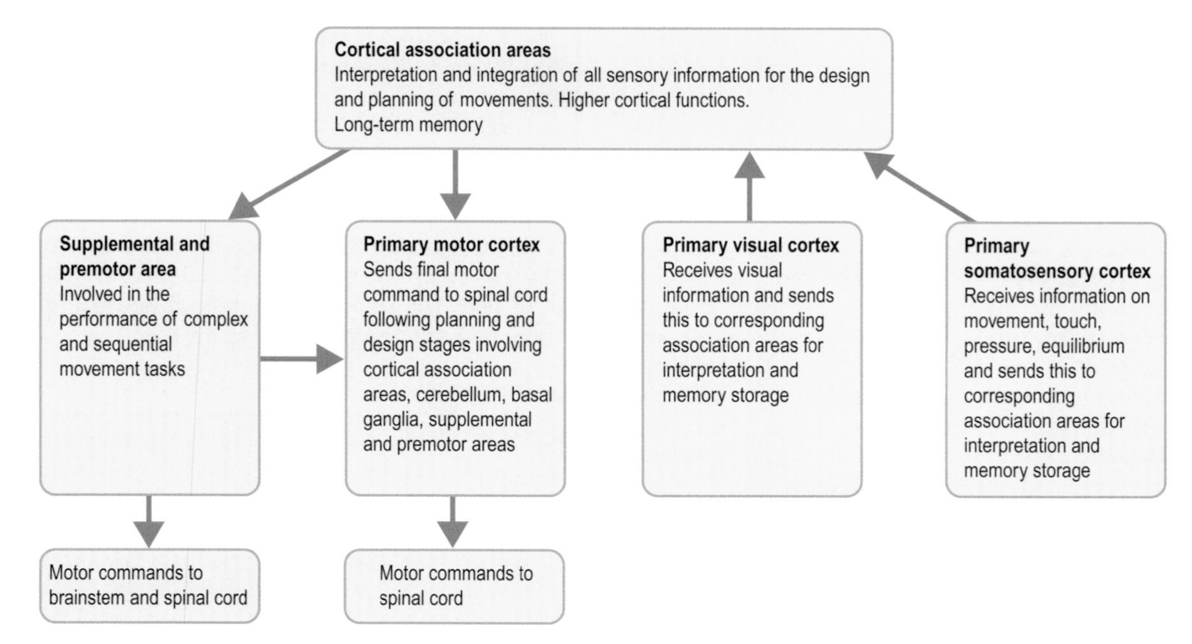

Figure 4.7 The cerebral cortex in movement control.

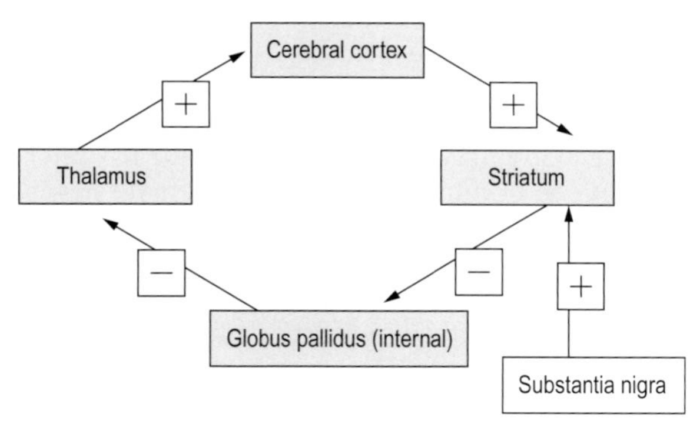

Figure 4.8 Hypothetical model of the direct loop in the basal ganglia.

motor and sensory information, pass through various brainstem nuclei before reaching the deep nuclei of the cerebellum. Likewise, output from the cerebellum exits via the deep nuclei to the cerebral cortex to help coordinate voluntary motor activity initiated there.

The cerebellum does not initiate motor activities but plays an important role in planning, mediating, correcting, coordinating and predicting motor activities, especially for rapid movements. It is vitally important for the control of posture and equilibrium, when it works in close relationship with the brainstem. Working with the basal ganglia and thalamus, the cerebellum helps to control voluntary movement by utilizing feedback circuits from the periphery and the brain. The distal parts of the limbs are controlled by information from the motor cortex and from the periphery, and this information is integrated in the cerebellum. This provides smooth, coordinated movements of agonists and antagonistic muscle groups, allowing the performance of accurate, purposeful intricate movements, which are especially required in the distal part of the limbs. This is achieved by comparing the intentions of the higher centres of the motor cortex with the performance of respective parts of the body. Overall, the cerebellum serves as an error-correcting device for goal-directed movements. It receives information on body position and movements in progress and then computes and delivers appropriate signals to the brainstem effector centres to correct posture and smooth out movements. The cerebellum is also important in the process of learning and acquisition of motor skills (Houk et al. 1997). Figure 4.5 showed the position of the comparator centre in the control of movement.

Brainstem

The principal role of the brainstem in control of motor function is to provide background contractions of the postural muscles of the trunk, neck and proximal parts of limb musculature, so providing support for the body against gravity. The relative degree of contraction of these individual antigravity muscles is determined by equilibrium mechanisms, with reactions being controlled by the vestibular apparatus, which is directly related to the brainstem region.

The brainstem connects the spinal cord to the cerebral cortex. It is comprised of the midbrain, pons and medulla oblongata. The central core of this region is often referred to as the reticular formation. This region of the CNS comprises all the major pathways connecting the brain to the spinal cord in a very compact, restricted space. It is also the exit point of the cranial nerves from the CNS.

It is through the integration of the information reaching the reticular formation that axial postural control and gross movements are controlled. Input to the reticular formation is from many sources, including the spinoreticular pathways, collaterals from spinothalamic pathways, vestibular nuclei, cerebellum, basal ganglia, cerebral cortex and hypothalamus. The smaller neurons make multiple connections within the area, whereas the larger neurons are passing through, being mainly motor in function.

The vestibular nuclei are very important for the functional control of eye movements, equilibrium, support of the body against gravity, and the gross stereotyped movements of the body. The direct connections to the vestibular apparatus of the inner ear and cerebellum, as well as the cerebral cortex, enable the use of preprogrammed, background attitudinal reactions to maintain equilibrium and posture. Working with the pontine portion of the reticular formation, the vestibular nuclei are intrinsically excitable; however, this is held in check by inhibitory signals from the basal ganglia (Hall, 2005). Overall, the motor-related functions of the brainstem are to support the body against gravity; generate gross, stereotyped movements of the body; and maintain equilibrium. This is achieved in association with the cerebellum, basal ganglia and cortical regions.

Spinal cord

The grey matter of the spinal cord is the integrative area for the spinal reflexes and other automatic motor functions. As the region for the peripheral execution of movements, it also contains the circuitry necessary for more sophisticated movements and postural adjustments.

Sensory signals enter the cord through the sensory nerve roots and then travel to two separate destinations:

1. same or nearby segments of the cord, where they terminate in the grey matter and elicit local segmental responses (excitatory, inhibitory, reflexes etc.)
2. higher centres of the CNS, i.e. higher in the cord, and brainstem cortices, where they provide conscious (and unconscious, i.e. cerebellum) sensory information and experiences.

Each segment of the cord has several million neurons in the grey matter, which include sensory relay neurons, anterior motor neurons and interneurons.

Interneurons are small and highly excitable, with many interconnections either with each other or with the anterior motor neurons. They have an integrative or processing function within the spinal cord, as few incoming sensory signals to the spinal cord or signals from the brain terminate directly on an anterior motor neuron. This is essential for the control of motor function. One specific type of interneuron is called the Renshaw cell, located in the anterior horn of the spinal cord. Collaterals from one motor neuron can pass to adjacent Renshaw cells, which then transmit inhibitory signals to nearby motor neurons. So stimulation of one motor neuron can also inhibit the surrounding motor neurons. This is termed *recurrent* or *lateral inhibition*. This allows the motor system to focus or sharpen its signal by allowing good transmission of the primary signal and suppressing the tendency for the signal to spread to other neurons (Rothwell, 1994). Together with the brainstem, the spinal cord contains a network of neurons that control walking. Figure 4.9 shows the basic components of a spinal reflex pathway.

ACTIVITY 4.3

Agnes tells Chris that she has a friend who has difficulties moving her left arm and also her left leg. She wonders if this may be similar to the problem Chris's friend has.

Work in small groups or on your own. Find a neurology textbook and identify the symptoms following a stroke. Explain why Agnes's friend has movement problems on the left side of her body. Which part of the nervous system has been affected?

Figure 4.9 The components of a basic spinal reflex pathway, using the stretch reflex as an example.

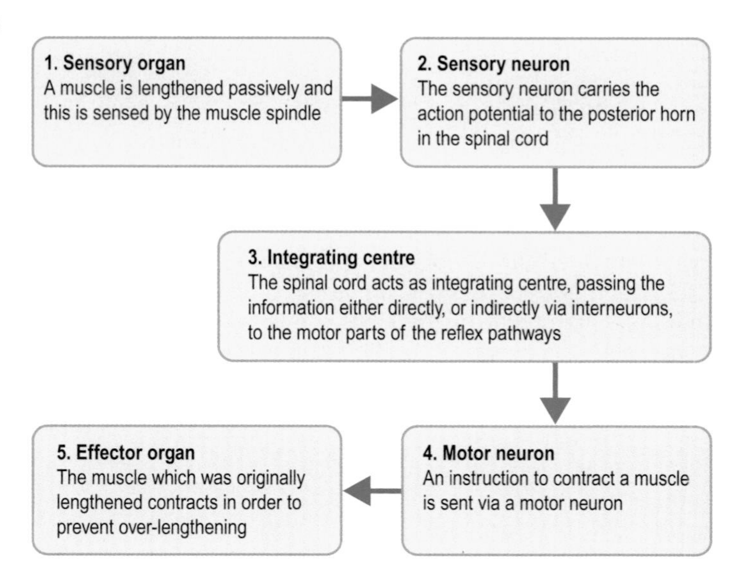

CONTROL PROCESSES OF VOLUNTARY MOVEMENT

Figure 4.10 summarizes in a simplistic diagram the control processes for voluntary movement. Follow the arrows and boxes from the design to the execution and then the return of feedback, which is finally stored as memory traces.

1. The cortical association areas play the key role in the design and planning of voluntary movements. Action potentials from the cortical association areas project to the basal ganglia for refinement and selective activation of movements and/or inhibition of unwanted movements.

2/3. The thalamus here is part of the basal ganglia loops and sends impulses to the motor cortex, which is seen as the final common pathway.

4. Impulses from the motor cortex are almost simultaneously sent to the cerebellum, the brainstem and the spinal cord. The cerebellum will use this information to compare it with the movement sensory information received from the periphery (6). The brainstem will play a role in maintaining background postural control, while impulses to the spinal cord are more of a focal nature for the activation of individual muscles or groups of muscles.
5. Alpha motor neurons cause muscle contraction.
6. The sensation of movement, together with other relevant feedback information, is sent towards the CNS. This sensory information is needed by various centres. The spinal cord will use it in its integration of spinal reflexes and the control of walking patterns. The brainstem utilizes sensory feedback mostly for postural control and balance. Sensory feedback is also sent to the thalamus. The cerebellum compares the movement as it occurs with the original movement instruction sent by the motor cortex.
7. If there is a discrepancy between the intended movement and the actual movement, correcting signals can be sent directly to the execution centres.
8. The thalamus distributes sensory feedback to its appropriate location on the sensory cortex.
9. Sensory experiences are interpreted by the cortical association areas, and memorized movements are stored for future use in the design and planning of movements.

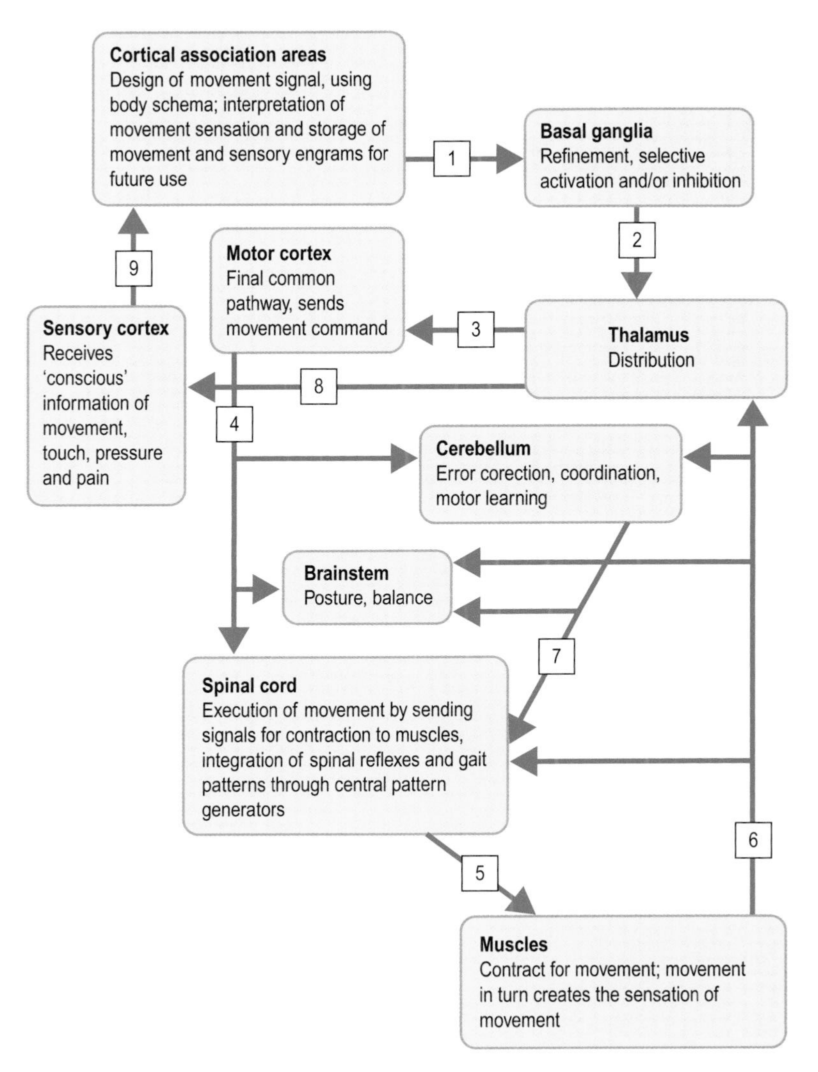

Figure 4.10 Control processes for voluntary movement.

References

Aruin, A., Ota, T., Latash, M.L., 2001. Anticipatory postural adjustments associated with lateral and rotational perturbations during standing. J. Electromyogr. Kinesiol. 11, 39–51.

Cohen, H., 1998. Neuroscience for Rehabilitation. JB Lippincott, Philadelphia.

Ganong, W.F., 2003. Review of Medical Physiology. Appleton & Lange, Norwalk.

Grillner, S., Deliagina, T., Ekeberg, O., 1995. Neural networks that co-ordinate locomotion and body orientation in lamprey. Trends. Neurosci. 18 (6), 270–279.

Guadagnoli, M.A., Etnyre, B., Rodrigue, M.L., 2000. A test of a dual central pattern generator hypothesis for subcortical control of locomotion. J. Electromyogr. Kinesiol. 10, 241–247.

Hall, J.E., 2005. Guyton & Hall Physiology Review. WB Saunders, Philadelphia.

Houk, J.C., Buckingham, J.T., Barto, A.G., 1997. Models of the cerebellum and motor learning. In: Cordo, P.J., Bell, C.C., Harnad, S. (Eds.), Motor Learning and Synaptic Plasticity in the Cerebellum. Cambridge University Press, Cambridge.

Kandel, E.R., Schwartz, J.H., Jessell, T.M., 2000. Principles of Neural Science. McGraw-Hill, New York.

Kiernan, J.A., 2008. Barr's The Human Nervous System: An Anatomical Viewpoint. Lippincott Williams & Wilkins, Philadelphia.

Latash, M.L., 2008. Neurophysiological Basis of Movement. Human Kinetics, Leeds.

Martini, F.H., Nath, J.L., 2008. Fundamentals of Anatomy and Physiology. Pearson Education, Upper Saddle River.

Massion, J., 1994. Postural control system. Curr. Opin. Neurobiol. 4, 877–887.

Morris, M.E., Iansek, R., 1996. Characteristics of motor disturbance in Parkinson's disease and strategies movement rehabilitation. Hum. Mov. Sci. 15, 649–669.

Rothwell, J.C., 1994. Control of Human Voluntary Movement. Chapman & Hall, London.

Shumway-Cook, A., Woollacott, M.H., 2007. Motor Control – Translating Research into Clinical Practice. Lippincott Williams & Wilkins, Philadelphia.

Stroeve, S., 1999. Analysis of the role of proprioceptive information during arm movements using a model of the human arm. Motor Control 3 (2), 158–185.

Tortora, G.J., Derrickson, B.H., 2008. Principles of Anatomy and Physiology. John Wiley, New York.

Vilensky, J.A., O'Connor, B.L., 1997. Stepping in humans with complete spinal cord transection: a phylogenetic evaluation. Motor Control 1, 284–292.

Chapter 5

Posture and balance

Clare Kell

CHAPTER CONTENTS

LEARNING OUTCOMES

When you have completed this chapter, you will be able to:
1. describe the ideal postures and the adaptations necessary for the journey through life
2. discuss the mechanisms for maintaining a functional posture
3. discuss the interdependence of control mechanisms during the maintenance of balance when performing everyday activities
4. discuss the therapeutic relevance of this chapter to practice.

CASE STUDY 5.1

This case study concerns challenges to balance during everyday activities. Our family have planned a trip to the cinema followed by a meal at a favourite restaurant to celebrate John's birthday. The film is popular, so, even though the tickets have been bought in advance, everyone has to queue to reach the theatre area. In this warm environment, with hands full of snacks, drinks and spare clothes, the family are jostled towards the appropriate screen. As they enter the dimly lit theatre, Jenny exclaims that she needs the toilet. Depositing her snacks with her parents, Jenny retraces her steps and the family members continue to their seats. Everyone finds the passage up the stairs inside the theatre tricky. With hands full and light flickering as the advertisements pass over the screen, everyone cautiously feels the step edges with their feet and adopts a slightly hunched posture. Barely have they sat down before Jenny joins them by bounding up the steps!

After the film, the family chatter about it on their way to the meal. Liz waits for Agnes, who is using the handrail to guide her down the steps. Agnes does not appear to be paying much attention to the conversations around her. Outside the cinema it is raining, so Liz and Agnes share an umbrella and Agnes gently rests on Liz's arm as they navigate the puddles, cars and excited children.

This story describes events with which many of us are familiar and postural responses that are varied but 'normal'. Throughout this chapter, we will be exploring why and how the family members acted in these ways. We will revisit this case at the end of the chapter, but for now keep the following question in mind as you read the text: if each family member is responding normally to their environment, why are they not all responding in the same way?

INTRODUCTION

Posture is variously defined as an attitude or position of the body (Cech and Martin, 2002); the maintenance for a period of time of a position in space as a prelude or background to movement (Bray et al. 1999); the intrinsic mechanisms of the human body that counteract gravity (Basmajian, 1964); or the position ordinarily held when locomotion ceases but distinct from positions of rest, feeding or other incidental poses (Morton, 1929), to state but a few. In some cases, while not stated in the same sentence as the definition, authors proceed to discuss the position of upright stance, giving the misleading impression that definitions apply and are of interest to this position only.

In this chapter, the term *posture* refers to the alignment of body segments such that the position of the body is ready for engagement in functional activity and responsive to both anticipated and unexpected perturbations or disturbances in balance. Such a definition acknowledges the active nature of posture and its maintenance, which, while essentially operated at a subconscious level, is a prerequisite to an efficient, effective, task-oriented existence.

Having briefly reviewed the requirements and challenges of humans' bipedal posture, the chapter will describe the body's segmental alignment in the positions of upright stance, unsupported sitting and prone lying, and consider the relevance of these 'standards' in the reality of human movement and function across the life- and work spans. The chapter will conclude with an in-depth discussion about the intrinsic and extrinsic requirements and systems essential for the maintenance of posture and postural control.

HUMAN POSTURE

THE SPINE AND ITS ROLE IN THE DEVELOPMENT OF BIPEDALISM

The adoption of the erect stance posture gives the human speed and agility but at the price of reduced stability resulting from the now raised centre of gravity (COG) and reduced base of support (BOS, outlined by the outer edges of the feet), through which the line of gravity (LOG) could act during balanced functional activities (see Ch. 9). With the body's weight now being transmitted to the supporting surface (through the spine, hind limbs and feet) from a significant height, the human needs a system for transmitting the load. The human spine is an amazing, biomechanically efficient structure designed as an S shape (Fig. 5.1) so that the curves can support the column while simultaneously offering a spring-like response to the loading. This elastic rod with its combination

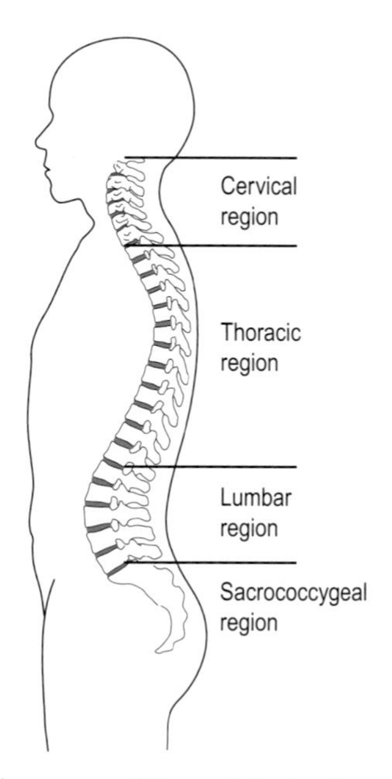

Figure 5.1 The curves of the vertebral column.

curves and arches spreads the body's weight while offering both support and flexibility to the body as a whole (Hamill and Knutzen, 2003). The load-bearing segments at the neck, hips, knees and ankles are balanced on and under the spine by the activity of sets of antagonist muscles that cross the joints 'like the springs of an anglepoise lamp' (Rothwell, 1994, p. 259). But such a many-linked chain produces its own problem: how are all the segments going to stay on top of each other as we move? We shall return to this issue later in the chapter.

THE HUMAN FOOT

As the foot is the human's contact point with the supporting surface while in upright stance, a basic understanding of foot anatomy and biomechanics is important in any discussion about human posture. Bipedal stance requires a structure that offers both a relative rigidity to support the weight of an upright posture against the force of gravity on a small base, and a foot capable of propulsion. To this end, the human is the only primate to have developed a foot characterized by a permanent arch (Reeser et al. 1983). As with the spine, the foot arch combines great strength with flexibility. Supported in quiet stance by the local bones and ligaments, the longitudinal arch offers stability and surface area contact, dissipating the weight-bearing load. System flexibility is provided by a mobile forefoot that is under intrinsic muscle control. This mobility permits weight distribution during gait, with the musculature becoming very active to support the arch when it is under most strain (when rising on to toes) so that the arch is still present as the gait cycle proceeds (Hicks JH, 1955, cited in Reeser et al. 1983).

In the next section, attention is turned to the reality of the upright posture in the modern human. The theories of force interaction on postural development (Basmajian, 1964) will be applied across the lifespan and then extended to discuss the self-imposed influences of our habits and environments. Readers are encouraged to also refer to Chapters 7, 8 and 13 to gain a more inclusive perspective.

DEVELOPMENT OF THE S-SHAPED SPINE ACROSS THE LIFESPAN

Cech and Martin (2002) described postural and functional development throughout life, the salient points of which are summarized below. At birth, the newborn infant spine is characterized by two concave forward-facing curves in the thoracic and pelvic regions. The latter curve is facilitated by the sacrum, being composed of individual sacral bones at this stage. At about 3–4 months of age, as the infant tries to raise their head from the prone lying position, gravitational interaction causes the convex forward development of the cervical spine. The final curve, the convex forward lumbar curve, begins to form as the baby sits up. Standing and walking increase the lumbar lordosis so that the COG lies over the pelvis in standing. At this stage, the child's feet are still flat. The location of COG differs from in adulthood, being higher (around T12) and anterior in the trunk, in response to the baby's proportionately larger head and liver. A relatively high COG challenges stability, which is commonly countered in toddlerhood by a wide BOS (Fig. 5.2). Cech and Martin suggest that foot arches and spinal curves approach adult form by the age of 6 years. It must be remembered, however, that the child has many more years of growth in front of them, when the spine formation can be challenged and adapted in response to environmental stressors. Nissinen et al. (1993) emphasize the specific vulnerability of the spine to formation stressors for the duration of puberty, when growth occurs as a spurt, leaving supporting structures temporarily challenged. These issues of spinal structure related to function are essential to both

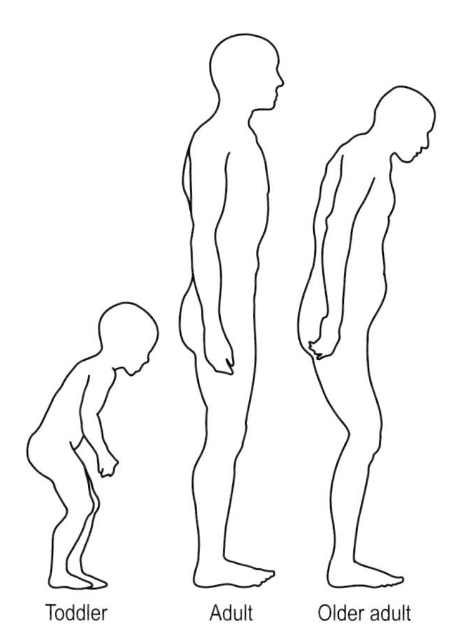

Figure 5.2 The curves of the vertebral column across the lifespan.

postural assessment and management and will be dealt with specifically later in the chapter.

By the age of 22 years, when growth has stopped (Nissinen et al. 2000), the spinal curves and the position of the body's COG within it will have matured. Ignoring for the moment all the factors that can influence individual posture, the position of upright stance will remain unchanged until a point in later life when ageing results in a changing interaction between physical ability and gravitational pull. Reducing physical strength and mobility is said to result in gravity exerting its influence to change the ageing stance posture, drawing it into flexion (lowering the COG) and reverting those curves developed during infancy (Cech and Martin, 2002). As a consequence, there is a tendency for the ageing population to display a posture of upright stance with a reduced cervical spine curve (Fig. 5.2), a flattened lumbar lordosis, and lowered and biomechanically inefficient longitudinal arches in their feet.

CASE STUDY 5.2

Let us think about the postures of our case study family introduced in Chapter 1. Sketch out your assumed postures for the family members. Acknowledging that this task is purely assumption-based (as you know very little about the individuals), why have you drawn them thus? Are there links between a person's posture and their movement? What clues have you picked up about the family and their movement from Chapter 1 to inform your thinking?

THE 'IDEAL' ALIGNMENT OF SEGMENTS

The word *ideal* is, in many ways, unhelpful, as it suggests a form of perfection against which we are all measured and towards which we are all trying to strive. Throughout this chapter, when reference is made to the term *ideal*, readers are requested to see it as a term used by researchers exploring biomechanical systems and applying that knowledge to the human body. The ideal thus becomes a biomechanically based framework from which real human posture can be described. Recognizing that this chapter focuses on the physical influences on human posture and balance, readers are urged to inform their posture descriptions with a more detailed understanding about the social, psychological and environmental influences on and consequences for a specific individual's postural alignment.

With that caveat, therefore, an ideal posture can be described as one that aligns the body segments so that the torques and stresses generated by gravity are minimized at each point in the chain (Pearsall and Reid, 1992). Such a posture requires the least internal expenditure of energy to maintain, i.e. the forces of gravity are neutralised by minimal internally generated counter-forces (active muscle work) (Basmajian, 1964). Consequently, one person's posture is not identical to another's; each is responding in a unique way with the gravitational environment given their unique combination of physical, muscular and soft tissue characteristics. Therefore care should be taken when labelling posture as poor or non-ideal. Until a careful history is taken and examination made of an individual, the environment and culture within which they exist, it would be quite wrong to start adjusting to the consensus ideal promoted by researchers in their investigations. Woodhull et al. (1985) and Pearsall and Reid (1992) concluded their studies with the observation that the most striking finding from their work was the postural variability between subjects. Pearsall and Reid suggested that each subject was displaying a posture that optimally compensated for their specific anthropometric differences (p. 84). Woodhull et al. cautioned that, while some subjects' postures are probably biomechanically 'better' than others', there is no reason to assume that the average is better than any one example (p. 115).

UPRIGHT STANCE

The location of the LOG (see Ch. 9) through the human body in upright stance has been investigated using various methodologies, including X-ray analysis, force platform analysis and staged immersion in water. The consensus is that, for the normal adult, the LOG ideally intersects the sagittal plane through the following points (Woodhull et al. 1985; Pearsall and Reid, 1992):

- the mastoid process or tragus of the ear
- just anterior to the shoulder joint
- just posterior to the hip joint
- just anterior to the knee joint
- just anterior to the ankle joint.

Woodhull et al. (1985) tried to locate the passage of the LOG about the hips and knees, and found that the line passes 1–2 cm behind the centre of the hip joint and 1–2 cm in front of the lateral epicondyle of the knee. The research suggested that this position of gravity generates torques (turning forces) that tend to keep both joints in slight extension while standing.

So what does the ideal upright stance posture look like if it adopts this segmental alignment in response to gravity? The most striking thing about truly ideal posture would be the left to right symmetry of the stacked segments, with no single muscle group seen to be heavily active except the erector spinae (Hamill and Knutzen, 2003) as they try to compensate for the continuous forward-bending action imposed on the trunk by gravity (p. 248). Viewing the posture from behind (posterior view), the observer would expect to see the following anatomical landmarks symmetrically aligned with the horizontal: ear lobes, breadth of shoulders, scapulae, waist creases, posterior iliac spines (dimples), buttock creases and knee creases (Fig. 5.3a). With the bare feet anteriorly placed and approximately 8 cm (hip distance) apart, the Achilles tendons would follow a perpendicular line to the floor. The spinous processes would follow a direct line from head to floor, with no deviation to left or right (scoliosis).

A front or anterior view (Fig. 5.3b) is good for noting central alignment, for example centrality of the chin above the sternum, a central and relaxed umbilicus, upward-facing anterior iliac spines and anterior-facing patellae. The observer would also notice a symmetrical distribution of weight through the feet. If you ran your hands around the edge of the feet, you would feel that they were equally accepting their BOS, rocking neither inwards or outwards; the longitudinal arches would be identical and run in line with the floor.

Observing the ideal upright stance posture from either side (lateral view) would show the head balanced on the neck, with no excessive chin protrusion or retraction. The spinal curves would demonstrate an open S shape such that the rib cage is anteriorly facing, and the abdomen would be flat but relaxed. A side view would show only one cheekbone, one shoulder, one scapula, one nipple, one buttock and one knee (Fig. 5.3c); if two of anything can be seen, segmental rotation is occurring. To clearly assess segmental rotation, the posture should be observed from both sides.

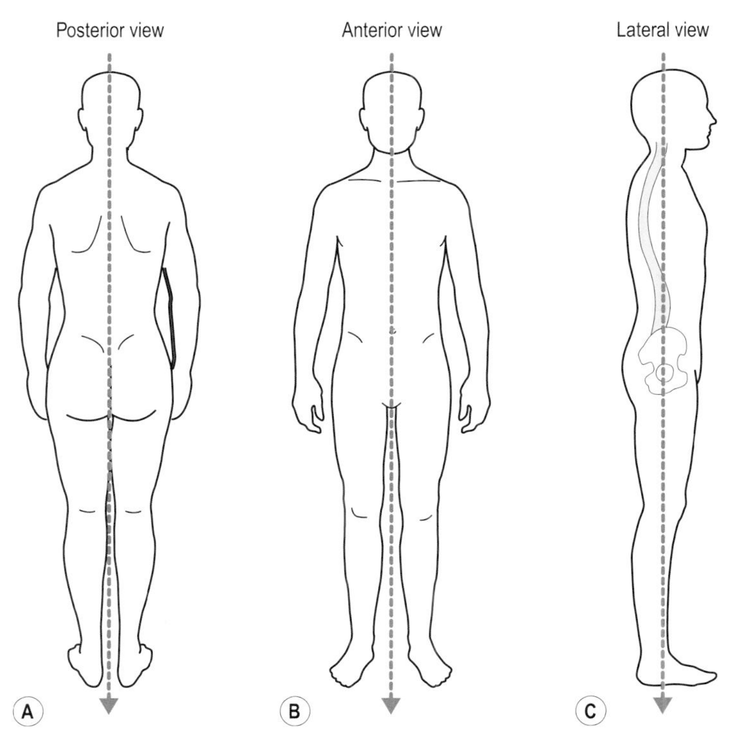

Figure 5.3 The 'ideal' alignment of segments in standing.

ACTIVITY 5.1

Looking at the postures drawn in Figure 5.3, what do you think you would see if you looked at these people from the front and back in terms of visible muscle activity, postural symmetry and alignment, etc.?

SITTING

When sitting, the height of the COG drops in relation to space but rises within the body (see Ch. 9), and the BOS is increased, with less load being placed on the lower extremities. In general terms, this posture requires less energy to maintain than that of upright stance, being inherently more stable. What has changed, however, is the angle of the pelvis on the lumbar spine, as the former is tilted backwards to allow the ischial tuberosities to be the focus of weight transference to the base (Hamill and Knutzen, 2003). The resulting flattening of the lumbar spine results in significant loading on the intervertebral discs and stretching of the posterior structures of that vertebral segment. Prolonged unsupported sitting can therefore have a detrimental effect on the lumbar spine. Supporting the lumbar spine with an appropriately measured and placed chair rest will dramatically reduce lumbar vertebral loading (please also see Ch. 8).

Ideally, therefore, the sitting posture should be supported by a chair whose backrest is high enough to support the upper thoracic spine (Fig. 5.4) and that inclines slightly backwards so that the lumbar spine is encouraged to stay in slight flexion (Hamill and Knutzen, 2003). The person's bottom should be placed to the back of the seat, with up to two-thirds of their thighs supported on the chair base to avoid compression of the posterior knee structures. Ideally, the chair height should be adaptable so that the knees and ankles will be at right angles as the feet are placed hip distance apart on the floor.

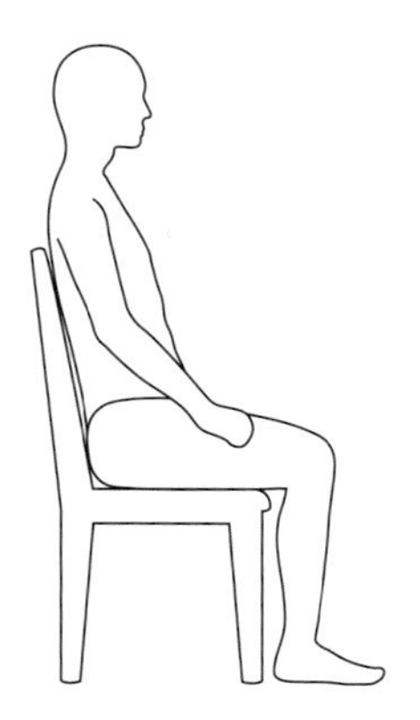

Figure 5.4 The 'ideal' alignment of segments in sitting.

ACTIVITY 5.2

Observe your colleagues or the people around you. How are they sitting? What factors could you change? How are you sitting at the moment? How long can you remain in this position? What is making you feel the need for change?

LYING

Lying is a posture of large stability, with both a low COG and a very large BOS. In this position, humans can achieve equilibrium with gravity with little if any energy expenditure (Basmajian, 1964). Basmajian stated that it is in lying that we spend most of our first year and about half of our lives thereafter! So it is important that we understand what is happening to the alignment of our segments in this position.

First, we must remember that the lying position is comprised of three distinct segmental orientations relating to the postures of supine (face up), prone (face down) and side lying. Each posture offers different contact points to the supporting surface and therefore exposes different areas to gravitational influence. The ideal supine position is one that mirrors the segmental alignment of upright stance (Basmajian, 1964). This position can be achieved only when the supporting surface is adaptable enough to permit indentation of the backward-protruding structures (e.g. the occiput, scapulae, thoracic spinous processes, sacral spines, ischial tuberosities and heels) (Fig. 5.5a). Too hard a surface will lead to these areas succumbing to pressure injuries (Fig. 5.5b), and too soft a support will result in mass spinal flexion (Fig. 5.5c). Pillows should be kept to a minimum (unless required for medical reasons) to avoid malaligning adjacent segments in the chain.

Because of the necessity to maximally rotate the cervical spine in order to breathe, and the accompanying hyperextension of the lumbar spine, the prone position is not advised for long periods.

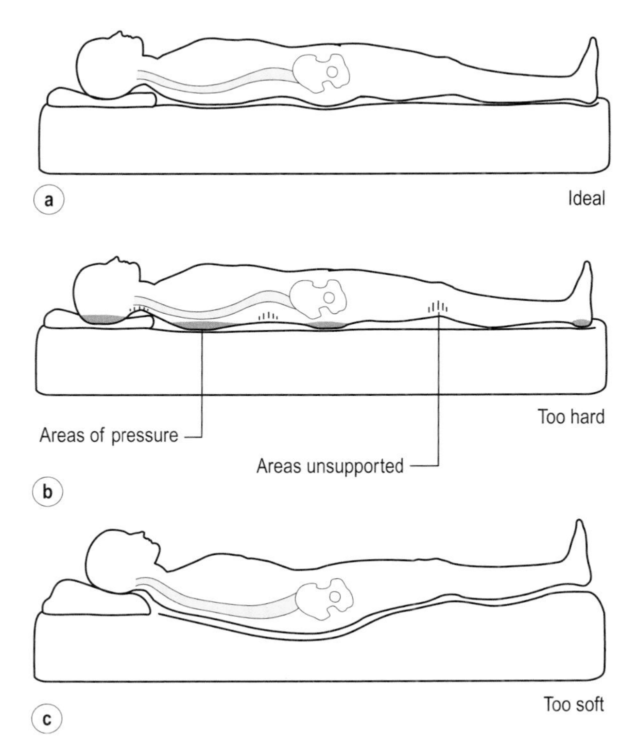

Figure 5.5 Possible mattress influences on segmental alignment in supine lying: (a) the 'ideal' alignment, (b) the possible effects of a very firm mattress, and (c) the possible effects of a very soft mattress.

The ideal side-lying posture again aims to mimic upright stance alignment. The influence of gravity in this position, however, necessitates support from external structures, most commonly pillows (Fig. 5.6). For the reasons outlined above, the supporting surface should allow indentation of prominent bony structures but not be too soft to encourage side flexion. In side lying, the cervical and lumbar curves require support because of their hanging free between the wider head and pelvis. As with supine lying, however, the use of external support must be targeted to the individual, supporting not exaggerating existing curves. If a pillow under the head is required, it should be accompanied by or include an extra support for the neck. The best option is to place a roll to support the neck; the worst is to use a man-made pillow that forces the head into side flexion and by morning supports only the ear! Similarly, a small roll may be needed to support the lumbar spine at the waist. Finally, in the side-lying position gravity draws the top leg towards the bottom, and the ideal hip distance separation of the lower limbs is lost, with resulting rotation of the lumbar spine towards the lower side. To counteract this rotation, it is advisable to lie with a pillow positioned between the knees. The above description provides the ideal posture for segmental alignment and reduction of stressful torque. While seemingly overpadded, this support is used widely throughout patient care environments.

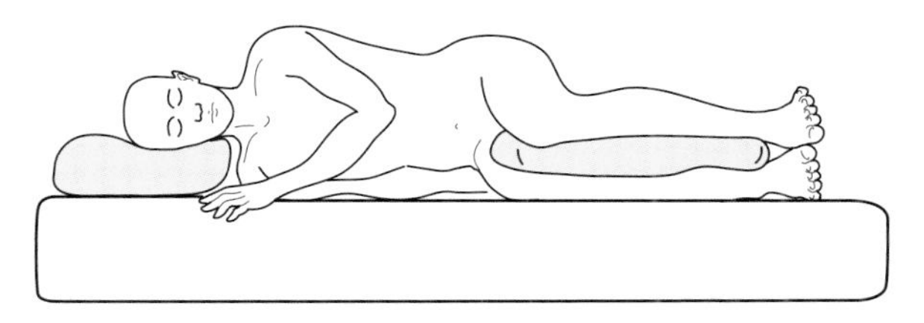

Figure 5.6 The 'ideal' alignment of segments in side lying.

Summary

This section has reviewed the ideal segmental alignment in three frequently described resting postures. The reader is urged to remember that the ideal is used as a frame of reference for postural assessment purposes and not as a guide to therapeutic outcome; the latter depends on an individual's functional environment. Postural analysis should be conducted in *each* position listed above, as poor alignments will emerge or become unmasked when gravitational and stability influences are changed. The most obvious change is seen when moving from standing to sitting, when trunk rotation disappears as legs of different length are removed from the supporting chain.

REQUIREMENTS FOR ACHIEVING THE 'IDEAL' POSTURE

A quick observation of our fellow humans (and the family as suggested above) will show us that very few people demonstrate an ideal posture. Why is this? Why do some people develop 'problem' or 'symptomatic' postures and not others?

The maintenance of any posture requires the following: an integrated system for detecting the posture, checking whether it is really what is wanted, and then bringing about any necessary change; effective and efficient muscles that can be appropriately recruited and then maintain the activity required; soft tissue structures of the appropriate length and flexibility to restrain and permit movement as designed; a psychological state that is motivated to engage with postural control; and an environment that permits postural alteration, for example women might have problems trying to stand upright with high-heeled shoes and short, tight skirts!

Chapters 7, 8 and 13 consider the environmental, psychosocial and ageing components, while this chapter explores the more physically based contributing factors. Here, we focus on things that can influence muscle strength and soft tissue flexibility, and therefore segmental alignment across the life- and work spans.

ACTIVITY 5.3

For each of the postural requirements listed above, note down a patient 'problem' that may result in them not fulfilling that requirement. How long is your list? How many people do you think have some form of postural deviation from the ideal?

THE REALITY OF POSTURAL ALIGNMENT

Several detailed studies have observed childhood postures in upright stance and observed those children until growth has stopped. Of interest to our discussion is the percentage of children who, while completely asymptomatic, display asymmetrical spinal postures. Juskeliene et al. (1996) found asymmetry in 46.9% of 6- to 7-year-olds, while Nissinen et al. reported spinal asymmetry in 21% of 10-year-olds (Nissinen et al. 1989). At 13 years old, the Nissinen population exhibited all sorts of postural combinations that were more common in girls than in boys (Nissinen et al. 1993), but when growth had stopped (age 22 years), equal numbers of boys and girls had asymptomatic, asymmetrical postures, with the most common presentation being a right scoliosis (Nissinen et al. 2000). So what sort of things could be causing these asymmetries?

FACTORS INFLUENCING SEGMENTAL ALIGNMENT

In childhood and adolescence

Juskeliene et al. (1996) reported that increased rates of asymmetry were found in two groups of children: those who had frequent childhood illnesses (defined in this study as having four or more acute illnesses in a year), and those who undertook little physical activity. The least asymmetry was found in the most active children. The researchers concluded by stressing the importance of muscle health and strength in the development of growing spinal alignment.

Their study of participants over a long period of time (from age 11 to 22 years) has enabled Nissinen and colleagues to make the following observations: left-handedness is a powerful determinant of hyperkyphosis (too much of a thoracic curve), probably resulting from working at desks designed for right-handed students (Nissinen et al. 1995); puberty is a very 'dangerous' time for spinal development, because of the imbalance of bone growth spurt and delayed growth and development of supporting and controlling soft tissue structures; post puberty, twice as many girls as boys had developed a scoliosis, because their sitting height (height while sitting at a desk) had grown faster than the boys and at a younger age – and they were still all using the same standard class desks (Nissinen et al. 1993); thoracic kyphosis is more prominent in males, while females are more lordotic (Poussa et al. 2005); and finally that, at the age of 22 years (i.e. during the slowing phases of bone growth), hyperkyphosis was significantly more prevalent in men than in women (Poussa et al. 2005). The messages from this extensive study suggest that we should be concerned about the environments in which children spend a considerable part of their day, and also acknowledge that postural development continues well past puberty and into young adulthood.

Gillespie (2002) reviewed the impact on childhood spinal development of computer (desk- and laptops) and electronic games (handheld and computer or screen-based) usage. Reporting staggering numbers of hours spent by children and young people in these activities, Gillespie noted that 'children and adolescents are increasingly engaged in activities that simulate work demands known to cause repetitive strain injury in working adults' (p. 249). During such activities, Gillespie (2002) suggests that young people are in deep concentration while adopting postures that are not suited to their needs (e.g. using environments where desk and chair heights are fixed, where there is a tendency for sharing computers but with only one mouse and keyboard, and where there is the tendency for children to use desktop computers at home set up for their parents – although she does make the astute observation that many units are not in fact set up appropriately for the adults using them either!). Drawing on this work, Gillespie et al. (2006) explored the impact of computer and electronic game use on self-reported episodes of musculoskeletal pain and/or discomfort. Their study suggests that girls were more likely than boys to report symptoms, but in general there was no overt link between hours spent engaged in these activities and reported episodes of musculoskeletal discomfort. Gillespie et al. (2006) did, however, note that discomfort was reported most when the young person was at risk of being overweight or needed vision correction (p. 4).

Concluding on a positive note, however, Gillespie (2002) suggests that, unlike adults, children, when uncomfortable, are happy to climb on to tables or chairs, lie on the floor, etc., and in so doing may be reducing the actual negative influence that they experience; she suggests that perhaps adults could learn from the unselfconscious attitudes of children (p. 256)! Figure 5.7 depicts the ideal computer station posture, with the screen height and angle adjusted so that a comfortable head position is maintained.

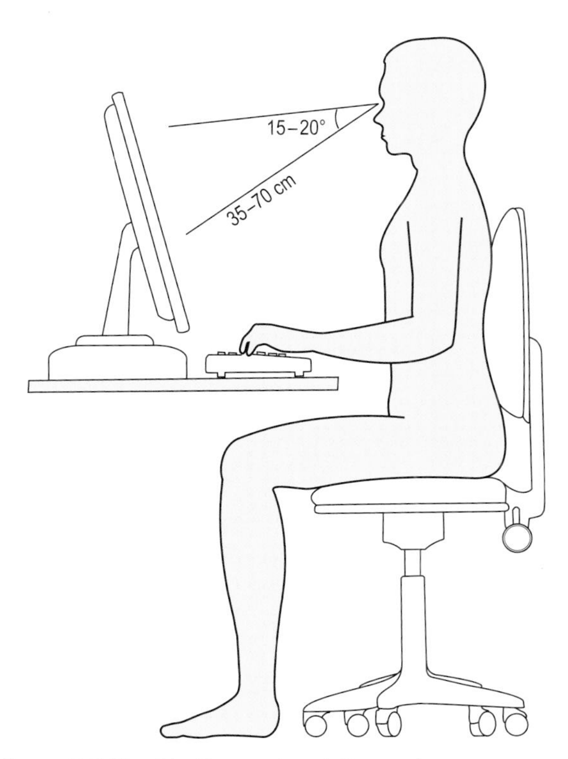

Figure 5.7 The 'ideal' computer station posture.

In adulthood

Maintaining a fixed posture

So if reduced physical activity and the adoption of awkward, sustained postures are risk factors for children adopting postural asymmetry, what factors influence adult posture? Grieco (1986) considered this issue while reviewing the likely impact on spinal evolution of taking millions of years to become *Homo erectus* and the relatively recent enforcement of *Homo sedens* in the execution of many of today's employment tasks (p. 347). Grieco reviewed the positions and ergonomics of workstations and working patterns and made the following observations.

- Enforced posture should be considered as a risk factor (to soft tissue and spinal injury) on the same level as the lifting of weights and vibration.
- Ischaemia in the paravertebral muscles occurs from prolonged isometric contraction, as seen in the maintenance of unsupported postures (and that can be any posture: sitting, standing, etc.).
- The problem, in terms of spinal injury, is one of postural fixity, not the loading of the spine at any one time; vertebral disc nourishment occurs via a pump or sponge mechanism, i.e. as a result of changing pressure resulting from change in posture.
- Ergonomic chairs may decrease the loading on the spine in general, but because the occupier is more comfortable, they do not move around, and then postural fixity becomes a problem.
- People who are uncomfortable move around.
- All employees should be educated in the use of their ergonomically sound devices and have 10-min 'postural pauses' (p. 359) built into their work pattern on an at least 50-min basis.

In conclusion, Grieco (1986) suggested that adults are susceptible to the same environmental stressors as children but now in the pursuit of employment. Postural education is a key factor in his argument.

Other causes of fixed postures

We must not forget that there are many adults in our population who – for reasons of poor stimulation, loneliness, weakness, illness, etc. – remain static or confined to one posture for substantial periods of time. Prolonged sitting for whatever reason will lead to the functional adaptation of a reduced lumbar spine lordosis, an increased thoracic kyphosis, and reduced range of movement at the hips and knees into extension (Cech and Martin, 2002). As the ideal thoracic alignment is designed to encourage maximal efficiency of lung function, people with static postures, especially with a thoracic hyperkyphosis, are at risk of developing pathologies associated with poor ventilation (e.g. reduced lung capacity and tidal volumes, leading to chest infections, and reduced efficiency of the thoracic venous pump, causing problems for lower limb venous return and fluid distribution) (Bray et al. 1999). These problems are compounded if the person is elderly and therefore their soft tissues are already reducing in flexibility and strength.

Hormone-related changes

Cutler et al. (1993) analysed the postures of 136 healthy women who were pre- and postmenopausal to see if there was a relationship between hormonal levels and postural hyperkyphosis (the so-called dowager's hump attributed to postmenopausal osteoporotic women). The results suggested that there was no link between degree of kyphosis and age, hormone levels or calcium intake. Again, there appeared to be a weak link with exercise levels: while many women displayed a kyphosis, those more active women were better able to straighten out of the kyphosis at will.

Temporary influences on postural alignment

The classic temporary influences on posture include situations in which weight distribution alters, as in pregnancy and when pain and/or pathological processes alter mood, tolerance levels, and general engagement with the environment (Clancy and McVicar, 2002). If the underlying cause of these postures is rectified promptly, no lasting adaptation to the new posture should occur, as a return to functional activity will restrengthen supporting muscles and soft tissues.

DEFINED POSTURAL DEVIATIONS FROM THE IDEAL

Classically, postural analysis has characterized non-ideal postures into six categories: scoliosis, kyphosis, hyperlordosis or hollow back, a combined kypholordotic posture, swayback and flat back postures (Fig. 5.8). Nominally (by title), all refer to deviations from the spinal ideal. Readers should note that this naming system rarely indicates that the defined deviation originated within the spine itself. The following section will briefly describe each of these postural types and discuss how the spinal deviation impacts on the segmental alignment of the whole body; a malalignment in one part of the body will have to be compensated for in another in order to keep the LOG falling within the BOS. Compensations seen will depend on the patient and context but would include structural realignment and/or increased muscle work in adjoining segments.

Scoliosis

- A static (fixed) or mobile (correctable) lateral curve usually seen in the thoracic and lumbar spinal regions.
- Soft tissues and muscles on the side of the concavity will be shortened and strong, while those on the convex side will be longer and weaker.
- Depending on the severity of the scoliosis, a compensatory side flexion may be seen further up the segmental chain.
- If the deviation is caused by a problem lower down the chain, the scoliosis may itself be the compensation.

Kyphosis

- Again, can be static or mobile. Many people adopt short-term slouched postures with no lasting effect.
- Always refers to an increased anterior curvature of the thoracic spine.
- A common posture in the older population.
- An increased kyphosis affects the higher segments such that the eyes focus on the ground unless the cervical spine compensates by increasing its lordosis, a position likely to cause muscle tension and joint compressions. Common symptoms of this effect are pain and headaches, as the neck is not biomechanically designed to cope with or adapt to this position.
- A long-standing kyphotic posture will result in shortening all musculature on the anterior chest wall (e.g. pectoral muscles) and the lengthening of the erector spinae to compensate.
- Severe kyphosis may result in impaired lung function caused by altered biomechanics for thoracic expansion.
- Occasionally, a kyphosis is the result of a compensation for shortened hip flexors.

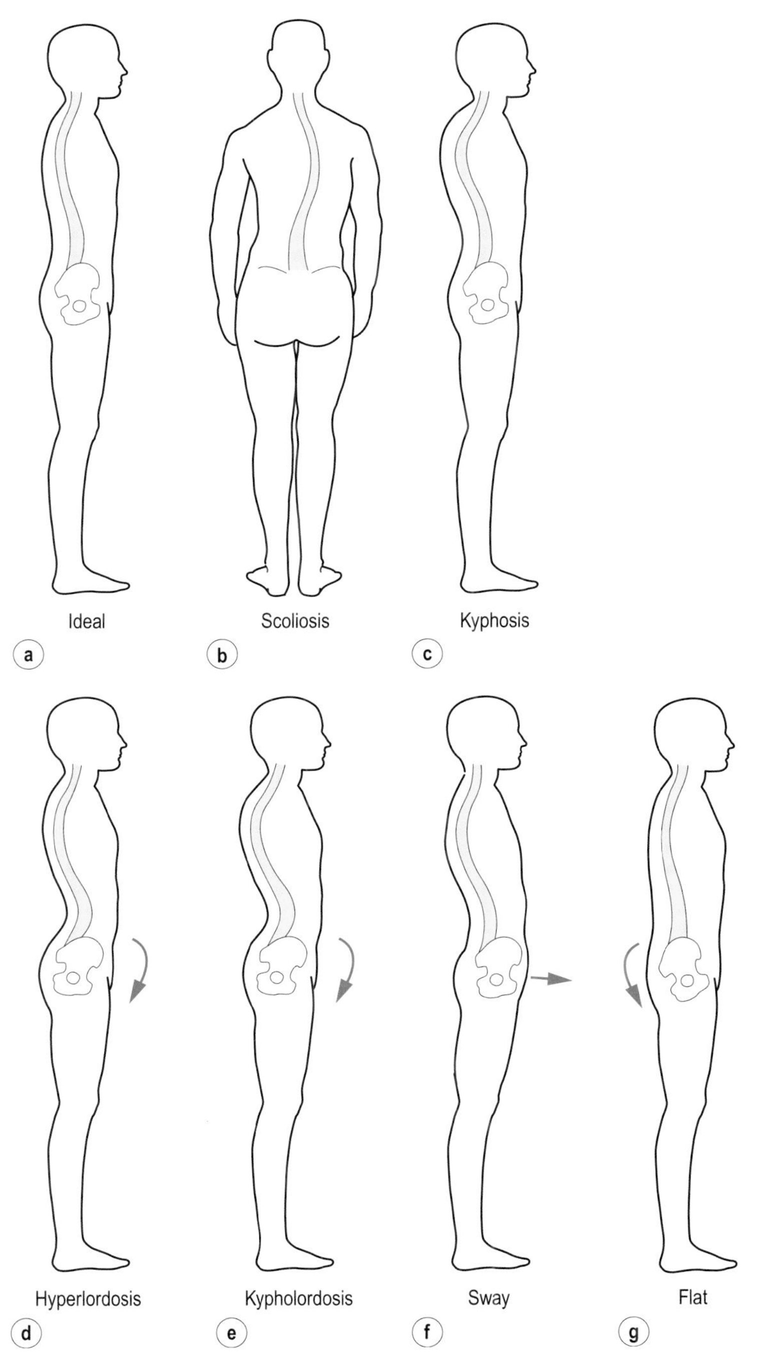

Figure 5.8 Diagrammatic representations of 'ideal' posture and deviations from it: (a) 'ideal', (b) scoliosis, (c) kyphosis, (d) hyperlordosis, (e) kypholordosis, (f) swayback, and (g) flat back.

Hyperlordosis

- A situation in which, in upright stance, the pelvis is held in anterior tilt.
- Classically seen after pregnancy and when abdominal musculature is weak.
- This position leads to an increased lumbar lordosis that shortens both the spinal extensors and the anterior muscles controlling the pelvis: iliopsoas, rectus femoris and tensor fasciae latae (Palastanga et al. 2006).

- The drawing forwards of the pelvis will lengthen and weaken the glutei and abdominal muscles.
- The posture is called *kypholordotic* when combined with a compensatory increased thoracic kyphosis.

Swayback

- Also known as *relaxed*, the characteristic profile of this posture is the slouch.
- Adopting this posture uses the least amount of active muscle work, as the posture is generally maintained by soft tissue and ligament length (e.g. pushing the pelvis forwards so that the hips go into extension allows the pelvis to hang on the anterior ligaments of the hip).
- This posture is commonly adopted by boys with Duchenne muscular dystrophy, for example.
- Working back up the segments, this anterior shift will need to be compensated for by an increased lumbar lordosis and thoracic kyphosis. The head will usually be held forwards of ideal.
- Depending on the extent of the pelvic shift, the knees may be locked into extension so that stance is maintained via bony approximation.
- A really mixed picture of altered joint and muscle biomechanics is presented by this posture, each deviation requiring careful assessment and consideration of the cause or compensation conundrum.

Flat back

- The key feature of this posture is a posterior tilting pelvis, which reduces the lumbar lordosis and gives the back its characteristic flat appearance.
- People adopting long-term slouched sitting positions commonly develop this posture.
- A posterior tilting pelvis may result in the person standing in hyperextension at both hips and knees, with lengthened hip flexors.
- Abdominal muscles are likely to be tight and strong, with the erector spinae reciprocally weak and long.
- Further up the segmental chain, the loss of lumbar lordosis will be compensated for by an anteriorly positioned head, which in turn may be compensated for by a slightly increased thoracic kyphosis.

Summary

This section has outlined the common features of a number of classically described postural deviations from the ideal. The reader should remember, however, that owning one of these postures does not automatically give the bearer pain, discomfort or any other problem. Kisner and Colby (2007) give detailed examples of when problems may arise for each postural type, but note that most people are symptom-free. It is also important to note at this stage that the existence of a pure flat back, for example, is as rare as the ideal alignment. As we know, each person's posture is the result of interaction with their environment. While the given labels are useful for initial consideration, postural assessment and intervention can occur only on a case by case basis.

MAINTAINING A FUNCTIONAL POSTURE

INTRODUCTION

Now that normal posture has been defined and the influences that affect segmental alignment described, the chapter will consider the concept of balance or postural control. How do we use posture to frame purposeful movement? Despite the potential instability of a relatively high COG above a small BOS, humans are able to maintain their posture and control the alignments of all segments so that purposeful movements occur about a stable base. In a nutshell, human postural control requires information from and integration between the systems responsible for visual, vestibular and somatosensory input and motor output. Despite the enormity of the task, we are only fleetingly aware of the systems' ongoing efforts to move and support our body in relation to gravity. Lackner and DiZio (2000) describe the perception of effortless ability to stand on one foot, our bodies seemingly unaware of the huge forces that are being transmitted through our supporting segments; it is only when we are ill or fatigued that we feel every gramme of load!

This section of the chapter will review the components of the postural control system and discuss how they are integrated to produce the background for functional activity. The chapter adopts the 'systems theory approach to motor control' described by Shumway-Cook and Woollacott (2007). The section will conclude with a discussion about the factors that influence normal postural control and consideration of the impact of abnormal control on effective and efficient movement.

SOME DEFINITIONS

To facilitate clear discussion, we will begin by explaining how we intend to use certain terms during the chapter. Lee (1988) and Williams et al. (1994) describe a system as a device or set of elements that transform input into a single or a selection of desired outputs. To function, a system requires inputs, outputs, and a means for the two to communicate together. In this context, the term *control system* refers to an ability of the system to set and then achieve its output, for example when reaching for an object, our postural control system is able to decide and then execute the correct amount of compensatory postural reactions (outputs) to maintain balance (COG over BOS).

Regulation refers, in this instance, to the ability of the system to do more than simply control for an output in a given situation. Regulation is the ability of the system to maintain the control–desired output responsively, i.e. in response to feedback gained from what is actually happening and in anticipation of what is to come. It is this ability of the postural control system to anticipate and regulate to react responsively to its environment that gives the human an exceptional array of highly skilled functional activities.

HOW DOES POSTURAL CONTROL WORK?

To answer this question, this section will consider each step in the postural control cycle: sensory information, integration and effector result. Please note again that this chapter is looking in detail only at the internal physical influences on posture. Readers are strongly advised to draw on Chapters 7, 8 and 13 when reading the following.

Input requirements

For any system to produce an effective product, it must receive good, up-to-date and relevant input data. Such is the intensity and diversity of the information required to frame functional movement that three input systems are used: the vestibular system, the visual system and the somatosensory system.

The vestibular system

Located in the inner ear are two sensory systems that send information down the eighth cranial nerve, the vestibulocochlear nerve. The cochlear portion of the cranial nerve carries information from the cochlear portion of the inner ear concerned with auditory stimulation; we hear with this part. Although not directly related to balance, over- or unexpected stimulation of this system can affect postural control, as we shall see later. The vestibular system comprises the other part of the inner ear complex and is responsible for our awareness of head orientation with respect to gravity and the head's linear and angular acceleration, i.e. any change in velocity and direction.

While this is not the text for detailed physiological discussion, it is important at this point to note that the sensory receptors of the vestibular system are constantly firing, and that it is the change in firing rate and pattern that are detected and interpreted by the interpretation centre for this system, the vestibular nuclei of the brainstem. The vestibular nuclei collect and redistribute information to other interpretation centres (e.g. the cerebellum, reticular formation, thalamus and cerebral cortex) but can also effect system output directly. Collectively, the nuclei will influence the antigravity muscles of the neck, trunk and limbs via the vestibulo-ocular and vestibulospinal tracts.

The visual system

The eyes provide the sense of sight but also play an essential role in giving us information about where our bodies are in space. This latter sense is sometimes called visual proprioception (Shumway-Cook and Woollacott, 2001). Visual proprioception informs the brain about body position in space, the relationship between one body part and another, and the motion of our body. This sense is the eyes' ability to distinguish the site of viewed movement: is it the object I am watching, is it my eye in my head or is it my head itself moving? It is suggested that information from the eyes passes to the superior colliculi (roof of the midbrain), where sensory maps compute how close to the body and how close to the midline of the body the movement is occurring. Information from the superior colliculi (visual system) passes to three main regions:

1. regions of the brainstem that control eye movements
2. the tectospinal tract that helps control neck and head movement.
3. the tectopontine tract that, through connections with the cerebellum, processes eye–head control.

In this way, the visual system is controlling output so that the postural control system can be sure about head position and movement in space with respect to surrounding objects (Shumway-Cook and Woollacott, 2007).

The somatosensory system

The awareness of joint position is provided by a complex interpretation of information collected from the unmyelinated Ruffini fibres in the joints themselves, the muscle spindles of those muscles passing over the joints, and the effector motor command signals (Lackner and DiZio, 2000). In the hand, where precise joint proprioception is crucial, a fourth input, from the cutaneous mechanoreceptors in the hand itself, adds to the information pool. In a similar way, the mechanoreceptors in the feet provide important information about the distribution of pressure through different areas of the soles of the feet. While we are able to think about the position of our limbs, proprioception is an essentially subconscious level sense. It is thought that information from the somatosensory receptors listed here passes directly to the motor neurons controlling postural stability at spinal level. In this way, sensory information can modulate movement that results from commands originating in higher centres of the nervous system (Shumway-Cook and Woollacott, 2001, p. 66). Such a direct communication system enables quick response times and is used to good effect when the sensors detect the likelihood of a fall. For example, if a person is standing, the pressure receptors in the feet will detect the perturbation and respond by effecting actions to widen the BOS (induce a stepping reaction) and/or lower the COG. We use this response every time we shift our COG over one leg (as in walking); detecting increased pressure under one foot, the somatosensory system responds by increasing extensor motor tone on the supporting side and simultaneously increasing flexor tone on the opposite side. Both these actions working together have the effect of transferring the COG over the new BOS (the supporting limb; Rothwell, 1994).

When do we need the input?

Lee (1988) offered a detailed review of the type of information that needs to be input into the postural control system. A summary of some key points is made here.

In order to form an output, it is important to know where you are at the start. Input data must therefore describe the *initial conditions*. We get to learn the patterns of setting up common moves (the postural reference) and can be caught out when something changes (e.g. when we have a leg in plaster). To avoid frequent falls every time something changes, information is also needed about alterations to both the internal and external environments that would challenge our balance. These *challenges to balance* are also called perturbations and are described as being mechanical in origin if they occur as a result of external or self-generated forces, or sensory if they are the result of an unexpected sensory input (e.g. an object moving quickly into our visual field or a loud crash occurs). Such is the complexity of the information received that the brain is able to determine the strength, duration, frequency and direction of the perturbation.

ACTIVITY 5.4

Why do we need to know all this information? Think about the perturbations you might experience when walking along a crowded pavement. How do you cope with postural challenges you can see coming? What happens if someone runs into you from behind? Why is there a difference?

Integrating input and managing postural orientation in different contexts

How the nervous system integrates and responds effectively and efficiently to the wealth of incoming sensory data about postural and environmental positioning is the focus of much current research. While, as we shall see, there is some disagreement about specific sensory integration processes, it is widely agreed that all three forms of sensory input (visual, vestibular and somatosensory) are used to maintain an upright posture when an individual is standing quietly (Shumway-Cook and Woollacott, 2007, p. 177). Similarly, as suggested above, it is acknowledged that, during short-lasting perturbations, the body relies more on somatosensory input than on visual or vestibular inputs.

But how does the nervous system work out which inputs to favour, which to downplay, etc.? Shumway-Cook and Woollacott (2007, p. 178) draw on more than three decades of research to suggest that the possible processes behind the sensory integration needed to effect postural control align with the following two hypotheses.

The intermodal theory of sensory organization hypothesis

Citing the work of Stoffregen and Riccio (1988), Shumway-Cook and Woollacott suggest that this hypothesis proposes that all three sensory inputs contribute equally at all times and it is their interaction, i.e. the relationship between the three senses at any one moment, that provides the information essential for establishing postural orientation. Shumway-Cook and Woollacott (2007) encourage readers to liken the relationship of the three senses to the relationship each side of a triangle has with its neighbours.

The sensory weighting hypothesis

Shumway-Cook and Woollacott draw together a wealth of research (citing the works of Nashner, 1976; Jeka and Lackner, 1995 and Oie et al. 2002) that suggests that, instead of all three senses contributing equally, the central nervous system modifies the weight or importance of each input depending on its perceived accuracy. For example, when we are wearing footwear that is impeding (or overloading) our foot somatosensory receptors, the input from this sense will be downplayed and input from visual and vestibular sources upgraded. Researchers supporting this hypothesis suggest that the advantage of such a weighting system is that we can maintain stability in a variety of environments. This hypothesis may also explain the different postural control strategies witnessed to come into play during different tasks and between people of different ages and in different environments (Shumway-Cook and Woollacott, 2007, p. 180).

Whichever (if either) hypothesis is in time proved to be correct, all researchers and practitioners agree that the sensory processing needed to effect postural control is highly complex!

Are things going as we planned? Regulation and anticipatory control

Once a movement has begun, the sensory systems cannot relax. The postural control system requires ongoing information about the effect of the planned response – or system's *feedback*. Again, Lee (1988) provides a very interesting discussion about feedback, defining the process as an active method for controlling error. Sensory inputs inform higher centres about the achieved or expected output, and then an error signal representing the difference between actual and desired output is detected. The response is a modification of signal output. We are normally unaware of this process, as feedback involving all the subsystems occurs at an automatic level.

Of course no system is perfect, and relying on a feedback system has its problems. The process just described has many steps to produce a movement, and for feedback to be effective the whole process has to be completed repeatedly as the movement or correction continues. Loop delay, or the time it takes to make a correction, is one of the big problems with feedback systems (Lee, 1988, p. 297). Additional conduction problems or difficulties with central processing will adversely delay anticipatory and feed forward information. Some individuals will therefore move with a high risk of error and thus have difficulty producing efficient and effective movement control.

To combat some of these problems, we also have another mechanism for making adjustments: anticipatory control (Massion, 1992; Rothwell, 1994; Shumway-Cook and Woollacott, 2007). Anticipatory control is a process in which, in a known or commonly experienced situation (when the likely perturbations have been learned), signals for postural compensation and modification are sent before (in anticipation of) receipt of sensory information that the intervention is actually required.

Anticipatory control occurs during most of our regular daily activities (e.g. writing and stepping), reducing movement execution times considerably and therefore increasing the efficiency of the task effected. Shumway-Cook and Woollacott (2007, p. 181) report the studies of Cordo and Nashner (1982), who discovered that anticipatory control is effected by muscle synergies and that these synergies are the same as those utilized by the postural control feedback systems. Building on these results, researchers have now confirmed that postural muscle synergies are preselected in advance of planned action when that action is serial, expected and/or practised. The process of this central preselection is know as *central set* (Shumway-Cook & Woollacott, 2007, p. 182) and is thought to be responsible for both our rapid postural responses in known environments and the skill with which postural control responses are modulated during practised activities. The central set therefore reduces the risk of our over- or under-recruiting postural control–related muscles, thus increasing our postural efficiency. While the processes of anticipatory control set out above appear to reflect current research consensus, there is much less agreement about where within the central nervous system these processes are coordinated.

ACTIVITY 5.5

Think about the following situation. You are walking down the stairs in a busy building, holding your books in front of you (so you cannot see your feet) and chatting to your companions; all is fine and you arrive safely at the bottom. How does the difficulty of the activity (walking down the stairs) change when you start to worry about where your feet are? Now what is going on? Several things are happening here. First, you are adding in extra communication loops and increasing the time delay from sensory input to motor output, but second, in order to see your feet you are probably going to have to lean forwards to see the steps – what does this do to your COG? What do you think is happening in your vestibular system when you look down at the step? How does your visual system cope with sensing the height you could fall? What happens if a step is suddenly a different height from the rest?

Is it a good idea to make patients' postural control responses conscious? Or should we try to build up segmental, automatic control by staging the complexity and difficulty (in terms of segments needing to be controlled against gravity) of the activities we ask them to practise?

Conclusion

This section has reviewed some basic physiology and discussed the systems used to maintain balance while the body is performing functional activities. We have seen that movement requires the dynamic and ongoing interplay of vision, proprioception, contact cues, efferent control and internal (learned) models. Such multisensory input is essential for body orientation, the apparent stability of our surroundings, and ultimately movement control (Lackner and Dizio, 2000, p. 286). Finally, it must be acknowledged that not all postural control occurs at a purely subconscious level. In their research, Schlesinger et al. (1998) suggest that increasing attention is needed to maintain balance when postural tasks and/or coexisting secondary tasks increase in complexity. In addition, there is a plethora of literature exploring the normal consequences of the human ageing process and the ensuing increase in conscious balance control (see Lacour et al. 2008 for a review of the recent literature). Thinking about your balance is not wrong but does slow down your response times (Jacobs and Horak, 2007), making you vulnerable, especially in standing. Elderly people who may be consciously thinking about their balance more frequently than most are thus particularly at risk of falls while dual-tasking (Lacour et al. 2008).

POSTURAL CONTROL SYNERGIES USED IN STANDING

Small perturbations to stable postures, such as upright, quiet standing, can be accommodated without any central control because of the intrinsic elastic properties of muscle (Rothwell, 1994). Together with reflex loop innervation, muscles are provided with a stiffness similar to that of a spring. During small perturbations, muscle acts as a stretching spring resisting the disturbance. In upright standing, with the COG projected anterior to the ankle joints, it is the calf muscle that is thought to play an important role in accommodating small perturbations. However, the body in standing is vulnerable to large and potentially multidirectional perturbations. To meet these challenges, the human body has developed an efficient but complex suite of responses. While we know this latter to be a fact from our personal experience of balancing, research is still evolving to help us unpick the system's complexity. What follows therefore is part of a developing field of research. You are advised to explore the recent original texts if this is a specific area of interest for you and your practice.

Managing anteroposterior challenges

When in upright stance, the body employs three main strategies to combat threats to balance and stability in the anteroposterior directions (Nashner, 1990, cited in Shumway-Cook and Woollacott, 2007): the ankle strategy, the hip strategy and the stepping strategy. The process of balance management is subconscious and reduces response times by treating effector muscles as groups or *muscle synergies*. A synergy describes the functional coupling of groups of muscles that are constrained to work together as a unit (Shumway-Cook and

Woollacott, 2007, p. 166). Sending one signal to a synergy will therefore get a wide-ranging, set response – 'cutting out the middleman' – reducing response times and increasing efficiency.

The ankle strategy

The ankle strategy is used when the foot is fully supported on the contact surface so that the dorsiflexors and plantar flexors about the ankle joint can exert their influence through *reverse action*, i.e. if we sway backwards the anterior tibialis muscles will contract to bring us back to midline, pulling the body over the foot. Similarly, a forward sway will cause gastrocnemius to become active (as the perturbation is now greater than can be accommodated by intrinsic properties alone) and draw the COG forwards over the BOS (Cech and Martin, 2002). The ankle strategy (also known as the *inverted pendulum*; Rothwell, 1994) is effective for correcting small perturbations when standing on a firm surface (so active when you stand quietly) and obviously requires intact joint range and muscle strength about the ankle (Shumway-Cook and Woollacott, 2007, p. 167).

Of course, humans can make the task of balancing in upright stance hard for themselves, and many people choose to do this by the footwear they use! Basmajian (1964; Fig. 5.9) described the results of an electromyographic study into the resting activity of lower limb muscles while wearing different footwear (citing Joseph and Nightingale, 1956). As early as 1956, it was known that high heels increase the resting activity of the calf muscles because the COG is shifted forwards. It appears that wearers of high-heeled shoes partially compensate for the threat to balance by combining this increased muscle work with a simultaneous increase in lumbar lordosis.

ACTIVITY 5.6

Think about the posture of wearers of high-heeled shoes (Fig. 5.9). Refer back to sections describing ideal posture and classic deviations in posture. Describe the likely label and compensations further up the body if the posture of the high-heeled shoe wearer is held and/or becomes the norm.

The hip strategy

When the perturbations are greater than a small sway, when the disturbance is quickly applied, or when the foot is on a small contact surface, the body uses another strategy: the hip strategy. Using the large muscles of the hip and knee together,

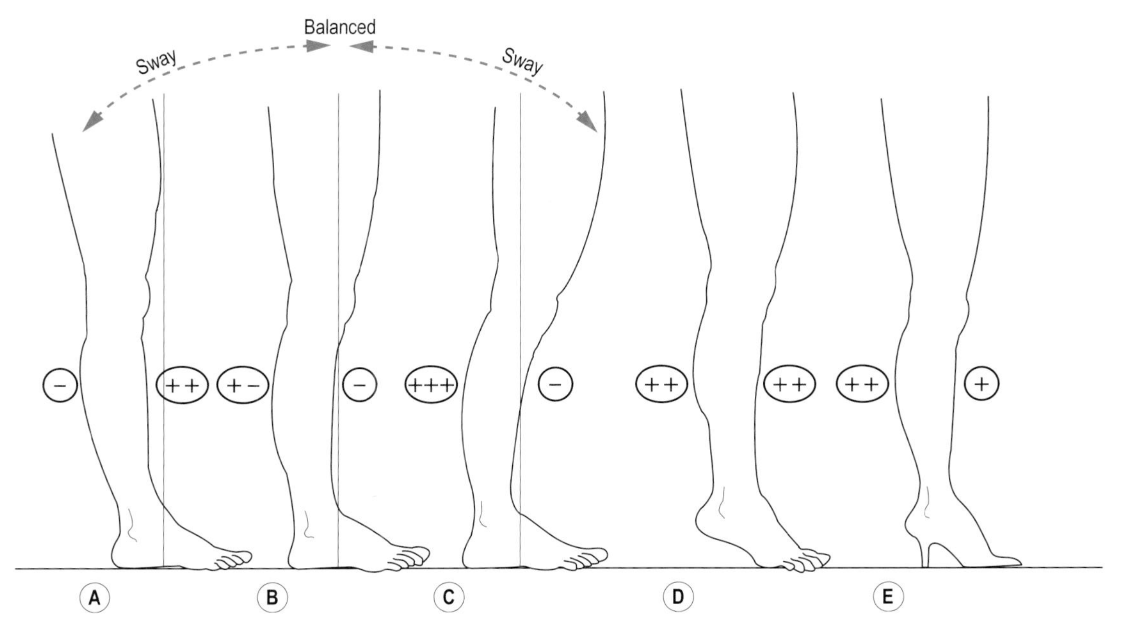

Figure 5.9 Muscular activity in leg muscles.

the hip strategy is better able to control the COG and prevent a fall (Shumway-Cook and Woollacott, 2007). Cech and Martin (2002) cited a study by Allum et al. (1998), who report that the hip strategy is evoked in preference to the ankle strategy for all perturbations that come from a mediolateral direction. Whatever the size of the perturbation, the study concluded that mediolateral forces activate muscle synergies in a proximal to distal order.

The stepping strategy

The third strategy, the stepping strategy, has been referred to above and is evoked when the perturbation is strong enough to really threaten stability. In this instance, a gross movement (e.g. a step or hop) is required to widen the BOS so that the COG can fall within it and balance be regained.

The diagrams (Fig. 5.10) from Shumway-Cook and Woollacott (2001, p. 179) summarize the boundaries to stability in upright stance and the sequence of strategy adoption.

Managing perturbations from other directions

Because of the limited mediolateral movements possible at the ankle and knee joints, these joints do not play a large or initial role in the adjustments needed to maintain mediolateral stability (Shumway-Cook and Woollacott, 2007). The primary mediolateral control of balance therefore occurs about the hip and trunk. As a consequence, the initiation of muscle patterns occurs in a proximal to distal direction.

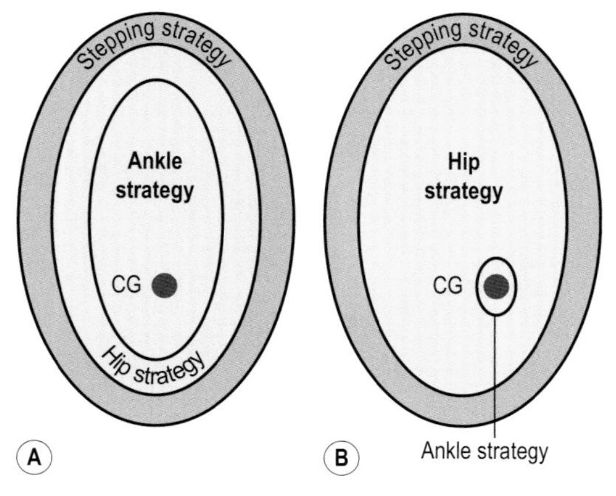

Figure 5.10 Boundaries for motor strategies. CG, centre of gravity. *(From Shumway-Cook and Woollacott, 2001, with permission.)*

When balance is challenged from multiple directions, the body draws on the complexity of muscle synergies, i.e. the notion that one muscle may be part of several synergies (Shumway-Cook and Woollacott, 2007, p. 170), and central organization to fine tune muscle activation to meet the needs of the challenge. Shumway-Cook and Woollacott suggest that, while research in this area is at an early stage, results propose that there may be only three 'robust' (p. 172) groupings of muscle. Potentially turning the earlier research on its head, they cite the work of Henry et al. (1998), who propose that two of the groupings are maximally active during diagonal and one during lateral balance challenges. Such research therefore suggests that anteroposterior synergies may not be the clear-cut strategies (ankle, hip and stepping) outlined above but rather a modification of a diagonal synergy. These are exciting times for postural control research, and we must remain alert to their findings and implications for practice.

Summary

This section has reviewed the postural control system and will conclude with a summary of the body systems and structures identified as essential for functional postural control (readers requiring more detailed information are encouraged to consult the original texts referred to above).

Requirements for an integrated postural control response

- Intact sensory system: sensors, nerves and communication pathways.
- Intact central nervous system: conduction pathways, internal communication networks, and framework for pattern reference or learned response patterns relevant to individual shape and body symmetry or alignment.
- Intact musculoskeletal system: effector pathways including nerves, neuromuscular junctions, strength and endurance of muscles, and range of movement about effector joints.
- Psychological engagement with surroundings: remember that the reticular formation is responsible for coordinating postural muscle tone activity in response to levels of higher centre arousal.
- An external environment that allows accurate postural changes to be made when perturbations occur.

ACTIVITY 5.7

How does your posture change when you are tired or in pain? What will be your response to perturbations in these conditions?

FACTORS THAT AFFECT POSTURAL CONTROL

Obviously, anything that impairs the functioning of the systems listed above will affect postural control responses, but there are some common situations that will challenge balance responses.

The effects of ageing

The effects of the ageing process on the postural control systems of healthy people can be summarized as follows (Woollacott and Shumway-Cook, 1990; Lord et al. 1991; Perrin et al. 1997, Kristinsdottir et al. 2001).

- Body sway increases with age. Only a small proportion (*c*.30–37%) of elderly people are able to maintain balance using the ankle strategy.
- The function of the sensory systems declines as part of the ageing process, but the visual system endures deterioration, becoming increasingly important as sensation and muscle strength reduce in the lower limbs and vestibular system function also declines.
- Both nerve conduction and central processing in the brainstem slow with age.

In practice, these challenges result in people having to think more about their posture during everyday activities, leaving little room for error or response to sudden change in circumstances, for example loud noises or sudden movements (Lacour et al. 2008). And one does not have to be of retirement age to experience the effects of ageing on balance. Poulain and Giraudet (2008) explored the balance responses of a range of individuals and suggested that participants aged between 44 and 60 years are relying on their visual sensory input, thus becoming very unsteady in darkened environments. Large numbers of our population will therefore find their postural control difficult, resulting in increased conscious control, slower response times and a reduced repertoire of correction responses. However, it is important to note that avoiding obesity and staying active during advancing years are important factors in maintaining the sensory integration needed for effective and efficient postural control (Prioli et al. 2005).

ACTIVITY 5.8

How do you think the changes seen with ageing will be noticed in terms of postural control? How might coexisting multipathology (e.g. cataracts) challenge postural control? What are the possible consequences on movement speed, energy efficiency, etc.?

The effects of coexisting multisystem activity

Many interesting studies have been conducted to look at the effects of adding activities (multitasking) to people trying to maintain increasingly difficult postures. As balance regulation requires increasing amounts of information-processing capacity (attention), greater demand is made of the available higher centre resources, in turn reducing the capacity for coexisting activity (Lajoie et al. 1993), for example memory and comprehension activities. While we have already explored the possible challenges for even healthy elderly people in this regard (Lacour et al. 2008), Schlesinger et al. (1998) warn that shift workers are also at risk from falls, work-related injuries, etc. because of the strong negative impact of sleep deprivation on postural control.

Hodges et al. (2002) and Dault et al. (2003) have investigated the effects of changing patterns of respiration and the muscle work of articulation on postural control. Both groups of researchers reported that these everyday activities have demonstrable effects on postural control. Hodges et al. suggested that the perturbations of respiration cause measurable counterbalancing activity in the trunk and lower limb muscles, a highly coordinated control system requiring multisegment synergies (p. 301).

The effects of high or low levels of exercise

Prolonged exercise (e.g. a 25-km run or an Ironman Triathlon) has been shown to reduce postural control when participants stop their activity. Lepers et al. (1997) and Nagy et al. (2004) suggest that this reduction is a result of the system's sensory receptors adjusting their levels of sensitivity in response to the hyperstimulation experienced during exercise and consequently needing to recalibrate on exercise cessation. Speers et al. (1998) also found that system recalibration occurred when they tested postural control responses in a sample of astronauts before and after space flight. The authors suggested that, in this instance, the recalibration was caused by hypostimulation experienced during weightlessness. Interestingly, when their sample of astronauts returned to land they were using hip correction strategies far more frequently than ankle strategies.

ACTIVITY 5.9

Why do you think astronauts demonstrate this postural control adaptation in the weightless environment?

One of the effects of prolonged low activity levels may be an increase in body weight. The effect of obesity on postural control has stimulated much research activity, with results suggesting that obesity negatively affects postural responses, especially if perturbations come from mediolateral directions (McGraw et al. 2000; Goulding et al. 2003). While not ignoring the influence of probable coexisting poor muscle endurance, both studies suggested that obese people have intact proprioceptive and sensory systems. Building on these studies, however, D'Hondt et al. (2008) explored the fine motor control of obese children in standing and found their motor function poorer than that of their non-obese peers. This observation suggests some underlying coordination difficulties and is currently challenging the earlier perception that postural control in obese people results from an altered muscle:body weight ratio (increasing body inertia so that more force generation is required for both the perturbation and the appropriate correction to occur).

Summary

In this section, we have thought about the systems required to make postural control efficient and effective and discussed some normal conditions under which predicted responses are challenged. Obviously, however, there are innumerable instances when internal and/or external events can unexpectedly prevent normal balance responses occurring (e.g. a limb amputation or lesions of the central nervous system).

RELEVANCE TO THERAPEUTIC PRACTICE

This final section of the chapter relates the given theoretical understanding to therapeutic practice. We shall review the possible precursors to abnormal, symptomatic posture and postural control and offer some examples of therapeutic options. Our goal is to facilitate understanding so that therapists consider some general routes to help manage and improve patients' safety and functional activity.

SYMPTOMATIC POSTURAL ALIGNMENT

Identifying causes for symptomatic postures

Earlier in this chapter, we described the ongoing interaction between the internal muscle and soft tissue forces and the external forces of gravity, friction and inertia. At any time, different force 'balance' will affect postural change. The task for the therapist is to identify the symptom's primary source and discuss with the patient the likely cause(s). We have discussed the importance of taking the patient through a detailed postural assessment in, and between, each of the three main resting postures to note how the patient's current posture deviates from the stated ideal. Causes of postural deviation may include muscles that are too weak, soft tissues too tight or long, and joint movements impeded by swelling or hampered by actual or anticipated pain.

Relieving the symptoms and preventing recurrence

It is essential that people suffering with symptomatic postures are given the opportunity to rest in positions of ease, in which their postures are supported, reducing the effort needed for their maintenance. Readers are referred back to the sections describing supported sitting and lying postures. Secondary intervention will manage symptom control and recurrence prevention through detailed ergonomic review, appropriate muscle strengthening and joint realignment disciplines. Postural rehabilitation is a long-duration activity requiring high patient motivation and patient–therapist interaction and support. At all times, therapists must be aware of the individual needs of the patient, and the context and environment in which that patient operates, helping to make therapeutic intervention relevant whenever possible.

PROBLEMS WITH POSTURAL CONTROL: BALANCE RE-EDUCATION

Identifying the causes

If a patient with a recent amputation or spinal cord injury is referred for balance re-education, then the reasons for balance problems may seem apparent, but therapists must also consider the impact of coexisting pathology (e.g. vision, vestibular processing and general system ageing). Whatever the cause of the balance problem, it is likely that the patient's soft tissues will have adapted quickly to the new environment and stimuli. It is essential that the musculoskeletal system is prepared (as above) so that it is in the best condition possible to respond to demands made on it by the central nervous system. For example, sensory endings cannot provide adequate information when surrounded by oedema, nor can the muscle function as required if its connective tissue has shortened or adhered.

When muscles, soft tissues, etc. have been prepared for action, it is important to help the patient integrate the changes made into functional activities. Remember that there are essentially two ways in which the postural control system operates: it is able to maintain our segmental alignment in quiet postures when no or minimal perturbations are expected, and it maintains balance when perturbations are great (e.g. while performing multitask or functional activities). This knowledge can be used to help stage patients' rehabilitation. Reflecting on the needs of a functional postural control system allows us to develop a tailored rehabilitation program. The points below reflect the range of questions that should be considered by the therapist when developing such a programme.

Some questions to ask

- Is there a need to try to increase existing sensory stimulation levels? Could we use a mirror or approximate the joints to increase mechanoreceptor input? Is the department too noisy?
- How good is the patient's segmental control? Should we reduce the difficulty of the starting position by lowering the COG and increasing the BOS?
- Is the patient's problem with segmental or intersegmental control? Should we work to achieve subconscious control segment by segment, or does the patient have problems initiating or controlling the muscle synergies?
- So should the patient be fairly static, or are they ready to have their balance challenged by expected and unexpected perturbations? Remember the existence of the feedback and feed forward systems: it is normal to have difficulty responding to perturbations from behind and those that come at high frequency – patients are not necessarily responding abnormally if they have problems with these!
- The patient is thinking very hard about their balance reactions; is this a problem? As we have seen, patients can be encouraged to think about preparing their segmental alignment for functional activity. Directly thinking about postural control during simple functional activities can cause more problems but is essential during advanced, difficult postural conditions and when performing multiple tasks simultaneously.

This list is incomplete but is intended to help the reader consider the issues that may be impacting on the patient's functional activity efficiency and effectiveness. The key is adequate musculoskeletal preparation followed immediately by integration into controlled functional activity. The nervous system is continually adapting, and we need to ensure that the components of the postural control system are challenged in such a way that compensations occur towards the normal.

CASE STUDY 5.3

We will revisit the challenges to balance during everyday activities. Let us go back to Case study 5.1 and our family enjoying a night out at the cinema and restaurant. Now that you have read the chapter, can you answer the following question? If each family member is responding normally to their environment, why are they not all responding in the same way?

What specific, normal, individual adjustments do you now notice in the case study story? Why did the adult members of the party find it difficult to navigate stairs when their arms were full and the light was dim and flickering? If you are not sure, look again at the section *How does postural control work?*

Thinking of Agnes, why might she have been a little bit quiet or seemed distracted and not engaging in the family's conversation as they walked to the restaurant? Look again at the section *Factors that affect postural control* if you are not sure.

Have you noticed these (and other) responses among your own family members but not really thought why they were acting in this way? Next time you are in a public space, sit quietly and have a look at:

- the potential challenges to people's balance in that environment
- how different people are responding to or navigating those challenges.

Think now about the traditional therapy department. What are the potential challenges to balance in these environments? How similar is this context to real life? If a person has been referred for investigations into falls, how relevant will any balance assessment you do in this environment be to their usual spaces? Is balance rehabilitation complete if a person can walk about safely within a therapy environment?

CONCLUSION

This chapter has explored the origins and norms of the upright human posture and the demands that are placed on whole body systems in order to produce effective functional activities. We have described the 'ideal' postural alignment and discussed the real life influences that make segmental alignment unique to the individual. We have acknowledged the complexity of postural control, the multisystem integration needed, and the frequency of balance problems in the patient and general populations. An understanding of posture and its control is essential for all involved in the rehabilitation of human movement.

References

Allum, J.H., Bloem, B.R., Carpenter, M.G., et al., 1998. Proprioceptive control of posture: a review of new concepts. Gait and Posture 8, 214–242.

Basmajian, J.V., 1964. Man's posture. Arch. Phys. Med. Rehabil. 45, 26–36.

Bray, J.J., MacKnight, A.D.C., Mills, R.G., 1999. Lecture Notes on Human Physiology, fourth ed. Blackwell Science, Oxford.

Cech, D.J., Martin, S., 2002. Functional Movement Development Across the Life Span, second ed. WB Saunders, Philadelphia.

Clancy, J., McVicar, A.J., 2002. Physiology and Anatomy: A Homeostatic Approach, second ed. Arnold, London.

Cordo, P., Nashner, L.M., 1982. Properties of postural adjustments associated with rapid arm movements. J. Neurophysiol. 47, 287–302.

Cutler, W.B., Friedmann, E., Genovese-Stone, E., 1993. Prevalence of kyphosis in a healthy sample of pre- and postmenopausal women. Am. J. Phys. Med. Rehabil. 72, 219–225.

Dault, M.C., Yardley, L., Frank, J.S., 2003. Does articulation contribute to modifications of postural control during dual-task paradigms? Cogn. Brain Res. 16, 434–440.

D'Hondt, E., Deforche, B., De Boudeaudhujj, I., Lenoir, M., 2008. Childhood obesity affects fine motor skill performance under different postural constraints. Neurosci. Lett. 440, 72–75.

Gillespie, R.M., 2002. The physical impact of computers and electronic game use on children and adolescents, a review of current literature. Work 18, 249–259.

Gillespie, R.M., Nordin, M., Halpern, M., Koenig, K., Warren, N., Kim, M., 2006. Musculoskeletal impact of computer and electronic game use on children and adolescents. Available at: http://www.iea.cc/ergonomics4children/pdfs/art0235.pdf (Accessed on November 13th, 2008).

Goulding, A., Jones, I.E., Taylor, R. W., Piggot, J.M., Taylor, D., 2003. Dynamic and static tests of balance and postural sway in boys: effects of previous wrist bone fractures and high adiposity. Gait Posture 17, 136–141.

Grieco, A., 1986. Sitting posture: an old problem and a new one. Ergonomics 29, 345–362.

Hamill, J., Knutzen, K.M., 2003. Biomechanical Basis of Human Movement, second ed. Lippincott Williams & Wilkins, Philadelphia.

Henry, S.M., Fung, J., Horak, F.B., 1998. EMG responses to maintain stance during multidirectional surface translations. The Journal of Neurophysiology 80, 1939–1950.

Hicks, J.H., 1955. The foot as a support. Acta Anatomica 25, 34–45.

Hodges, P.W., Gurfinkel, V.S., Brumagne, S., Smith, T.C., Cordo, P.C., 2002. Coexistence of stability and mobility in postural control: evidence from postural compensation for respiration. Exp. Brain Res. 144, 293–302.

Jacobs, J.V., Horak, F.B., 2007. Cortical control of postural responses. J. Neural. Transm. 114, 1339–1348.

Jeka, J.J., Lackner, J.R., 1995. The role of haptic cures from rough and slippery surfaces in human postural control. Experimental Brain Research 103, 267–276.

Joseph, J., Nightingale, A., 1956. Electromyography of muscles of posture: leg and thigh muscles in women, including the effects of high heels. The Journal of Physiology 132, 465–468.

Juskeliene, V., Magnus, P., Bakketteig, L.S., Dailidiene, N., Jurkuvenas, V., 1996. Prevalence and risk factors for asymmetric posture in preschool children aged 6–7 years. Int. J. Epidemiol. 25, 1053–1059.

Kisner, C., Colby, L.A., 2007. Therapeutic Exercise: Foundations and Techniques, fifth ed. FA Davis, Philadelphia.

Kristinsdottir, E.K., Fransson, P.A., Magnusson, M., 2001. Changes in postural control in healthy elderly subjects are related to vibration sensation, vision and vestibular asymmetry. Acta Otolaryngol. 121, 700–706.

Lackner, J.R., DiZio, P.A., 2000. Aspects of body self-calibration. Trends. Cogn. Sci. 4, 279–288.

Lacour, M., Bernard-Demanze, L., Dumitrescu, M., 2008. Posture control, aging, and attention resources: models and posture-analysis methods. Clin. Neurophysiol. 38 (6), 411–421.

Lajoie, Y., Teasdale, N., Bard, C., Fleury, M., 1993. Attentional demands for static and dynamic equilibrium. Exp. Brain Res. 97, 139–144.

Lee, W.A., 1988. A control systems framework for understanding normal and abnormal posture. Am. J. Occup. Ther. 43, 291–301.

Lepers, R., Bigard, A.X., Diard, J.P., Gouteyron, J.F., Guezennec, C.Y., 1997. Posture control after prolonged exercise. Eur. J. Appl. Physiol. 76, 55–61.

Lord, S.R., Clark, R.D., Webster, I.W., 1991. Postural stability and associated physiological factors in a population of aged persons. J. Gerontol. 46, 69–76.

Massion, J., 1992. Movement, posture and equilibrium: interaction and coordination. Prog. Neurobiol. 38, 35–56.

McGraw, B., McClenagha, B.A., Williams, H.G., Dickerson, J., 2000. Gait and postural stability in obese and non-obese prepubertal boys. Archives Phys. Med. Rehabil. 81, 484–489.

Morton, D.J., 1929. Evolution of man's erect posture. J. Morphol. Physiol. 43, 147–179.

Nagy, E., Toth, K., Janositz, G., Kovacs, G., Feher-Kiss, A., Angyan, L., Horvath, G., 2004. Postural control in athletes participating in an iron man triathlon. Eur. J. Appl. Physiol. 92, 407–413.

Nashner, L.M., 1976. Adapting reflexes controlling the human posture. Experimental Brain Research 26, 59–72.

Nashner, L.M., 1990. Dynamic posturography in the diagnosis and management of dizziness and balance disorders. Neurol. Clin. 8, 331–349.

Nissinen, M., Heliovaara, M., Seitsamo, J., Poussa, M., 1989. Trunk asymmetry and scoliosis. Anthropometric measurements in prepubertal school children. Acta Paediatr. Scand. 78, 747–753.

Nissinen, M., Heliovaara, M., Seitsamo, M., Poussa, M., 1993. Trunk asymmetry, posture, growth and risk of scoliosis. Spine 18, 8–13.

Nissinen, M., Heliovaara, M., Seitsamo, J., Poussa, M., 1995. Left handedness and risk of thoracic hyperkyphosis in prepubertal schoolchildren. Int. J. Epidemiol. 24, 1178–1181.

Nissinen, M., Heliovaara, M., Seitsamo, J., Kononen, M.H., Hurmerinta, K.A., Poussa, M., 2000. Development of trunk asymmetry in a cohort of children ages 11 to 22 years. Spine 25, 570–574.

Oie, K., Kiemel, T., Jeka, J.J., 2002. Multisensory fusion: Simultaneous re-weighting of vision and touch for the control of human posture. Cognitive Brain Research 14, 164–176.

Palastanga, N., Field, D., Soames, R., 2006. Anatomy and Human Movement: Structure and Function, sixth ed. Butterworth-Heinemann, Edinburgh.

Pearsall, D.J., Reid, J.G., 1992. Line of gravity relative to upright vertebral posture. Clin. Biomech. 7, 80–86.

Perrin, P.P., Jeandel, C., Perrin, C.A., Bene, M.C., 1997. Influence of visual control, conduction and central integration on static and dynamic balance in healthy older adults. Gerontology 43, 223–231.

Poulain, I., Giraudet, G., 2008. Age-related changes of visual contribution in posture control. Gait Posture 27, 1–7.

Poussa, M.S., Heliovaara, M.M., Seitsamo, J.T., Kononen, M.H., Hurmerinta, K.A., Nissinen, M.J., 2005. Development of spinal posture in a cohort of children from the age of 11 to 22 years. Eur. Spine J. 14, 738–742.

Prioli, A.C., Freitas, P.B., Barela, J.A., 2005. Physical activity and postural control in the elderly: coupling between visual information and body sway. Gerontology 51, 145–148.

Reeser, L.A., Susman, R.L., Stern, J.T., 1983. Electromyographic studies of the human foot: experimental approaches to hominid evolution. Foot Ankle 3, 391–407.

Rothwell, J., 1994. Control of Human Voluntary Movement. Chapman & Hall, London.

Schlesinger, A., Redfern, M.S., Dahl, R.E., Jennings, J.R., 1998. Postural control, attention and sleep deprivation. Neuroreport 9, 49–52.

Shumway-Cook, A., Woollacott, M., 2001. Motor Control: Theory and Practical Applications, second ed. Lippincott Williams & Wilkins, Baltimore.

Shumway-Cook, A., Woollacott, M., 2007. Motor Control: Translating Research into Clinical Practice, third ed. Lippincott Williams & Wilkins, Philadelphia.

Speers, R.A., Paloski, W.H., Kuo, A. D., 1998. Multivariate changes in coordination of postural control following spaceflight. J. Biomech. 31, 883–889.

Stoffregen, T.A., Riccio, G.E., 1988. An ecological theory of orientation and the vestibular system. Phychological Review 95, 3–14.

Williams, L.R.T., Caswell, P., Wagner, I., Walmsley, , Handcock, 1994. Regulation of standing posture. N. Z. J. Phys. 22, 15–18.

Woodhull, A.M., Maltrud, K., Mello, B.L., 1985. Alignment of the human body in standing. Eur. J. Appl. Physiol. 54, 109–115.

Woollacott, M.H., Shumway-Cook, A., 1990. Changes in posture control across the life span – a systems approach. Phys. Ther. 70, 799–807.

Chapter 6

Motor learning

Nicola Phillips

CHAPTER CONTENTS

LEARNING OUTCOMES

At the end of this chapter, you should be able to:
1. define the term *motor learning* and describe the limitations to motor control
2. demonstrate an understanding of models of motor control
3. define the term *skill*
4. discuss the main components involved in the skill acquisition process
5. discuss how long- and short-term memory are involved in learning and performing a motor skill.

INTRODUCTION

Earlier chapters have explained some of the biomechanical principles of movement and the musculoskeletal and neurological bases of movement production and control. This chapter aims to explain how these human movements are learned and become the skilled coordinated patterns of activity necessary for function.

MOTOR LEARNING

DEFINITION

Motor learning has been defined as a set process associated with practice or experience leading to relatively permanent changes in skilled behaviour (Schmidt, 1988).

The set process mentioned in the above definition will be explained in this chapter. The other important term to note in the definition is *relatively permanent changes*. A change in technique of carrying out a particular task is considered learned only if it can be performed consistently rather than merely by chance. Vereijken et al. (1992) described motor learning as the process of adjusting movement characteristics to a new task or challenge. This concept has a very similar meaning to Schmidt's definition but has an emphasis on observation of the movement characteristics as opposed to models of what might be happening in the brain. Again, this concept will also be explained later in the chapter.

HISTORICAL PERSPECTIVE

The field of movement control has historically been studied from two entirely different areas: those that provide models considering motor control as a top down process, and those that describe it as a bottom up process. The psychology and neurophysiology researchers tend to describe a central nervous system (CNS) control of learning movement patterns. In contrast, biomechanists have adopted engineering principles to describe models of motor learning that are adapted to changes at the peripheries. It is very likely that motor learning is a combination of both schools of thought, but unfortunately the language used by all these different groups can make comparison and integration of principles confusing. Sherrington (1906) was an important early influence in neural control theories, and his concepts of reflex responses to stimuli causing movement of the extremities is still a foundation of many treatment approaches today. It was this formative work that highlighted some of the sensory receptors involved in proprioception and introduced the concept of reciprocal innervation of agonist and antagonist muscle. The term *final common pathway*, when the influences from reflexes, sensory sources and cognitive sources converge at spinal level, was also introduced at this time.

Weiner first developed the information-processing model in 1948 and likened the brain to a computer in which information is received and processed, leading to an output to muscles, creating movement. This approach was termed *cybernetics*. Since then, there has been a gradual progression, in this line of thinking, towards a model of cognitive information processing (Schmidt, 1975).

Bernstein (1967) was one of the formative authors on motor control and learning, and was one of the first authors to attempt to integrate biomechanical, neural and psychological models of motor control. Much of his work has underpinned some of the current thinking of this subject.

Winstein (1991) stated that the acquisition of motor learning is fundamental to human life and consists of neural, physical and behavioural components. This statement conveniently encompasses many of the aspects covered in this chapter and serves as a reminder of the different areas that need to be considered to understand the basic concepts of motor learning.

TYPES OF MOVEMENT

Movement can be broadly divided into two types:

1. reflex (these movements are usually inherited)
2. learned (these do not appear to be inherited and therefore need practice).

Both these types of movement can be either simple or complex. For example, a simple reflex task would be blinking in response to an object near the eye. We might not be consciously aware that this has happened until after the response, and quite possibly then only because the object, such as dust, might have caused some irritation. We certainly would not remember ever having to learn how to blink at the right time.

Conversely, a simple learned task would be clapping your hands or reaching for a toy. Learning to clap requires conscious control and a number of attempts to bring the hands into contact at the right time to make a noise; the concentration on a young child's face is testament to this!

Similarly, breathing is a complex motor task subject to a great deal of variation but is under reflex control. We are not consciously aware of the general rate or depth of our breathing unless it changes dramatically, such as after a fairly intense physical effort, but there is a high degree of coordination required to produce the optimum rate and depth of breathing for every circumstance. On the other hand, a gymnastic tumbling routine is a complex learned task that requires hours of practice and takes a great deal of conscious control throughout the learning process before it can be performed accurately, consistently and efficiently.

The learned movements described above are all tasks that have been refined through trial and error practice until they produce a successful outcome. This might be a baby managing to clap their hands or grasp a toy after a few months of waving their arms around in a seemingly haphazard way, or it might be the gymnast perfecting a tumbling routine, often after many painful failed attempts at the component skills. How these movements are learned is where the experts have differing views. Some believe that movements develop and become more refined as the CNS develops, and that developing more complex movements is not possible until the CNS has developed enough. On the other hand, others hypothesize that movement control develops as a response to the requirements made in the limbs, in which case the CNS develops as an adaptive mechanism and will therefore adapt only if those demands are made on the body.

These different theories will be described separately, but some of the points made will overlap and relate to each other despite the differing language used.

MOTOR CONTROL

THE CLOSED LOOP THEORY

Coordination of movement, whether reflex or learned, can be termed *motor control*. The different areas of the CNS that deal with reflex or learned responses have been highlighted in the previous chapter and include spinal cord, brainstem, motor cortex and cerebellum. The study of this field is about how movements are selected in response to sensory information obtained from the environment and/or within the body, based on previous experience. The process, termed the *schema theory* by Schmidt (1975), is thought to be controlled by the long-term memory and then modified by other centres in the CNS. Figure 6.1 provides an overview of this concept, and despite coming from an information-processing model bias, it also acknowledges the importance of environmental constraints on the execution of a task and developing motor control.

This process can be divided into three main parts for clearer explanation.

Stimulus identification

Input via interoceptors or exteroceptors is identified as stimuli at CNS level. This stage can be subdivided into three stages.

Stimulus detection

A myriad of different stimuli could be detected via different sensory receptors at any one time. For example, when walking downstairs the CNS is receiving information from the visual field; auditory input of the sound of each step, background noises and possibly continuing a conversation; and proprioceptive input from joint, muscles and tendons about the depth and width of the step. A filtering process is therefore adopted to avoid information overload. Consequently, there is not usually conscious awareness of this level of perception unless the brain receives input it does not recognize or is not expecting in comparison with previous similar functional tasks. For instance, during that normally familiar task of walking downstairs you do not notice the width and depth of each step, allowing you to continue your conversation – unless one step is suddenly different. You then become very aware of the steps beneath your feet. Your brain takes in the visual input and proprioceptive and tactile feedback but alerts your conscious thought only if there is a problem. How many times have you walked down a flight of steps and almost tripped because one step is slightly different to the others? You turn and look at the step and, for a while, watch each step carefully to avoid stumbling again. Normally, stimuli that have been received for a while are ignored, whereas any new stimuli are passed on to the next stage of the process. This is a type of internal feedback and is termed *knowledge of performance*.

Stimulus interpretation

How you interpret information will depend on what sort of stimulation is expected and any prior experience of similar situations. The accuracy of this interpretation will depend on how efficiently the individual can retrieve information about previous experience, stored through various memory mechanisms discussed later in the chapter. A typical example in the sporting world would be two badminton players, one experienced player and one novice. The experienced player will recognize the body position and foot movements of an opponent who is about to make a shot and react accordingly, successfully returning the shot in an attacking position. The novice may miss the telltale visual signs allowing earlier reaction and not be able to get into position quickly enough, either missing the shot entirely or only able to return defensively, providing the opponent with another attacking opportunity.

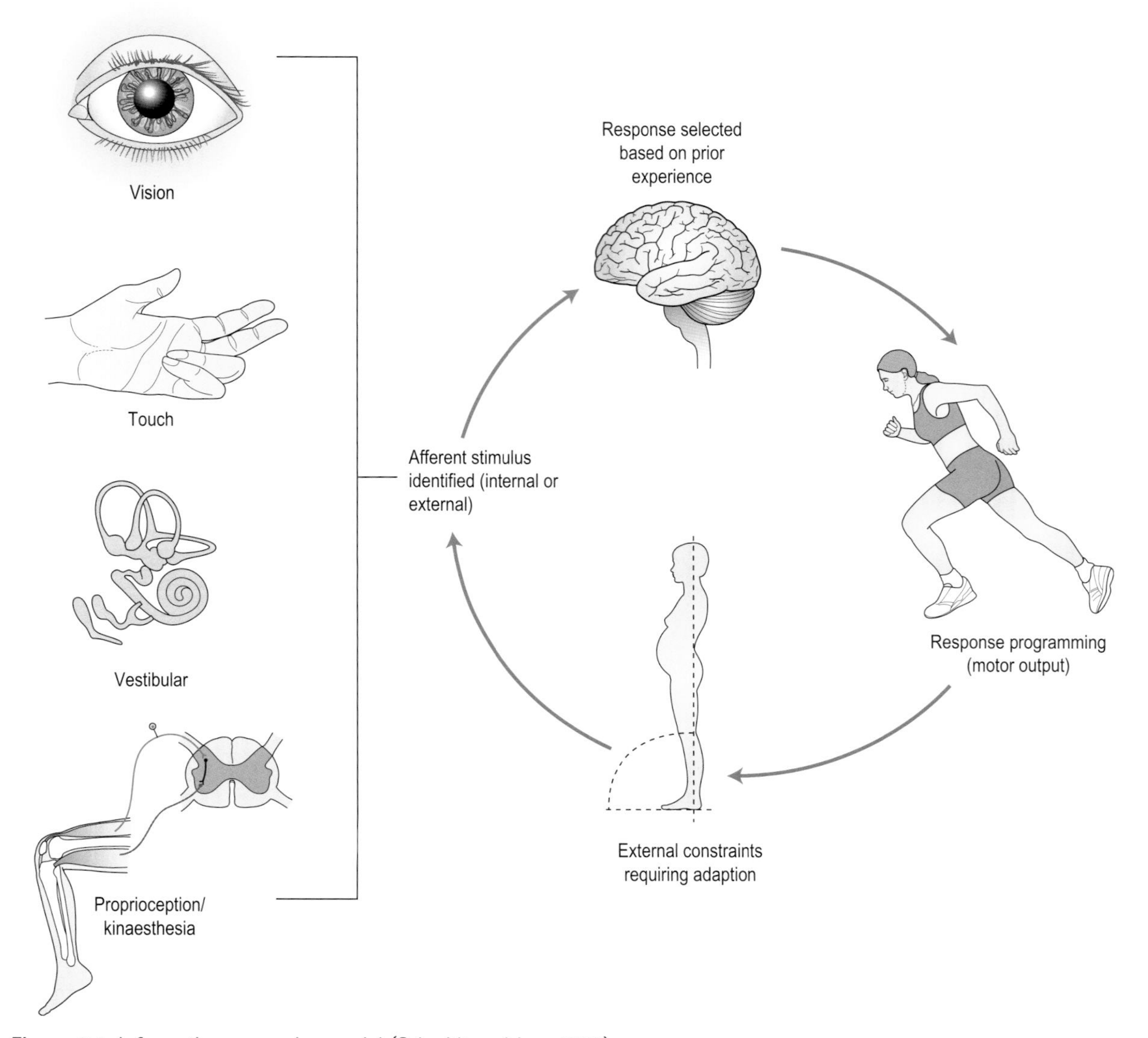

Figure 6.1 Information processing model (Schmidt and Lee, 1999).

Stimulus selection

How much or what type of stimuli selected to be passed on to the next stage of the decision-making process will depend on how much concentration or attention is devoted to this particular function at the time. Selection of the most useful stimuli for that particular task requires the correct allocation of attention to ensure that the appropriate information is passed on. The balance of the breadth of attention needs to match the focus of attention for the correct selection of movement. For example, compare attention to a beam of light: it can be a tightly focused spotlight or a diffused beam covering a large area. The focus needs to be tight enough to increase concentration, but concentrated too tightly it will create a tunnel vision therefore missing important cues around the periphery. Someone learning to ski has to concentrate on controlling the skis and is probably not able to take in the alpine scenery and stay on their feet at the same time. Conversely, they may not take in peripheral cues such as other skiers or trees in time to do anything about it!

Response selection

The appropriate movement pattern in response to the identified stimuli is likely to be chosen based on prior experience. Once the afferent stimuli have been accepted and recognized, the CNS must then decide what sort of response to make. A movement plan is assembled based on identical or similar previous performances. Imagine you are about to assess a patient in an outpatient clinic, having been given a referral with a brief diagnosis of a condition you have not treated before. First, you look up the condition and different methods of treatment in a textbook (long-term or declarative memory). Then, you make a rough plan of what you are going to do in the assessment (short-term memory). During your assessment, you modify what you do depending on what information you receive from your patient. Once you finish your assessment, you organize the subjective and objective information you have collected and formulate a treatment plan. This whole process is similar to the decision-making process involved in a motor task. From this example, you will see that if this cognitive process was the only strategy, then there is likely to be a substantial delay, which would be too slow for most functional tasks; even more so when considering complex skills. This will be discussed further as different concepts are described.

Response programming

The relevant motor experience is carried out with feedback to decide if it was the right choice. This is the stage in the process at which the selected response is coordinated by recruiting the appropriate muscles with the right amount of effort, in the right direction, at the right time. These patterns of movements are thought to be stored as action memory traces in the motor cortex once they have been refined through feedback mechanisms (Vansant, 1995). Any environmental changes (external) that happen during the execution of the task would then be relayed back through the loop (internal) and further modifications made to movement as required.

Neural control of the motor system is covered in more detail in the chapter on neural control of movement.

At first consideration, this whole process would seem to be an insurmountable task to be performed within the time limits needed for almost all skilled movements. If every stage of all our movements had to be controlled by cognitive processes, how could we possibly be able to think of other things or be able to speak or listen at the same time?

Information processing theorists suggest that central programming in the motor cortex allows unconscious performance of some skills once they have been learned, allowing an individual to be able to divert cognitive processes to other areas that might be necessary at the time. Findings of some studies would certainly seem to support this thinking (Doyon and Ungerleider, 2003). For example, Milton et al. (2007), using functional magnetic resonance imaging, found very different patterns of CNS activity in novice versus expert golfers. The posterior cingulate gyrus, amygdala and basal ganglia were active only in novices, whereas expert activation was mainly in the parietal lobe. These findings suggest that novices may well use a more cognitive approach to movement control (basal ganglia activity) and also have more difficulty filtering out sensory information (posterior cingulate gyrus activity).

Relating back to the example of the badminton player, the experienced player will be able to use a more parallel style of taking in external cues, possibly through a more efficient CNS control of all the information, as described by Milton et al. (2007) in golfers. It means that instead of dealing with one stimulus at a time, which would result in quite a delay in some responses, the expert player could notice and react to a few things at once, allowing cues to be taken in at a glance. This would in turn allow concentration on other skills, such as tactical decision making.

ACTIVITY 6.1

If you can drive a car, think back to when you were learning.

- How many different things did the instructor tell you to pay attention to at once?
- How many different things did you have to do with your feet and your hands at the same time?
- How many lessons did it take you to be able to cope with all those tasks at the same time?

Constant monitoring of the movement through proprioceptive *feedback* allows knowledge of performance before knowledge of results. This means that we are able to have an awareness of how the movement is progressing by utilizing sensory information from muscle spindles and joint and tendon receptors without waiting for information from visual or auditory receptors on completion of the task. Continued monitoring each time a motor task is performed means that an expert in a particular skill will know if that movement was successful before viewing the outcome. For example, an experienced weightlifter will know whether they have made a successful clean and jerk attempt before the referees pass or fail the lift. The weightlifter can relate the current performance to previous attempts and compare the feel of that particular movement with previous successful and unsuccessful attempts as the movement is happening rather than waiting to view the outcome. Consistent afferent information from joints and muscles is vital for the success of this function, and this has been investigated in individuals with known highly skilled activities, such as expert pianists (Lim et al. 2001).

Feed forward

It should be noted that some motor tasks are much too fast for such a cognitive control, which takes approximately 200 ms. A boxer's punch has a reaction time of 90 ms (Schmidt, 1991). This is well below the reaction time necessary for a cognitive response. A *feed forward* mechanism is thought to control this type of task. The memory traces in the motor cortex are thought to instigate preparatory muscle stimulation in order to respond at this speed. Although this mechanism allows a faster response, the boxer does not have time to think about altering the movement once the response has been initiated. This is why a boxer has to practise each different type of punch in their repertoire until the feed forward mechanism is developed sufficiently for each response. This is likely to be stored as procedural memory (see later in the section on memory).

Realistically, the brain would never have the capacity to store individual movement programmes for each and every functional task. An expert at such a task is more likely to have an overall picture of the task, requiring some variation depending on the circumstances, rather than a separate prepared pattern for each component of the task. The less predictable the task, as in boxing, the more practice is required. It is thought to take at least 10 000 h of deliberate practice for an elite athlete to reach an appropriate level of expertise.

Limitations of motor control

Having discussed how this process of top down motor control works, there are obviously going to be limitations within the whole mechanism. These limitations are as follows.

- Capacity: how much information the system can process at any one time. Every individual has a limit to how much information they can cope with all at once.
- Speed: how fast it can process the information. This will depend on whether the individual has to deal with the information in series (i.e. one at a time if it is unfamiliar) or in parallel (i.e. at the same time if it is very familiar).
- Distortion: the extent to which information is lost or distorted during the process. This limitation will often depend on concentration or attention and will also be affected by memory (emotional or semantic) associated with previous attempts at the same task. The different types of memory are explained later in the chapter.

The relatively lengthy process of a closed loop model described above was the major criticism of the dynamical systems theorists, who proposed that centrally driven processes were far too slow for most functional tasks and therefore developed a model based on externally driven processes.

DYNAMIC SYSTEMS THEORY

As suggested by its title, this approach attempts to explain motor control from a more mathematical point of view by observing and predicting patterns of multisegmental movement. The definition provided by Vereijken et al. (1992) earlier in the chapter works on the principle of dynamic systems theory. The initial stimulus for development of this theoretical concept came from the observation of natural phenomena such as hurricanes, in which complex patterns of air can emerge from simple initial conditions, through self-organization of air particles, without the need for an overriding central control system.

Some of the reasoning behind these theories is that individuals have been observed to perform specific tasks in similar ways, despite the opportunity to get to the end point in a variety of routes. This suggests that, for many tasks, there is likely to be an optimum way of moving that requires the least energy for that length and weight of limb as well as the sort of movement required. The resulting movement could then theoretically be achieved through a combination of multiple subsystems rather than a single central command centre.

A joint will have a certain number of degrees of freedom of movement. For example, a metacarpophalangeal joint will have one degree of freedom (flexion or extension), while the shoulder has three (flexion or extension, abduction or adduction, and rotation). In addition to that, there are more muscles going over most joints than there are degrees of freedom, which gives us the choice of using different muscles to produce variations in joint movement in any particular direction. Finally, each muscle has a number of different fibre types as well as different nerve endings that produce different qualities of muscle action and performance, such as power generators or stabilizers.

With all these parameters, we can perform a single movement in a host of different ways, yet we usually do familiar things in a similar way. For instance, a footballer will have a particular way of striking a ball that works best for them. That pattern still has to be flexible enough to be able to change for slightly different circumstances but is essentially similar for that skill. How much hip flexion compared with knee extension power used to kick the ball will depend on the individual. Some players may include a different proportion of ankle movement or trunk rotation to put a spin on the ball, while others will seek a straighter trajectory with more power.

The more expert a person is at a skill, despite having well-rehearsed, predictable movements, the more they are able to free up some of these degrees of freedom when needed so that they can subtly alter movement. This compares with the psychological models explained above, in which an expert can change their movement more easily in response to the environment, when the initial basic movement pattern becomes more automatic.

Thelen (1998) describes how babies learn to reach for objects depending on their initial, individual styles of early movement. The babies that tended to move quickly had to deal with trying to be more accurate and avoid missing the toy they were aiming at. On the other hand, the babies who moved more slowly were accurate but had to develop better antigravity control, as their arms were held in the air for longer. They have to do all this with large heads compared with the rest of the body and narrow shoulders with weak muscles. Added to that, they have just spent 9 months floating around in an aquatic environment that did not prepare them for dealing with the effects of gravity. Any decision about which combinations of muscles and movement to use has to be done through trial and error. For instance, do they use the closest arm and abduct their shoulder or the other arm and adduct? Do they move the shoulder more and keep the elbow bent or vice versa? Do they need to have palm up or down? The list could go on forever, and that is just reaching for a toy!

ACTIVITY 6.2

Think of a task such as reaching for a glass of water or throwing a ball.

- How many different ways could you do this?
- How would you do the same task if your shoulder movement was restricted?
- How would you do the task if your elbow movement was restricted?
- What alterations did the restricted movement attempts involve?

All the adaptation you would have tried in the task would have required changing the degrees of freedom you used at different joints in the upper limb segments and probably in your trunk and lower limb as well. You increased the variety of movement at some joints while restricting the choice in others. This is called recruitment or suppression of biomechanical degrees of freedom (Kelso, 1998).

Table 6.1 summarizes the main concepts of the closed loop (information-processing) and dynamic systems theories.

SKILL

The term *skill* has been mentioned frequently when discussing the process of motor learning. Before explaining how skills are learned and developed, it might help to define the word.

DEFINITION

Skill is the accuracy, consistency and efficiency of movement deployment (Higgins, 1991).

Table 6.1 Main concepts of the information processing and dynamic systems theories

CLOSED LOOP (INFORMATION PROCESSING) THEORY	DYNAMIC SYSTEMS THEORY
Uses a computer analogy (central programming and control) • Commands controlled by central nervous system • Memory-based learning (i.e. motor programmes)	Uses a biological analogy (adaptation to environment) • Commands initiated by the environment • Adaptation to environmental constraints (perception–action coupling) • Coordination (synergies) • Stable, efficient movement
Explaining mechanisms of control from inside the system	Explaining the relationship between environment and behaviour from outside the system

ACCURACY, CONSISTENCY AND EFFICIENCY

The desired outcome of the motor task has to be achieved (accuracy) in a high proportion of the attempts (consistency or precision) and with the minimal amount of physical effort (efficiency). For example, a beginner throwing a dart at a dartboard might well hit the bull's-eye. If they did, most people watching would put this success down to beginner's luck. There would be a very slim chance of the lucky beginner reproducing their success consistently until they have learned which components of the combination of movements produced the first successful attempt. Magill (2003) uses this sort of concept to explain the difference between accuracy and precision, using target shooters as an example. Figure 6.2 illustrates the differences; the target on the left (A) shows a widespread array of shots, some of which were over the centre, whereas the one on the right (B) shows a tight cluster but all of them a little way off the centre. Although both shooters might have achieved the same score in that particular attempt, which shooter do you think might be able to improve their scores more easily?

The answer is shooter B on the right. A simple adjustment of their sights would mean that the tight cluster achieved could be moved over the centre of the target. This tight cluster reflects *precision*. Shooter A was equally as *accurate* as shooter B, because they had the same score, but not as *precise*, because they could not reproduce the same performance consistently.

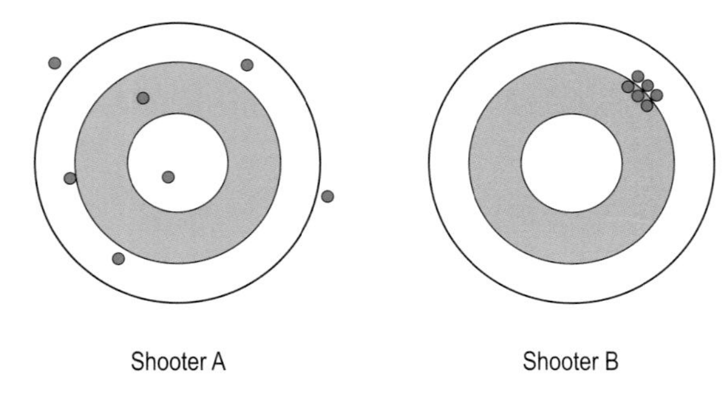

Figure 6.2 Accuracy versus precision.

In addition to learning the optimal coordination of movements for a task, the beginner also has to learn the optimal amount of muscle work necessary. Usually, someone learning a new task will hold themselves very stiffly, only moving the limb sections absolutely necessary for the task in hand. Anyone who has ever tried out a new sport will remember the feeling of aching all over despite needing to work only certain areas for that activity. A student who is learning a new manual technique will have to consciously relax the shoulders as they gradually get closer to the ears and their arms begin to ache! This is known as freezing degrees of freedom (Vereijken et al. 1992), and the principle was discussed earlier in the chapter. As the individual becomes more proficient, the limb and trunk segments are given a little more freedom of movement, making the functional task appear more fluid but requiring a higher skill level to do so.

There are, once again, a variety of theories about how this learning process takes place. Despite a large variety of nomenclature, these fall into two main schools of thought: the maturation approach (Gessell et al. 1974) and the perceptual cognitive approach (Bressan and Woollacott, 1982).

Maturation approach

This approach describes alternating periods of stability and instability during maturation. The periods of instability are thought to be the times when new patterns of control are being learned and hence result in some instability or lack of control until the pattern becomes skilled. For example, a toddler learning to walk will take a number of attempts over a few months until they can control a walk without falling over. During this time, the child's skill level in walking will be unpredictable; some days almost perfect and other days disastrous. Following this approach, each functional ability would develop in the same way throughout childhood, adolescence and adulthood.

Perceptual cognitive approach

The perceptual cognitive approach suggests that intellect may have a bearing on how well a motor skill is learned. Sensory input is regarded as having a significant impact on motor performance and feedback of the outcome of a particular movement. With this approach, trial and error plays an important part in the learning process. The learner consciously discards the motor patterns that produced an unsuccessful outcome and retains those that produced an outcome closer to the ideal.

MEMORY

Whichever the approach favoured, once a motor pattern has been established, it then takes repeated practice to develop that pattern into a skilled movement. Different aspects of memory are thought to play a part in various phases of learning. It would therefore be useful at this stage to look at an overview of the different aspects of memory in relation to their role in motor learning. A summary of this text can be seen in Figure 6.3.

Short-term (working) memory

This function can be subdivided into three areas as follows.

Phonological loop

This functional involves remembering sequences of numbers and/or letters such as telephone numbers or e-mail addresses. It is useful for very short-term repetition but is easily displaced by distraction.

Visual–spatial

This function involves tasks such as scanning text while reading, for example in order to remember where you were on a page if reading a novel and you have to put the book down to answer the telephone. Again, this is useful only over relatively short periods of time.

Central executive

This is the least understood part of short-term memory and involves tasks such as reasoning, mental arithmetic, organizing processing for storage, and filtering. The ability could be considered as similar to the random access memory on a computer, which also allows you to deal with a few tasks at once. Ironically, despite this area being the least researched, it is probably the most important area in terms of rehabilitation of functional ability.

Non-declarative memory

This function is thought to be controlled at the midbrain and brainstem areas of the CNS and can also be subdivided.

Procedural memory

This is the part of memory that allows us to do the many automatic tasks, such as driving a car or riding a bicycle, that were mentioned earlier in the chapter. They are the motor skills we are able to perform without necessarily being able to consciously describe how. They are also the skills that, once learned, can usually be recalled with relative ease, even after long periods without practice, hence the expression 'like riding a bike'. It would therefore seem that the components of this type of skill must be laid down somewhere in the long-term memory. Acquisition of these sorts of fairly complex skills are thought to be controlled in the cerebellum and putamen, while the ingrained habits such as driving are probably controlled in the caudate nucleus.

Classic conditioning

These are also deeply ingrained habits that are thought to be stored somewhere in long-term memory, such as having to eat a chocolate biscuit with a morning coffee, regardless of whether you are hungry or not. This function is the typically quoted research on dogs and rats regarding responses to repeated stimuli through reward or an adverse stimulus.

Emotional responses

This includes areas such as fear memory. It is thought that these memories appear over time in response to adverse stimuli linked to specific situations. Phobias are thought to fit into this category, when sometimes very extreme physical responses can be observed in response to a specific visual stimulus or environment.

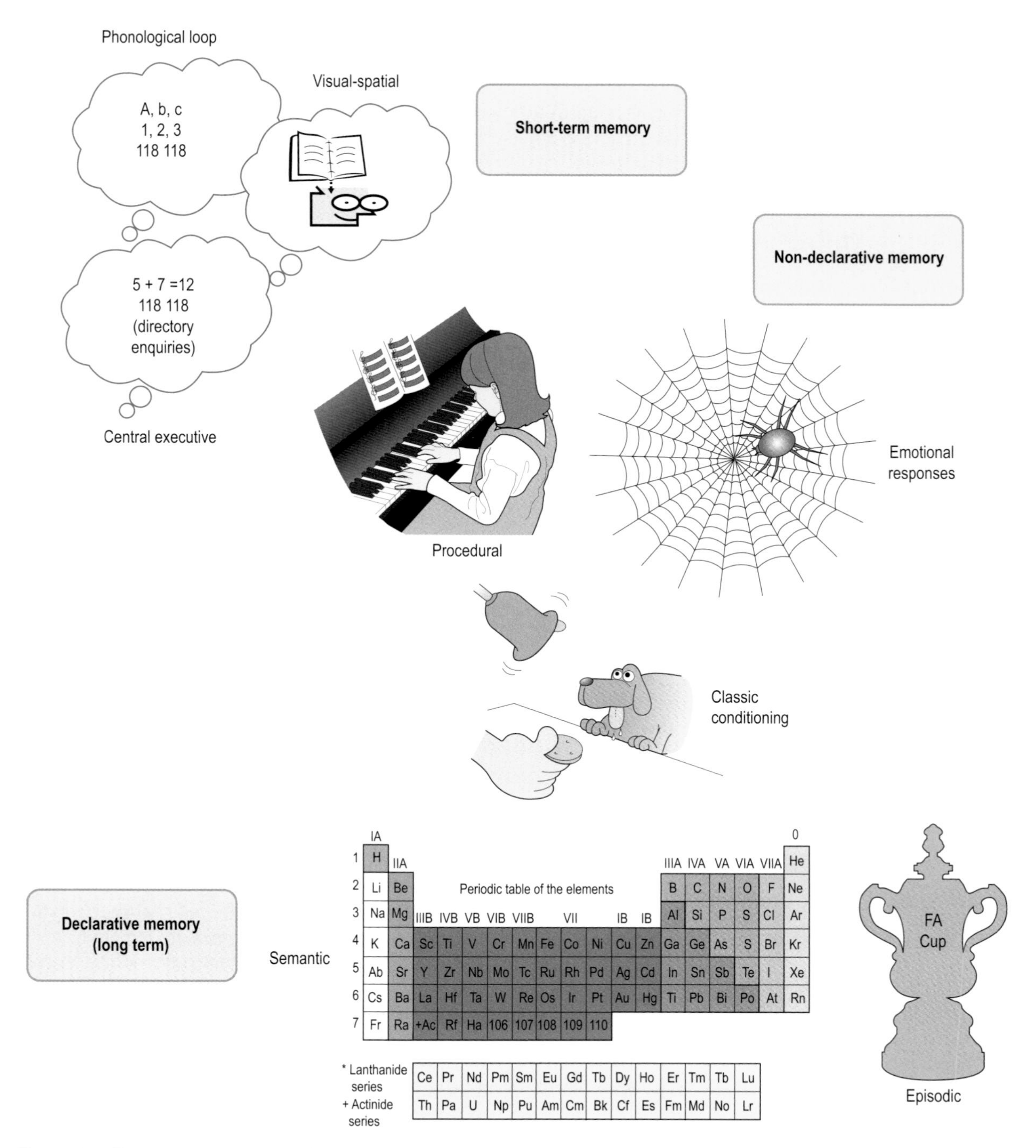

Figure 6.3 The memory process controlling learning of a new skill.

Declarative memory

This is the section of memory that is formally considered as long-term memory and, once again, can be subdivided.

Semantic memory

Semantic memory is the ability to retain and recall facts. This function would be used when initially learning a skill, when information about the new skill needs to be stored in order to develop a movement strategy in response to the environment constraints within that activity. This might be related to the rules of a particular sport or information about any equipment to be used.

Episodic memory

This relates to recall of specific events, both personal and historical. They tend to be highly individual and based on personal responses when initially experiencing the event, either directly or indirectly. They would often be historical events for which typically many people can remember what they were doing when it happened, such as the election of Barack Obama to the US presidency.

These various memory functions will be referred to in the section covering phases of skill acquisition. First, it would be useful to look at the different types of skills we use every day, as the type of skill will have a bearing on how long it takes to achieve expert level in that activity.

CLOSED AND OPEN SKILLS

For the purposes of this chapter, skills are split into two broad categories:

1. closed skills
2. open skills.

A closed skill is movement that is repeated in the same way for each performance of the one task. Using the example of the weightlifter again, the technique for the clean and jerk will be identical for each attempt at the same weight. This skill requires a very high degree of spatial control, with relatively little time limit in comparison with something such as catching a ball. To demonstrate what happens when a greater temporal (time) control is applied, observe what happens in a competition when a lifter has to rush out to make their attempt because time is running out. It quite often fails, because the weightlifter has practised the movement skill many times but at the same speed. The environment that this type of skill is performed in is said to be a stable environment. If any part of that environment is changed, the ability to perform that skill will be compromised. The fairly rigid sequencing that occurs in a movement such as this limits the performer to a relatively narrow choice of movements (degrees of freedom) compared with an open skill but allows very efficient performance (Browenstein, 1997).

An open skill requires a combination of both spatial and temporal control. The movement has to be practised many times, as in a closed skill, but variation in speed and effort has to be applied to be able to adapt to a changing environment. This time, refer back to the badminton player as the example. Changes in the intended movement may well have to take place depending on the actions of the opposing player plus possibly the actions of their doubles partner. This type of environment is said to be an *unstable environment*.

When re-educating a motor skill in a rehabilitation environment, both these principles need to be considered. For example, to start progressing through a new skill, such as balancing on one leg following an ankle injury, a high degree of cognitive attention is required, using visual cues as well as input from mechanoreceptors in the limb. At this stage, balancing can be maintained only as a closed skill, and the individual will usually attempt to freeze some degrees of freedom at selected joints in order to reduce the variable elements that have to be controlled for that activity. If any part of the environment changes, such as introducing an unstable base of support, balance reactions are challenged once more. Consequently, the patient will have to take a step on to the unaffected leg much earlier than an uninjured person, as they will have restricted themselves to a limited repertoire of balance reactions to be able to have better control. As balance improves, secondary activities are introduced. If balance has to be maintained while catching a ball or being pulled in one direction by elastic tubing, the balance strategies have to change with each different circumstance. Exercises such as this encourage skills learned as a conscious effort to become automatic and to adapt to a changing environment. They also encourage a gradual freeing up of the variation in degrees of freedom of movement. This same balance activity has now become an open skill in preparation for the functional mobility required of the ankle in everyday life.

Taylor et al. (1998) highlighted the effects of diverting attention on motor performance in their study investigating joint position error detection during concurrent cognitive distraction. Joint position sense was reduced in subjects who received an auditory distraction while performing the task. The researchers point out that this can have a positive or negative effect depending on the timing of the distraction. For someone concentrating on mastering a very new task, distraction could reduce the safety of the activity. Conversely, appropriate distraction as a progression in later stages would supplement a rehabilitation programme.

Both types of skill need a significant amount of repetition to allow efficient retrieval from long-term memory and the development of efficient motor control for a specific task, for example (Kottke et al. 1978):

- learning to walk (to age 6) – 3 million steps
- parade ground marching (end of army basic training) – 0.8 million steps
- hand knitting – 1.5 million stitches
- violin playing to a professional level – 2.5 million notes (4500 h of practice)
- baseball throwing (pitcher) – 1.6 million throws
- basketball shot from any angle – 1 million throws.

SKILL ACQUISITION

It is now generally accepted that the development of a skill towards expert level can be divided into three main stages. As with all other topics covered in this chapter, there is a myriad of different terms used to describe the phases. Therefore, for simplicity in this section, they will merely be termed phases 1, 2 and 3 (Fig. 6.4).

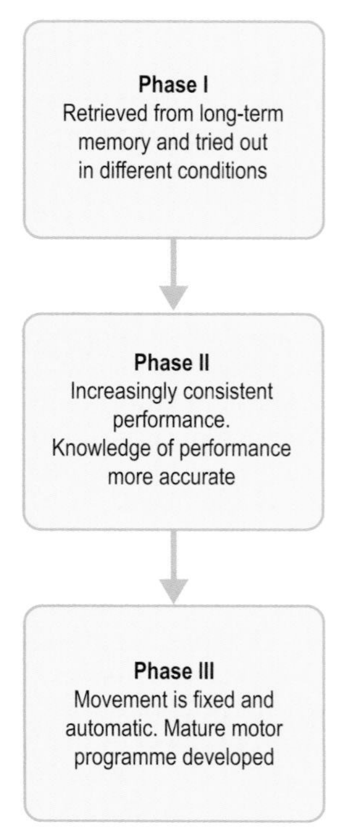

Figure 6.4 A model of the stages of skill acquisition. *(From Bressan and Woollacott, 1982, with permission.)*

Phase 1

The first stage involves organization of the skill when an approximate estimation of the movement is constructed, which will vary depending on prior experience of similar activity. Memory is thought to play an important part in this stage of learning, when facts on the requirements of the new activity are taken in and compared with facts stored of similar activities (declarative or semantic memory). Similar movement patterns used in other motor tasks can be transferred to the new skill and adapted as appropriate (procedural memory). For example, a good basketball player could transfer their ball-handling skills to other sports such as rugby. This would be called *positive transfer*. In this instance, hand–eye coordination, wide visual field and manual dexterity skills would be similar enough in both to allow modification between the two activities.

However, a squash player would have problems on initially trying to play tennis as, despite the improved hand–eye coordination, the racquet action required in squash has a detrimental effect on the technique used in tennis. This is called *negative transfer*.

A fair amount of trial and error will also be involved for the learner, who will need to devote a relatively high level of cognitive attention to the task to appreciate an overall picture of the required movement task. This would explain why a learner cannot always listen to a teacher or coach at the same time as attempting to perform the new skill being taught. Watching a child learning to write illustrates this point quite well.

Phase 2

As learning continues and the skill begins to stabilize, successful attempts will become more frequent and movement patterns will become more consistent. Feedback providing knowledge of results is important at this stage to correct minor faults as the motor pattern is becoming established, which will be difficult to change once the movement becomes automatic. At this stage, the individual is committing a possibly quite complex movement pattern to procedural memory. There is a great amount of discussion around this aspect in published literature, much of it still conflicting. However, in rehabilitation-related literature the use of structured feedback is generally supported. For example, Prapavessis and McNair (1999) found that subjects who received verbal instruction and correction improved subsequent attempts at a jumping and landing task more quickly than those who used their own sensory feedback from each attempt. Hewett et al. (1996) had similar findings, in which subjects taken through a programme designed to improve neuromuscular control in addition to strength in the lower limb were able to reduce landing forces on the knee more efficiently than those who progressed through a strengthening programme alone.

Phase 3

By the final stage, the movement pattern has been established. Knowledge of performance is now being monitored, and the proprioceptive feedback is routinely being filtered out from conscious awareness unless there is any significant change. This change in allocation of attention means that more can be devoted to higher processes, within the individual's capacity, during the task. Think back to one of the earlier examples of the experienced badminton player who is able to time a shot more appropriately and also be more aware of the tactical considerations happening around them. This ability has become possible because the actual movement skill has become automatic, freeing up more attention for anticipation or planning the next shot while completing the first one.

Think back to one of the earlier tasks about driving a car. Once you have learned to drive, you can probably listen to music, talk to a passenger, read road signs, and be aware of other traffic and pedestrians while performing the task of driving automatically. What happens if you suddenly have to drive in another country on the other side of the road? What sort of things do you do to be able to perform the task?

Suddenly, the task changes and cannot be dealt with automatically. The sorts of things you might do would be to turn the music off, stop talking, and ask people to look for road signs and landmarks.

What you have done with these actions is to reduce some of the sensory input and additional cognitive tasks being performed at the same time in order to make more capacity for a motor task that now requires cognitive attention rather than being automatic.

Any change in sensory input involved with a learned stabilized skill will have a similar effect. Injury is no exception to this. For instance, if a joint has less range of movement after an injury, the kinaesthetic input from joint and muscle proprioceptors will be different. The cognitive process of relearning a motor skill, whether it be walking, running or climbing stairs, will need to be repeated, albeit with some prior experience for comparison, until the task becomes automatic again.

ACTIVITY 6.3

Think of ways in which the experience of some pain or discomfort following injury would affect the control process in a functional movement that had been automatic prior to injury.

VARIABLE PRACTICE

Changing an activity during the learning process of a skill will have the effect of distracting from information stored in short-term memory. This will encourage retrieval from long-term memory to facilitate retention of the new skill. Repeated retention and retrieval will strengthen the memory trace. In addition to this, if the second activity performed is similar to the first, it is thought that some learning is transferred between skills, thus speeding up the learning process. This type of approach is called variable or random practice, as opposed to block practice, when the same activity is repeatedly drilled until performance is achieved and there is some evidence demonstrating its efficacy (Lee et al. 1985; Shea and Kohl, 1990; Eidson and Stadulis, 1991).

LOCUS OF CONTROL

In addition to considering the order of how learning activities are structured for a task performed, how you give instructions is also an area that warrants consideration in order to optimize the learning opportunity.

ACTIVITY 6.4

Think about how you would teach someone an exercise as part of a rehabilitation programme and how you could improve the skill acquisition process. Think of an example in each of the following cases:

- improve understanding for more efficient storage in semantic or procedural memory
- provide information on knowledge of performance that encourages a more external focus of attention
- progress to a more automatic movement.

An internal locus of control, or focus of attention, involves concentrating on how you achieve a particular body position in order to perform a task, such as bending your knees on landing from a jump. An external focus in the same task would be thinking about landing as quietly as possible. The focus has now become much more related to the task itself rather than the conscious thought process around limb movement and control. Relating this back to earlier in the chapter, you should be able to see the link between performance of a task being very cognitive as opposed to automatic movement. Not surprisingly, therefore, the general consensus in coaching and sports science teaching is that a cue or feedback that has an external focus is more effective for performance. However, as with most areas of research, there is some discrepancy in what is actually termed internal versus external, which has led to some conflicting recommendations. Wulf et al. (1999) is an example of a number of papers by the same author strongly advocating use of external cues when teaching new skills, whereas Perkins-Ceccato et al. (2003) suggest that novices might be better served with more internally focused cues, with experts performing better with external cues. Both studies looked at golf, with Wulf et al. (1999) comparing control or arm swing as internal and golf club swing as external focus. Perkins-Ceccato et al. (2003), on the other hand, regarded club swing as more internal and golf ball trajectory relative to the hole as external. It may be that novices, or patients in the case of a rehabilitation setting, may at first require some more internally focused attention to start developing an awareness of the movement but at some point will need to be helped towards responding to more external cues in order to make the movement more automatic. This thinking is supported by the functional magnetic resonance imaging findings mentioned very early in this chapter, indicating that novices have more difficulty in filtering out important sensory input and also, at least initially, have to think about the activity more than experts do.

CONCLUSION

Motor learning requires a combination of factors to be successful, and evidence for the exact mechanism remains conflicting in parts. This chapter has outlined many of the principles involved, but the reader is directed to some of the texts solely devoted to this subject for more in-depth information. These texts are listed in the reference section. The main factors to consider in the field of motor learning are short- and long-term memory, transfer of skills, modifying degrees of freedom of a movement, knowledge of both performance and results, and finally practice. How we use these principles in the clinical setting will have a significant impact on the success of a rehabilitation programme, whether the ultimate goal is to walk unaided, transfer from bed to chair, or return to a highly skilled activity.

You should now have an understanding about how an individual controls movement for both simpler repetitive movements and more complex skilled tasks. You should understand how these skills are learned and how the learning environment can be manipulated to facilitate this process.

References

Bernstein, N.A., 1967. The Co-ordination and Regulation of Movements. Pergamon Press, New York.

Bressan, E.S., Woollacott, M.H., 1982. A prescriptive paradigm for sequencing instruction in physical education. Hum. Mov. Sci. 1, 155–175.

Browenstein, B., 1997. In Browenstein, B., Shaw, B. (Eds.),

Functional Movement in Orthopaedic and Sports Physical Therapy: Evaluation, Treatment and Outcomes. Churchill Livingstone, Edinburgh.

Doyon, J., Ungerleider, L.G., 2003. Functional anatomy of motor skill learning. In: Squire, L.R., Shacter, D.L. (Eds.), Neuropsychology of Memory. Guilford Press, New York.

Eidson, T.A., Stadulis, R.E., 1991. Effects of variability of practice on the transfer and performance of open and closed motor skills. Adapted Phys. Act. Q. 8 (4), 342–356.

Gessell, A., Ilg, F.L., Ames, L.B., 1974. Infant and Child in the Culture of Today. Harper & Row, New York.

Hewett, T.E., Stroupe, A.L., Nance, T.A., Noyes, F.R., 1996. Plyometric training in female athletes. Decreased impact forces and increased hamstring torques. Am. J. Sports Med. 24 (6), 765–773.

Higgins, S., 1991. Motor control acquisition. Phys. Ther. 71 (2), 123–129.

Kelso, J.A.S., 1998. Co-ordination dynamics. In: Latash, M.L. (Ed.), Progress in Motor Control, vol. 1. Human Kinetics, Leeds.

Kottke, F.J., Halpern, D., Easton, J.K.M., Ozel, A.T., Burrill, C.A., 1978. The training of co-ordination. Arch. Phys. Med. Rehabil. 59, 567–572.

Lee, T.D., Magill, R.A., Weeks, D.J., 1985. Influence of practice schedule on testing schema theory predictions in adults. J. Mot. Behav. 17 (3), 283–299.

Lim, V.K., Altenmuller, E., Bradshaw, J.L., 2001. Focal dystonia: current theories. Hum. Mov. Sci. 20, 875–914.

Magill, R.A., 2003. Motor Learning: Concepts and Applications. McGraw-Hill, Columbus.

Milton, J., Solodkin, A., Hluštík, P., Small, S.L., 2007. The mind of the expert performer is cool and focused. Neuroimage 35, 804–813.

Perkins-Ceccato, N., Passmore, S.R., Lee, T.R., 2003. Effects of focus of attentions depend on golfer's skill. J. Sports Sci. 21 (8), 593–600.

Prapavessis, H., McNair, P.J., 1999. Effects of instruction in jumping technique and experience jumping on ground reaction forces. J. Orthop. Sports Phys. Ther. 29, 352–356.

Schmidt, R.A., 1975. A schema theory of discrete motor skill learning. Psychol. Rev. 82, 225–260.

Schmidt, R.A., 1988. Motor Control and Learning. A Behavioural Emphasis. second ed. Human Kinetics, Champaign.

Schmidt, R.A., 1991. Motor Learning and Performance. Human Kinetics, Champaign.

Schmidt, R.A., Lee, T.L., 1999. Motor Control and Learning. third ed. Human Kinetics, Champaign.

Shea, C.H., Kohl, R.M., 1990. Specificity and variability of practice. Res. Q. Exerc. Sport 61 (12), 169–177.

Sherrington, C.S., 1906. On the proprioceptive system, particularly in its reflex aspect. Brain 29, 467–482.

Taylor, R.A., Marshall, P.H., Dunlap, R.D., Gable, C.D., Sizer, P.S., 1998. Knee position error detection in closed and open kinetic chain tasks during concurrent cognitive distraction. J. Orthop. Sports Phys. Ther. 28 (2), 81–87.

Thelen, E., 1998. How infants learn to reach. In: Latash, M.L. (Ed.), Progress in Motor Control, vol. 1. Human Kinetics, Leeds.

Vansant, A., 1995. Motor control and motor learning. In: Cech, D., Martin, S. (Eds.), Functional Movement Development Across the Lifespan. WB Saunders, Philadelphia.

Vereijken, B., van Emmerik, R.E., Whiting, H.T.A., Newell, K.M., 1992. Free(z)ing degrees of freedom in skill acquisition. J. Mot. Behav. 24 (1), 133–142.

Winstein, C.J., 1991. Knowledge of results in motor learning: implications for physiotherapy. Phys. Ther. 71 (2), 140–149.

Wulf, G., Lauterbach, B., Toole, T., 1999. The learning advantages of an external focus of attention in golf. Res. Q. Exerc. Sport 70 (2), 120–126.

Chapter 7

Psychosocial influences on human movement

Sally Scott-Roberts

CHAPTER CONTENTS

LEARNING OUTCOMES

When you have completed this chapter, you should be able to:

1. recognize primary and secondary socializing agents and understand the influence they have on an individual's movement and motor behaviour
2. identify how social factors such as gender, race and ethnicity can influence motor participation
3. recognize how the social environment can impact on an individual's psychological well-being
4. appreciate the role the expert plays in supporting the acquisition of motor behaviours.

INTRODUCTION

This chapter considers the influence the social environment has on movement. It will look at how individuals learn to use movement to establish motor behaviours that will be repeated as they regularly participate in a range of daily activities. As discussed earlier in this book, we know that movement is the result of a dynamic interplay between both internal (individual attributes) and external (environmental) factors. This chapter will focus on one of the external factors, the social environment, encouraging you to think about how movement is influenced by those around us. It will begin by considering how others and the institutes we work and play in shape the motor patterns we learn and continue to use. It will conclude by considering how these social interactions affect our perceptions of competency, motivation and sense of self, all of which impact on our desire to continue to participate and our psychological well-being.

Individuals, by watching and interacting with others in the world around them, learn quickly what motor behaviours are appropriate to use and when. A small child, often long before they can verbally greet you, learns to wave hello. They not only learn how to do the hand gesture but also quickly associate it with its intended meaning. Children's motor skills are developed, adapted and refined in response to repeatedly watching others but also in reaction to the feedback they get when they use such skills. This motor behaviour will continue to be used in the correct social situation throughout the child's life, only changing if it is not the gesture expected in a certain setting (e.g. when entering formal settings such as interviews).

Goffman (1971) suggests that we learn to use an array of outward expressions to make the right impression in the right situation, for example the correct dress, behaviour (including motor) or speech. He, along with other sociologists, has recognized that individuals have a tendency to repeat or do the same things over and over again. Much of sociologists' work is spent looking at these regularities, the meanings attached to them, and the ways in which they are shaped by social forces. If we are going to fully understand why people use certain movements, or sets of movements, in specific situations, we also must consider them in the context of the social environment that has and will continue to shape and give the movements meaning.

ACTIVITY 7.1

Consider the motor behaviours you have adopted to read this book. Are you using familiar movements? How did you know which ones to use? Would the movements you use differ if you had been brought up in another culture (e.g. the direction you scan the text or turn the page)? Who has played a part in shaping your motor knowledge?

SOCIALIZATION

Valued behaviours are learned through a process called socialization. Socialization is an active process of learning and social development, which occurs as we interact with one another and become acquainted with the social world in which we live (Coakley, 2001).

As social beings, humans live and interact with a number of groups of people. The social atmosphere, or the culture of each group, is created by its shared values and norms that are adhered to and acted out by its members. New members learn to be part of a group by watching significant others' behaviour. Behaviours considered necessary for acceptance and membership are copied. This process of socializing an individual to a group not only ensures that the individual understands the cultural expectations but also plays a significant part in establishing their self-identity.

THE FAMILY AS PRIMARY SOCIALIZERS

The family provides children with their first role models. Family members play an important role in the primary socialization of their children, introducing them to the values and norms of not only the family but also the community and society at large. This may include the reinforcement of socially formed traits deemed to be associated with specific groups of people (e.g. characteristics associated with gender).

When we consider the impact early socialization has on motor behaviour, it can be seen that the actions and behaviour of family members will have a positive or negative influence. For example, research has indicated that there is a positive impact on the future motor behaviour of children who are brought up in households in which their parents or carers take part regularly in physical activity (Davison et al. 2003). The sociologist Bandura (1977), who proposed that all learning occurs within the social context, hypothesized that the actions of the adult members of a group are viewed by the children as behaviours required for family membership, and as such the children endeavour to replicate them.

Not only do family members model the accepted behaviours of the group, but they also provide children with the opportunity and ongoing support to engage and pursue certain motor activities. This includes offering opportunities and support to develop skills that will later enable them to take part in activities that are considered valuable.

Often, inadvertently, engagement in these activities is rewarded because they are valued and form part of a family's identity. For example, the family that skis each year will be keen to involve and support their children in the sport from an early age. In later years, they may continue this encouragement by providing financial support for the child to continue their development, perhaps on an organized school trip. Vygotsky (1980) would suggest that to aid the child's acquisition of the skills necessary to become an expert skier the family offers *scaffolding*. Learning requires some risk taking, and initially the family helps manage this by employing expert tuition. As the child progresses, the scaffolding becomes less overt and direct, i.e. by offering the opportunity to practise, skills are consolidated, and by altering the context (a new ski resort or slopes), understanding and retention of skills are increased.

THE INFLUENCE OF A FAMILY'S BELIEFS AND VALUES

Observation of family life has revealed that the types of activities pursued and encouraged can be influenced by the beliefs and values that families hold about certain social factors such as gender. There may be a division of labour within a household, with observable differences in the motor behaviours used to complete activities by the male and female members. This division of labour may be based on preference or skill but may have been influenced by strong socially formed notions of what women and men can or should do. Children will assimilate this association with male or female identity, and if reinforced over time, it will form the basis of their own social values about gender.

ACTIVITY 7.2

What divisions of labour are noticeable in your family or household? What influences this? What motor behaviours have been refined by individuals to enable them to complete their allotted activities?

Play opportunities offered to girls and boys could be seen to differ depending on parents' notion of gender suitability. For example, rough and tumble games may be more readily accepted as appropriate for young boys, while negative attitudes about girls taking part in contact sports may restrict their opportunity to participate in team games. Stereotypical influences may lead to differences in skill development between the sexes and affect individuals' choice of activity as they grow older. The popular film *Billy Elliot* portrays the struggles of a young man fighting gender stereotyping to pursue his dream of becoming a ballet dancer. The boys in his community were expected to join the boxing club, and only the girls joined the dance class. Billy, even as a 12-year-old, was very aware that his desire to dance did not fit with the norms of his family or those of the community in which he lived. He tried extremely hard to keep his dancing hidden from his father and brother, and when his activities were discovered, not only was his masculinity questioned but the family name was also called into disrepute. Both his family and the local community's beliefs had to be challenged and altered (temporarily at least) for him to pursue an occupation deemed unsuitable for a boy.

Gender differences in participation may be augmented by religious beliefs. For some, religion is an integral part of family culture. Their lifestyle choices, including their willingness to participate or encourage their children to take part in certain activities, will be influenced by their beliefs. For example, dress codes as well as gender labelling of activities could see some physical activities not pursued or undertaken only in same sex groups.

This notion of labelling, unless successfully challenged, will potentially be passed down from generation to generation. It is not uncommon to observe parents buying toys or playing games with their offspring that had been their own childhood favourites. Many of the dads who stand watching their sons playing football during the weekend will have and may continue to play football. By introducing their child to a familiar ball game at an early age and supporting their development, they introduce them to motor activities they have or had enjoyed, continuing the cycle or pattern of behaviour. It is hoped that those same fathers will encourage their daughters to also take up the 'beautiful game'; it is only with a steady increase in the number of people challenging the stereotypes that future generations will be encouraged to do the same, with less and less thought or attention paid to their social factors, such as gender.

SECONDARY SOCIALIZATION

Socialization continues beyond early childhood. As children venture beyond the immediate family, they will enter new groups with potentially different values and norms.

FRIENDS AS SECONDARY SOCIALIZERS IN ADOLESCENCE

As children move into adolescence, their motor skills continue to adapt to accommodate the activities that are considered valuable among another group: their peers. In adolescence, the influence of friends on motor behaviour becomes greater than the influence of family. Self-exploration becomes the preoccupation of adolescence, and peers act as a social mirror. For peer acceptance, individuals will need to demonstrate that they are willing to take on board the values and norms of their group of friends.

A common stereotype of a UK teenager portrays them as physically inactive, happy to stay in bed for long periods of the day, or as uncommunicative individuals who adopt poses and postures that make it difficult for them to give eye contact. However, realistically, a large proportion of teenagers will seek acceptance into a group in which belonging will predominantly require some element of participation and appropriate motor behaviour (e.g. the netball or rugby team, local dancing troupe or school rock band). Most will have developed the necessary motor skills and now spend time adapting and refining them to ensure that their performance improves (e.g. a violinist or tennis player will continue to monitor and adapt their bow hold or racket grip to improve their playing but also ensure their place, and often status, within a chosen group).

Acceptance into some activity-based groups does not, however, have to be dependent on motor performance. Donnely and Young (1999) suggest that newcomers can gain membership by adhering to the wider expectations and behaviours of the others. In formalized groups, such as the local rugby or football club, this may include formally agreeing to attend non-sporting events together, and to wear a stated uniform in and out of the sporting arena. Such socialization is also visible in informal groups in which there are unwritten expectations made of members about the dress code and non-motor participation, for example the music people choose to listen to. For some, therefore, a poverty of the required motor skill that would allow them to achieve success may not be a deterrent or exclude them from their preferred social group. Time spent with others who do perform well, acquiring knowledge about the activity, and ensuring adherence to the expected behaviours may be sufficient to ensure acceptance. This is something that practitioners can draw on when working with young people who may have difficulties with motor skills.

ACTIVITY 7.3

The middle sibling in our family, Dan, has recently purchased a skateboard. What group norms, motor or otherwise, might he need to adhere to to become an accepted member of the skateboard group that meet in the local park?

FRIENDS AS SECONDARY SOCIALIZERS IN ADULTHOOD

The need for social acceptance and inclusion does not stop after our teens. Continued motor participation through involvement in sport remains one of the predominant ways in which adults spend their leisure time and adapt and maintain their motor skills. Membership of a club or team, along with its adherence to group behaviours and expectations, offers social opportunities throughout the lifespan. Participation can be solely motivated by the human need for company. Agnes's membership of the lawn bowls club means that she remains in contact with her peers. John's membership of the badminton club ensures that each week he meets with long-term friends, plays a number of friendly games, and enjoys a drink at the local pub on the way home. Neither appears oriented towards the potential competitive aspect of the sporting event that they engage in but rather see the membership as a medium through which to socialize. However, for others the desire to refine their motor skills or to boost their psychological well-being through more competitive involvement may be the intrinsic driver.

Like many adults, the adults Agnes and John in our case-study family, will, when questioned, identify that their participation in sporting activities is also part of a larger personal campaign to remain fit. They do wish to improve their individual motor performance to fulfil an intrinsic desire to feel fitter, stronger, slimmer and potentially younger. While both will have monitored an internal sense of well-being associated with their participation, they will also have been socialized to believe that exercise taken on a regular basis is a vital component of a healthy lifestyle. A national campaign to raise awareness about the benefits of a healthy lifestyle has had impact on our family members. It is also evidence of other social influences beyond family and friendship groups shaping individuals' beliefs.

THE WIDER SOCIETAL INFLUENCES ON SOCIALIZATION

While the norms and values held by our significant others offer the initial template for our socialization, it is unrealistic to think that they remain an isolated agent of change. They are themselves influenced by the larger systems that they function within. From quite early on in our lives, it becomes obvious that the behaviours we use in and outside the home will have been shaped by the values and norms of our nation, community, and institute and organization.

Children will step out of the home into schools, which will have a set of beliefs not only about how their pupils will be educated but about the behaviours that they will and will not accept. Some of these beliefs will have been shaped by national norms and values set down and reinforced through government. For example, the UK government, in the National Curriculum guidelines, lays down guidance about the nature of and time spent on physical education in secondary schools. Early in their school career, children learn which motor behaviours the school values. You will all be able to remember that school expected you to walk, not run, in the corridors, and that in assembly you were expected to sit on the floor with your legs crossed. These behaviours are reinforced by a combination of praise and sanctions, depending on an individual's willingness to adhere to the norm.

ACTIVITY 7.4

Schools are not the only institutions that establish norms that demand certain motor behaviours. With reference to the institutions that you are part of, identify the motor behaviours that are valued. How are they reinforced?

RACE AND ETHNICITY

Sociologically, when considering the bigger systems or factors that influence motor behaviour, we cannot ignore that individuals are ascribed to groups because of their nationality and/or race. *Ethnicity* is the term used to describe an individual's cultural heritage; an ethnic group will have a shared cultural legacy, set of traditions and set of customs. *Race* is identified by physical or biological traits: visible characteristics. As with gender, stereotypical notions associated with race and ethnicity are prevalent, often fuelled by media discussion. For example, there have been unsubstantiated claims made about the apparent physiological make-up of certain races, particularly when a group excels at international events. The excellent performance of a number of African sportsmen and women in world athletics has led to the suggestion that, as a group, individuals of African origin have heightened stamina and strength. Very little consideration has been paid to sociological factors that may have increased their chances of taking up athletics (e.g. a lack of opportunities to participate in other sports because of cost implications, lack of facilities or lack of positive role models). National origin can also be seen to reinforce associations with certain sporting activities by heightening the status offered to certain sports personnel. While in the UK youngsters may aspire to be a footballer, their peers in Pakistan may want to be cricketers, and in Japan they may strive to be sumo wrestlers. The sport of boxing has a lengthy history within the Roma and Traveller cultures. There is a strong tradition of young men being encouraged to participate in the sport. In 2008, for example, there was much publicity surrounding an 18-year-old member of the British Olympic boxing squad who was known to be from the Traveller community. Billy Joe Saunders successfully reached the Olympics and, in doing so, reinforced a cultural tradition. He offers a positive role model to the boys and young men of the community, encouraging them to participate in an activity that is strongly supported and made accessible to them by their elders.

CELEBRITY ROLE MODELS

Individuals, as we have previously discussed, learn about their role in society by observing others. It is the motor behaviours that are considered valuable that will be replicated. Access to media coverage of sport events offers young people visible role models who they can relate to but may never meet. Many individuals are inspired to take up jogging following the coverage of the ordinary men or women competing in the London Marathon. Others will be inspired to take part because of the successes of sportsmen and women who they feel they can identify with. For example, individuals such as Lewis Hamilton (the Formula One racing driver), Tiger Woods (the golfer) and the Williams sisters (champion tennis players) have all been cited as being exciting new role models for black young people. They, with considerable parental support, have achieved success in sports that in the past have produced few role models. Their success has challenged any socially constructed notion that these sports are not accessible to black youngsters.

GOVERNMENT AND THE MEDIA

The government or ruling party of any nation can always be viewed as a socializing agent. It can, through law or statutes, reinforce the values and norms of the nation or respond to changes in the values of the population it serves. For example, the recent implementation in the UK of a smoking ban in public places is an example of a law that has responded to current research and changes in public opinion.

The government can use the media to reinforce changes in expectations and shape the thinking of the nation. For years, road safety education has been reinforced by media campaigns and advertising. The acquisition of the correct motor behaviours at zebra and pedestrian crossings as well as among parked traffic have not been left to chance, and the skills needed to use these crossings are taught and reinforced through cartoons and advertisements appropriate across the age spectrum.

The recent focus on increasing children's fitness, in response to medical concerns about the incidence of childhood obesity in the UK, has been accompanied by a plethora of local initiatives (e.g. free swimming sessions), all promoted through multiformat media. The success of preschool children's programming such as *LazyTown* in getting children up and moving, and the impact (across the world) of *Strictly Come Dancing* on the number of adults seeking dance lessons, demonstrate the power of the media to motivate and potentially socialize the nation to take up more physically active interests.

Unfortunately, the media can have a negative impact too. For example, in the same struggle to encourage children to be more active, researchers have found that there is a reluctance by parents to allow their children to play out of doors, (Preston, 1995). Perceptions of danger have increased over time, fuelled by the often graphic media coverage of the sad, but few, cases of children coming to serious harm when playing outside.

THE IMPACT OF OTHERS ON OUR PSYCHOLOGICAL WELL-BEING

We have so far considered how the development of motor behaviour is influenced by the social milieu. There is clear indication that the need to be accepted by others shapes not only what we choose to do but aids the development and refinement of motor skills. We have also suggested that those around us can form our attitudes about activities. What we have not yet focused on is how these interactions shape individuals' attitudes about their own motor ability and the consequential impact on their participation.

SELF-ESTEEM AND PARTICIPATION

Self-esteem is an individual's self-evaluation of their abilities and difficulties. It is, in part, the worth that we all attach to those opinions that others hold about us. For example, there is evidence in western society that children favour being good at physical skills over being good at classroom or academic skills. As a consequence, this consensus of opinion increases the value that children attach to physical competency (Cratty, 1997). It is not surprising then to learn that children who have poor motor skills, because of such conditions as developmental coordination disorder, can have lowered self-esteem (Schoemaker and Kalverboer, 1994).

James (1890), in his attempt to explain self-esteem, suggested that self-evaluation could be explained as a fraction. He stated that self-esteem is the product of the number of successes someone experiences divided by the individual's perception of his or her potential. He attributes low self-esteem to a discrepancy between an individual's achievement and what they believe they can achieve. So, to alter self-esteem he suggests that either the number of successes needs to be altered or an individual needs to change their expectations. A study undertaken by Coopersmith in 1967 with a group of 10-year-old boys identified that, while the group were from similar socio-economic backgrounds with the same academic potential, when measured, the boys had differences in their self-esteem ratings. The boys who had high self-esteem were observed to be confident, positive and realistic, and happy to try new activities. Those with low self-esteem scores were self-conscious and oversensitive to criticism. These latter boys perceived themselves as less competent than their peers and so were reluctant to try new activities. Coopersmith deduced that the social environment had played a part in shaping these boys' evaluation of themselves. He identified that the boys who had high self-esteem had been offered consistent parental support and their views and actions were valued, whereas the others had experienced inconsistent support and an erratic approach to matters such as discipline.

Other psychologists, such as Harter (1982), have continued to look at the impact of the social environment and also found that perceived support from a significant adult has an impact on children's self-esteem. She also suggested that individuals have multiple self-representations across different contexts. She saw self-esteem as domain-based, so individuals may have different self-evaluations across a range of skill areas (e.g. social and physical).

Why is this important to our consideration of motor behaviour? First, it is safe to assume that if someone has a low self-esteem in the physical domain, there will be an increased likelihood that they will avoid participating in motor or physical activities. The adult who has negative memories of poor performances on school sports days will still recall their lack of success. The child who is regularly left until last when peers pick their rounders or baseball team, will perceive that others value their skills as poor or less useful than those of their friends who were chosen before them. This may lower the child's expectations of their ability to play not only that team game but also all team sports. Often, the messages that we receive from others are not direct or overt, nevertheless these implicit messages can have an impact on an individual's self-evaluation of physical competency and ultimately their willingness to try new activities or continue with familiar ones.

THE INFLUENCE OF OTHERS ON THE CHOICES WE MAKE

There is no doubt that our past experiences influence our future motor choices. As part of the decision-making process, we appear to evaluate the personal cost and/or gain that taking part will offer. While this decision-making is often at a subconscious level, we assess what we will gain and have gained from doing an activity in the past and what was or could be the cost to us if we pursue it now.

Grandmother, Agnes, was reluctant to join the family on the ice on their last skating trip. There could be several reasons why. Agnes may have noticed that she is not as well balanced as she used to be and that the dimmed lights at the rink impede her deteriorating eyesight further. Her evaluation then would not be based on comparison with others' values and beliefs but would have been made through her own assessment of the biological changes she has and is experiencing. Alternatively, Agnes could have been influenced by others' reactions and opinions. For example, she may have perceived that others think she is too old to be skating or could have actually been told that she should not take part as she would be placing herself at unnecessary risk. Agnes's beliefs about her capacity to skate may have been shaped by wider societal opinions about age and ageing not necessarily pitched at her but at the older generation, who are often stereotypically portrayed as less able and more at risk. The impact of such social typecasting could have left her feeling anxious about participating in a range of activities that may be considered risky.

ANXIETY AND ITS IMPACT ON PARTICIPATION

Potentially, Agnes's increase in anxiety in the scenario above will have more of an impact on her ability to safely complete an activity than her age. Biologically, the limbic system, midbrain, is geared to prepare the body to respond when it perceives it is under attack. The fight or flight response is activated not only when an individual is confronted by a snarling dog or an oncoming car but also when individuals, such as Agnes, become anxious because of negative thinking about an event or activity. The body's release of adrenaline (epinephrine) and redirection of the blood supply, to ensure that her body is ready to react, will leave both the individual who needs to run from the dog and Agnes, who has become anxious about her safety on ice, with the same physical sensations (i.e. nausea, breathlessness and palpations). We know that for some this bodily reaction provides the drive and often the thrill to take part in extreme sports or motor activities such as parkour (free running). For others, however, it can and will prevent them participating. The positive correlation between the strength of the negative emotion felt and the restriction on rational thinking does decrease an individual's ability to gain control of the situation by problem solving or being flexible and receptive to learning, increasing their chances of an incident occurring (Fredrickson, 2004).

THE IMPACT OF SUPPORT WHEN LEARNING MOTOR BEHAVIOURS

When we are introduced to anything new, we quickly become aware that learning is a process that requires a period of practice to consolidate skills. For example, being able to drive a car safely, and without constant conscious thought about the position of your feet or hands, is achieved only after a period of practice, often long beyond passing a formal test of competency. This autonomy can be achieved only by practising in context, out on the roads, and becomes an automatic process once it is laid to long-term memory. This learning process will be enhanced by certain conditions, for example the support you receive from others, the learning styles adopted to teach you, and access to suitable materials (a suitable car in this case). In light of what we now know about self-esteem, we also need the individual to be motivated to take part with sufficient esteem to take a risk and have go.

ACTIVITY 7.5

Spend some time to reflect back on how you learned a complex motor activity such as riding a bicycle or learning cursive, joined up writing. How did others help you? Did they scaffold your learning?

Consider how you may have supported someone else's learning. Did you grade your support or scaffolding as they became more proficient?

At any stage of learning, there is a difference between what the learner can achieve alone or with the support of others. The learner will be able to develop only if they are at a stage of readiness to take on a new level. Vygotsky (1980) describes this as the *zone of proximal development*. Individuals need to have the necessary foundations in place on to which they can build new skills. If we return to our analogy of learning to drive a car, it is considered that by the age of 17 years (in the UK), most individuals will have sufficient motor and intellectual skills to master driving. This new stage will, however, need to be facilitated by another more experienced and skilled person, a driving instructor or parent. For other motor activities, this might be a coach, a teacher, a therapist or a more skilled peer or sibling. This facilitation by another, as briefly discussed earlier, is described by Vygotsky as scaffolding. The learning is dependent on an exchange of knowledge, through language and often demonstration, between the expert and the novice. As the learner becomes more proficient, the level of support needs to be decreased until we are able to complete a task or an activity alone. The teaching approach adopted by the expert, as well as their ability to judge when to moderate, increase or reduce the support, is essential to help maintain the learner's motivation to continue.

SUMMARY

This chapter aimed to encourage you to begin to think about how movements, and in particular motor behaviour, can be influenced by the social environment in which we perform them. The desire to participate in motor activities appears, in part, to be driven by a need to be accepted, initially into the family and later into friendship groups. However, copying valued motor behaviours does not only help to ensure our inclusion but also provides us with the opportunity to develop and improve our motor skills. Additionally, our psychological well-being can be affected by a perception of acceptance and by the feedback we receive from others as we learn and continue to participate.

Each one of us, at some time in our lives, will have or will be responsible for encouraging others to participate in motor-based activities. For some, on a daily basis, we will be using participation as the medium through which we will aim to remediate or develop motor skills. By having an understanding of how individuals are influenced by others around them, we can begin to understand their relationship with motor behaviours and activity participation. This understanding also ensures that we remain alert to the part we might play in the shaping of that affiliation with movement and how we might influence people's sense of competency and/or level of motivation to participate in motor activities.

Glossary of terms (taken from Giddens 2006)

Culture: The life characteristics, values and norms of a given group.

Norms: Rules, or behaviour, which reflect or embody a culture's values. They either prescribe a certain type of behaviour or forbid it. Norms are always backed by sanctions of one kind or another (e.g. disapproval or punishment).

Self-identity: An ongoing process of personal discovery. The development and establishment of a unique sense of who we are and 'our relationship to the world around us' (p. 1033).

Values: Ideas held by individuals or groups about what is desirable, proper, good or bad. Values will vary between cultures.

References

Bandura, A., 1977. Social Learning Theory. Prentice Hall, Upper Saddle River.

Coakley, J., 2001. Sport in Society: Issues and Controversies. McGraw-Hill, Hightstown.

Coopersmith, S., 1967. Antecedents of Self-Esteem. Freeman, San Francisco.

Cratty, B., 1997. Coordination Problems Among Learning Disabled Children. Harwood Academic Publishers, Amsterdam.

Davison, K., Cutting, T., Birch, L., 2003. Parents' activity-related parenting practices predict girls' physical activity. Med. Sci. Sports Exerc. 35 (9), 1589–1595.

Donnely, P., Young, J., 1999. Rock climbers and rugby players: identity construction and confirmation. In: Coakley, J., Donnelly, P. (Eds.), Inside Sports. Routledge, London, pp. 67–76.

Fredrickson, B.L., 2004. The broaden and build theory of positive emotions. Philos. Trans. R. Soc. 359 (1449), 1367–1371.

Giddens, A., 2006. Sociology, fifth ed. Polity Press, Cambridge.

Goffman, E., 1971. The Presentation of Self in Everyday Life. Pelican, London.

Harter, S., 1982. The perceived competence scale for children. Child Dev. 53, 87–97.

James, W., 1890. The Principles of Psychology, vol. 1. Holt, New York.

Preston, B., 1995. Cost effective ways to make walking safer for children and adolescents. Inj. Prev. 1, 187–190.

Schoemaker, M.M., Kalverboer, A.F., 1994. Social and affective problems of children who are clumsy: how early do they begin? Adapted Phys. Act. Q. 11, 130–140.

Vygotsky, L., 1980. Mind in Society. Harvard University Press, Cambridge.

Chapter 8

The influence of the environment on human movement

Philippa Coales

CHAPTER CONTENTS

LEARNING OUTCOMES

At the end of reading this chapter and completing the activities, you should be able to:

1. describe a systems approach to human activity
2. apply aspects of the effects of the physical environment on human movement
3. apply aspects of task design to human movement
4. consider the impact of technology on human movement
5. consider the impact of organizational design and psychosocial environment on human movement
6. discuss the impact of anthropometrics on human movement
7. consider how anthropometric factors act on individuals and the consequences to individuals' movement.

INTRODUCTION

Throughout this book, you will see that there are many factors that can influence the production and execution of human movement. Many of these have been involved with the physiological processes within the body and the human's psychological situation and could be termed as *intrinsic factors*. Factors extrinsic to the individual human's physiology can also have an impact on their movement; these can be described as environmental factors, and in this chapter we shall discuss what these factors are and how they may influence human movement.

WHAT IS THE ENVIRONMENT?

The dictionary definition of environment is 'the surrounding objects or circumstances in which a person, animal or plant lives or operates' (Oxford English Dictionary, 2009). In this day and age, the term *environment* has taken on a rather different connotation, and you are most likely to hear the word used in reference to issues related to climate change, rain forest destruction, sustainability and other green factors.

However, if we consider the environment in relation to human movement we need to understand which environmental factors may have an impact on it. If we go back to the original dictionary definition, we can break it down into two parts: surrounding objects and surrounding circumstances.

ACTIVITY 8.1

Make a list of some surrounding objects that you think might influence human movement.

From Activity 8.1 above, you may have mentioned factors such as clothing, desk height, carpets, furniture, pavements, street furniture (such as lamp-posts) and seating that could be considered as surrounding objects. Factors such as the weather, level of lighting, temperature and noise can be considered as physical circumstances surrounding a person. These are simple physical factors; however, there is a further dimension that could best be categorized under the division of surrounding circumstances. This factor is more about the psychosocial situation of an individual, rather than the physical, and needs to be considered for a complete picture. An example of this may be the impact on a person who is required to work in an open-plan office environment when they prefer to work in a single office environment.

WHY DO WE NEED TO CONSIDER THE ENVIRONMENT?

Human movement can be affected, either positively or negatively, by the environment within which the movement takes place. Consider a sprint athlete who runs the 100 m while wearing training shoes. He is unlikely to achieve as good a time wearing these shoes as he would if he wore specifically designed spiked running shoes. During athletic competition, the wind speed is always measured, as it is recognized as having an impact, either positively or negatively, on times achieved. If our runner was running into a low headwind, his speed would be reduced, as some of his force would be needed to overcome the additional obstacle of the wind. On the contrary, if he had a high-speed tailwind his performance would be enhanced and movement assisted by the wind (see Ch. 9). Let us also consider other surrounding circumstances, such as what is motivating the athlete to run. Is he there because he really wants to be, or is he there just because his brother is a runner and so his parents bring him along as well? If he really wants to run, his performance will be better than if he is running under duress or with no motivation, as his psychological profile will affect his movement production either consciously or subconsciously.

Thinking about one activity such as running 100 m, it is clear that it is impacted on by many aspects of the environment, surrounding objects, surrounding physical circumstances and surrounding psychosocial circumstances, all of which have an influence on the human movement performance. It is important to consider the environment in which any movement takes place when considering that movement's achievement for a full and clear understanding of it.

It is clear that human movement can be affected by numerous aspects of the environment in which the movement takes place, and it would be helpful to have a strategy by which the environmental impact on the movement could be considered.

A SYSTEMATIC APPROACH TO CONSIDERING THE ENVIRONMENTAL IMPACT ON MOVEMENT

Considering the environmental impacts on movement is very desirable to ensure all influences on movement are addressed. Professionals who are involved in analysing human movement – including physiotherapists, occupational therapists and ergonomists, as well as others involved in some specific aspects of human movement, such as sports coaches and trainers – aim to maximize a person's safety, comfort and productivity. A logical, thorough process using a system approach can attain the best outcome. This approach considers all aspects that may impact on an individual's ability to perform their function, and the systems can be identified in all walks of a person's life – home, work, leisure, etc.

A *system* is any interaction that a human may have in any aspect of their life that involves cognitive thought; initiation of movement and/or activity; interaction with physical, human and psychosocial influences; and a desired outcome (Arnold et al. 1998).

An example of a system would be a person loading a dishwasher. The individual identifies that the dishwasher needs loading; initiates the activity to load it; interacts with the plates, cups, dishwasher, etc.; and loads the dishwasher, closes the door and starts the machine, achieving the goal. What movement was used to achieve this may be influenced by the position of the dishwasher, the size and weight of the plates, and the psychological state of the individual (see the discussion of the balance model below).

Another example of a system would be someone working at a conveyor belt in a factory, when he is required to load components that are in front of him into boxes that come along the belt. In this case, he initiates the repeated movements; interacts with the components, boxes and conveyor belt, and possibly his colleague working on the opposite side of the belt; and achieves the desired pile of filled boxes. How this is achieved may depend on the height of the individual in relation to the conveyor belt, the duration that the activity had been done for, and the relationship the individual has with his colleague (see the discussion of the balance model below).

The system approach can be applied in all aspects of life – for example having a shower, playing a game of badminton, sitting in a lecture, eating a meal, watching the television and putting on a coat – and in that way all aspects of a person's life can be considered in a logical way.

ACTIVITY 8.2

Make a list of some of the systems in which you have been involved today.

If we think about this book's case study family, we can identify some systems that the individuals will be performing. For example, Agnes may get herself up from bed and dress herself. John may cut the grass, and Liz may work at a computer terminal. Chris may drive the car, Dan may play chess after school and Jenny may walk the dog.

It is hoped that now you can identify systems within which people move.

ACTIVITY 8.3

- Consider two members of the family and create a list of systems within which each may operate regularly.
- Consider the differences between the two family members and generate some ideas about these differences.

CASE STUDY 8.1

Let us consider Agnes, who, at 85, spends most of her time in the house on her own. She is self-caring and able to do small domestic jobs around the house to assist the family. She uses the bathroom along the landing from her bedroom. In the morning, Agnes is able to get herself out of bed (system 1), get to the bathroom and wash herself (system 2) and return to her bedroom and dress herself (system 3). Once downstairs (system 4), she can prepare her breakfast (system 5) and eat it and wash up (systems 6 and 7). Having eaten her breakfast, she tends to wrap herself in a light blanket and sit in her armchair in the sitting room to read (system 8). Although she is able to stand up from sitting, she tends not to get up until it is time for lunch (system 9), as it is warmer to stay sitting and wrapped in the blanket. She is able to make herself a sandwich (system 10) and cup of tea (system 11).

Jenny is 9 years old and is able to mirror her grandmother in the morning's activities as systems 1–6 are all completed, though possibly in a different fashion because of the age difference (see Ch. 13). Jenny then collects her school-bag and walks to the bus stop (system 7); travels to school by bus (system 8); attends classes, in which she works as part of a group (system 9); and moves between classrooms, which can be on different floors (system 10), to attend a science class in a laboratory (system 11). At lunchtime, Jenny eats her packed lunch (system 12), chats with her friends (system 13) and plays outside with a skipping rope (system 14).

You can see from the above case study that there are some core systems that are common to both people and are mostly to do with basic self-care, such as washing, eating and dressing. While they may be a common activity, the way in which they are carried out will differ between them because of factors that are intrinsic and extrinsic, such as environmental factors.

In addition to the differences in performing similar activities, there are differences in the actual type and number of systems in which each individual is involved. Jenny has many more systems within which she interacts, and these are of a more active nature, requiring movement, strength and flexibility. It is also clear that Jenny has more interaction with other people in her systems, whereas Agnes spends much of her time alone. This interaction with others may have an impact on each individual's psychological factor. If Agnes was previously a gregarious person, this isolation may demotivate her and reduce her desire to move. Jenny's attitude towards schoolwork will impact on whether she runs to the bus stop or walks dragging her feet, and whether it is raining or not may impact on whether she is able to go outside to play at break time.

Having completed Activity 8.2, you will be beginning to understand how many and varied are the activities within which human movement may occur. However, you may be wondering what impact the environment may have on these systems of movement.

THE ENVIRONMENT'S IMPACT ON SYSTEMS OF MOVEMENT

As we have already seen, there are several components in the environment that may affect a person's movement, which include their immediate physical surroundings, their immediate surrounding circumstances and their surrounding psychosocial situation.

Some of the obvious immediate physical surroundings that may impact on a person's movement include clothing, which may restrict movement, and physical stature in relation to objects they are using, which may make the activity uncomfortable or difficult if there is a mismatch. Some aspects of a person's surrounding circumstances include temperature, which may make movement difficult if at extremes of physical comfort; light levels, which can either enhance or detract from easy movement; noise, which may alter ways in which movement is achieved; and the duration of the activity, which may alter movement patterns because of fatigue.

As you can see, there are many aspects of the environment that need to be considered when thinking in relation to human movement, and it is important to address them all to achieve the best quality of movement possible.

Therefore wide thinking is needed if you are to consider the impact of all aspects of environmental factors. An aid to this type of wide thinking can be used that will help you consider as many aspects as possible in a logical way. A model, the system model based on balance theory, is shown in Figure 8.1. This aims to concentrate on the individual while considering as many types of factors as possible and the impact of the various factors on that individual, giving a considered balanced approach. As you can see, the multiple factors are grouped together into five areas: the individual, technology, task design, task environment and organizational design. The use of this type of model is very helpful when used in conjunction with a system approach to movement, as described above, as it can be applied to any identified system and allows logical and thorough consideration of factors that may impact on human movement.

Using this model, the individual is the most important factor, and it allows factors that impact on the individual to be considered. In this case, the task can be considered to be any system in which

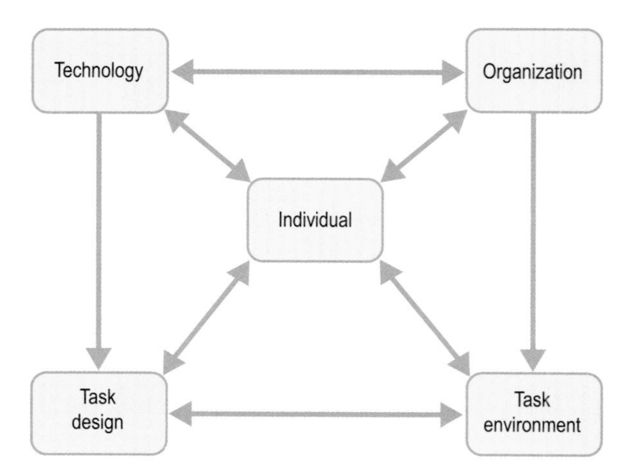

Figure 8.1 A system model based on balance theory. *(From Smith & Sainfort 1989, with permission.)*

From Case study 8.5, you can see that the organization within which movement systems occur can have a profound impact on the movement that an individual person may achieve or perform. In this case, a policy that was implemented to encourage workers has led to John carrying out heavy physical activity despite the availability of assistive devices to help him.

ACTIVITY 8.11

Think about the organization within which you are currently functioning, whether at home, at work or for leisure. Make a list of some aspects of that organization that may impact on your movement activities and identify if these are positive or negative impacts.

It is interesting to identify how some innocuous, or even historical, situations actually impact on human movement without us being aware. In some cases, the organization can be manipulated by altering the use of technology (John's assistive devices in Case study 8.11), for example, to alter a person's movement. The ability to adjust an organization, in the same way as you may adjust a task design, may need careful negotiating skills but may ultimately allow safer human movement.

THE INDIVIDUAL

Using the balance theory approach to human movement, the main component is the individual. How the other components impact on that individual is important, as we have seen, but individuals have their own characteristics, which will have an impact on their movement and will impact on the other components.

We have already mentioned human anthropometrics in the section on task design and looked at how body stature and physique may impact on the movement it can achieve. We identified that each individual's unique anthropometrics (e.g. height, chest girth and overhead reach) will impact on how similar activities are carried out and whether or not an individual is actually able to achieve the goal. Let us revisit the shopper in the supermarket whose height is 155 cm, who is able to reach objects on lower and middle height shelves but is unable to reach objects on the top shelf and requires assistance. As we know, the majority of systems are designed for the 'average' person, which may cause considerable difficulty for those at extremes of this norm. Take, for example, a man of 190 cm who is trying to buy a new car that is comfortable for him to drive. It is very difficult, as many of his anthropometric measurements limit his ability to achieve a comfortable position. For example, his lower leg length means it is difficult to comfortably reach the pedals, while his foot size may make safe use of the pedals impossible. His thigh length means that he has to sit with his hips flexed to more than 110°, and it is difficult for him to get his knees under the steering wheel, while his trunk length means that his head is rubbing on the ceiling. Someone who is at the shorter end of the scale may have difficulty actually reaching the pedals and/or the steering wheel and may not be able to see adequately over the steering wheel. Car manufacturers design features such as adjustable seat height, steering wheel position and seat position, which mean that the majority can achieve comfort, but there will always be the extremes who are not so easily accommodated. If a person is too far from the norm to achieve the goal of a movement system, they may require adaptation of the systems to enable them to achieve the goals without further outside assistance.

ACTIVITY 8.12

Think about one movement system that you perform regularly (e.g. cleaning your teeth, making a cup of tea or working at a computer terminal). Consider how different ranges of standard anthropometric measures may influence how that activity is achieved by different sized people.

In addition to human anthropometrics, other personal characteristics may influence how a person moves. For example, a person's strength may allow them to achieve an activity or movement that a weaker person may achieve differently or even not at all. Let us go back to John's work, in which he moves furniture from a warehouse on to a van and from the van into people's homes. We already know that to speed the process up, and get home early, he avoid using the devices in situ to help him, the trolley and the hydraulic lift, and physically lifts and carries the items. He is able to do this as he is muscularly strong enough to do so, whereas a work colleague with less strength may not be able to or, in trying to, may harm themselves.

ACTIVITY 8.13

Think about one movement system that you perform regularly (e.g. playing a sport or doing household chores). Consider how different individual strength may influence how that activity is achieved by different people.

Another characteristic of an individual that could influence how movement is performed or achieved is the individual's present or previous health. Agnes, who at 85, may have arthritis in her hips and knees may climb the stairs one step at a time using the banisters on both sides, while 15-year-old Dan may run up the stairs taking two steps at a time and not using the banister at all. Long-term conditions may influence movement, for example someone with a chronic low back pain may walk in a more flexed posture and at a slower speed, while a person with chronic bronchitis may have a hunched upper body posture that may limit shoulder and neck movements. Acute illness or injury may also impact on movement in a variety of ways from the person who is acutely ill and bed-bound temporarily to someone who must use aids to achieve movement, such as the individual with a fractured ankle who needs to use crutches to walk.

However, it is not only an individual's physical health that may impact on movement quality and quantity. Someone suffering from clinical depression may limit their movement in all aspects, and when they do move it may be achieved slowly and possibly in a flexed posture. It is not only physiological clinical mental health issues that can influence movement. An individual's psychological profile and personal motivation may influence the quality of their movement. For example, Jenny, who at 9 years old is very enthusiastic about school, rushes along the pavement running and skipping to get the bus; however, Dan, at 15 years old, is bored with school and disinterested in it so walks along the pavement with his head down, shoulders drooped and dragging his feet.

CASE STUDY 8.6

John and Liz are keen gardeners. They enjoy working together in their flower and vegetable garden and are out in the garden in all weathers (the weather may have an impact on their movement, as already discussed in the section on the task environment above). Their enthusiasm for their hobby means that, whatever the weather, they are happy to be gardening, and this influences their movement activity, ensuring smooth and coordinated activity. John is 180 cm and quite stockily built, while Liz is 155 cm and petite in physique. Over the years that they have created and managed their garden, they have realized that each is more suited to some of the regular jobs than others and they work to their individual strengths. John digs and fertilizes the vegetable garden, as he is strong enough to achieve good soil turnover and is able to lift and manoeuvre the fertilizer bags. Liz works in the greenhouse bringing on the seedlings, as John finds the height of the greenhouse shelving forces him to bend his back and he becomes uncomfortable. Although the height of the shelving is appropriate for Liz, she is not strong enough to lift the bags of potting compost, so John decants the big bags into several smaller bags so that she can lift them. When planting time comes, it is Liz's job to plant out the young plants into furrows that John has created. During the growing season, the garden is kept tidy by Liz using a hoe regularly to keep the weeds under control. Although she is weaker than John, he is hampered in this job as his height in relation to the length of the hoe handle means he is uncomfortable working in a flexed posture. The lawn is kept neat by John doing the actual mowing and Liz trimming the edges, so each does activities suited to their individual characteristics so they are safe and effective. Their different characteristics come in very useful when it is time for pruning, as their different statures mean that they can each reach areas that the other cannot and they complement each other's activities.

As can be seen from Case study 8.6, a person's individual characteristics mean that they are more suited to some activities than to others. It is clear that both John and Liz could achieve each of the jobs in the garden, but they have divided the activities to allow each to achieve them using easily achieved and safe movement. For one to achieve a job the other does may require exertion and potential harm.

SUMMARY

In this chapter, we have discovered how human movement can be considered in relation to movement systems and how environmental factors, which may influence this movement, can be logically addressed using a balance theory model. This model applies factors such as task environment, task design, the influence of technology, and the organization within which the movement system occurs to the individual person doing the movement, and considers how each factor, separately and collectively, may influence movement achieved. Additionally, it considers individuals' unique personal characteristics and how they may influence the movement achieved. This approach allows both physical and psychosocial aspects of each factor to be applied to the movement system (please see Ch 7 for a comprehensive discussion about the psychosocial influence on human movement).

It is hoped that, having read this chapter and completed the activities, you will have a clearer understanding of physical and psychosocial environmental factors that may influence human movement.

References

Alexander, N.B., 1994. Postural control in Older Adults. J. Am. Geriatr. Soc. 42 (10), 93–108.

Arnold, J., Cooper, C.L., Robertson, I.T., 1998. Work Psychology. Understanding Human Behaviour in the Workplace, third ed. Prentice Hall, Harlow.

Berg, K., 1989. Balance and its measurement in the elderly: a review. Physiother. Can. 41 (5), 240–246.

Oxford English Dictionary, 1989. http://www.askoxford.com accessed 12.1.09

Parker, S., Wall, T., 1998. Job and Work Design. Sage Publications, London.

Pheasant, S., Haselgrave, C.M., 2006. Bodyspace: Anthropometry, Ergonomics and the Design of Work. Taylor & Francis, London.

Smith, M.J., Sainfort, P.C., 1989. A balance theory of job design for stress reduction. Int. J. Ind. Ergon. 4 (Part 1), 67–79.

Chapter 9

Biomechanics of human movement

Robert W.M. van Deursen and Tony Everett

CHAPTER CONTENTS

LEARNING OUTCOMES

At the end of this chapter, you will be able to:

1. describe the concept of force and its relevance for human movement
2. discuss the general kinematics and kinetics of human movement
3. apply your knowledge of kinematics and kinetics in specified situations
4. describe the graphical and mathematical analysis of force systems and use this knowledge in the analysis of solid and fluid mechanics
5. integrate the above knowledge to perform simple biomechanical movement analyses.

INTRODUCTION

Biomechanics can be defined as the study of the structure and function of biological systems, such as the human musculoskeletal system, by application of (Newtonian) mechanics. Newtonian mechanics refers to Sir Isaac Newton, who laid the foundations for the field of mechanics as currently used. The scope of biomechanics is both wide and deep. Therefore a choice has been made to highlight the aspects considered most relevant for the study of human movement. The main focus of this chapter is the understanding of forces and their effects. Principles of kinematics (the description of motion) and kinetics (the description of motion including the consideration of forces as the cause of motion) will also be discussed.

A first step in understanding human movement is knowledge of the anatomy and physiology of the human body, the musculoskeletal system in particular. However, anatomy books describe movement as though it occurred in a state of weightlessness and only using the anatomical position as the starting point. It is important to realize that movements cannot be considered out of context but occur in an infinite number of body configurations and within a mechanical environment. A biomechanical analysis is required to unravel these influences to determine which muscles (if any) are required to produce a given movement. In addition, such an analysis can quantify the amount of loading occurring on different anatomical structures. Although biomechanical analyses can be very powerful in providing information about movement, it should be noted that the environment within which the movement occurs is generally a lot richer than mechanics alone. Therefore it is best to consider this field of study, as with the other chapters of this book, as a set of tools in a toolbox that contains various other methods used to learn more about human movement.

BASIC CONCEPTS

It is important that before any calculations are attempted, the basic elements of trigonometry are revised and some basic mathematical conventions are explained.

SI units

By convention, the international system of units is used for the dimensions of biomechanical quantities. Table 9.1 gives the basic SI (Système International d'Unités) units and their equivalent values in other systems.

Table 9.1 SI units, their names and imperial equivalents

QUANTITY	SI UNIT	CONVERSIONS
Mass	Kilogram (kg)	1 kg = 9.807 N 1 kg = 2.2 lbs 1 stone = 6.35 kg
Time	Seconds (s)	
Length	Metre (m)	1 m = approximately 39 inches 1 ft = 0.305 m
Angle	Degrees (deg)	1 rad = 57.3 deg 1 deg = 0.0175 rad

Trigonometry

The most basic concept of trigonometry is that of the triangle, with its three sides and three angles. The three angles of any triangle will always add up to 180°. If one of these angles is 90° (the right-angled triangle) and the dimensions of the sides of the triangle are known, then the value of the angles can be calculated. If an angle and a side or two sides are known, then it is possible to calculate all the remaining unknown parameters. This is possible by the use of the *sine*, *cosine* and *tangent* rule. Figure 9.1 shows a right-angled triangle in which angle ABC is the right angle. Side b is known as the hypotenuse, side c as the base or adjacent, and side a as the perpendicular or opposite. Mathematically, the value of the angle CAB (ϕ) can be found using the following formulae:

$$\sin\phi = \text{opposite}/\text{hypotenuse} \text{ or } \sin\phi = a/b$$

$$\cos\phi = \text{adjacent}/\text{hypotenuse} \text{ or } \cos\phi = c/b$$

$$\tan\phi = \text{opposite}/\text{adjacent} \text{ or } \tan\phi = a/c.$$

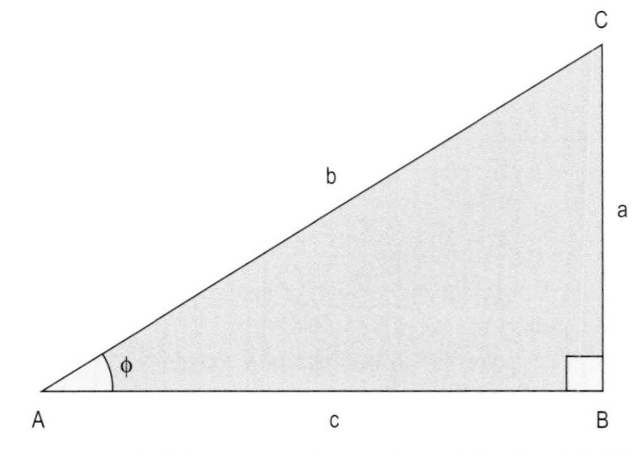

Figure 9.1 A right-angled triangle is used for the definition of the sine, cosine and tangent of the angle ϕ.

It is also possible to calculate the lengths of the sides of the triangle by using Pythagoras's theorem, which states that the square of the length of the hypotenuse is equal to the sum of the squares of the lengths of the other sides. In this case,

$$b^2 = a^2 + c^2 \text{ or } b = \sqrt{(a^2 + c^2)}.$$

Scalar and vector quantities

In biomechanics, two types of quantities are discussed. The first are those that have magnitude only, such as mass, temperature, work and energy. These are called scalar quantities. The second are those that have both magnitude and direction, such as force. These are called vector quantities. Scalar quantities can be added, subtracted, multiplied or divided without any problems. To manipulate vector quantities, trigonometry is used because those quantities may have varying directions.

Rotations

Figure 9.2 shows that rotations in biomechanical terms can occur in one of two directions: anticlockwise, as illustrated in Figure 9.2a, which is defined as positive rotation, and clockwise, as illustrated in Figure 9.2b, which is defined as negative rotation. In this example, the thigh is fixed and the leg is moving. The arrow indicates the direction of movement, and it is important that this direction arrow is added to any diagrammatic representation of movement.

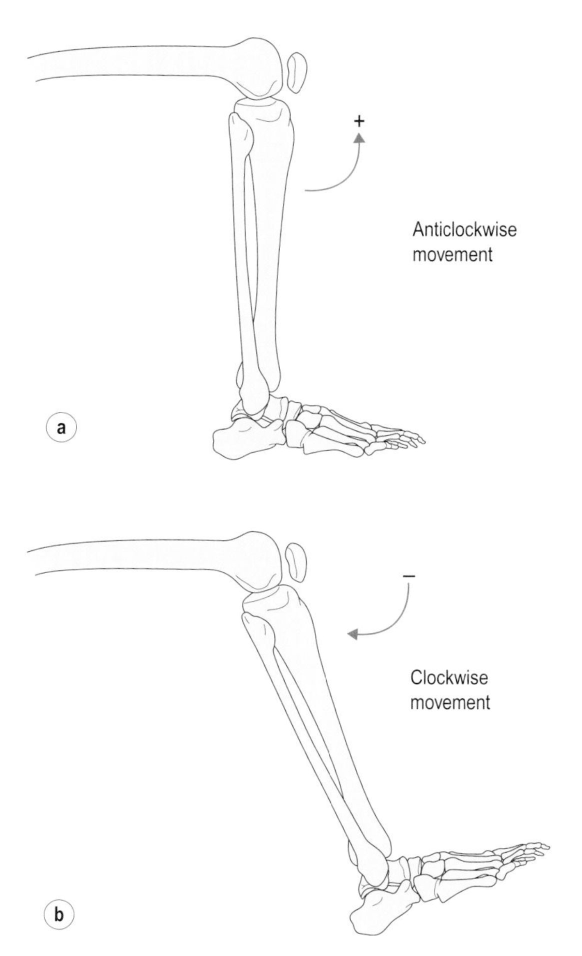

Figure 9.2 In describing the direction of a rotation, the convention is to call anticlockwise rotation positive and clockwise rotation negative.

NEWTON'S LAWS OF MOTION

Although not the first to study movement, Sir Isaac Newton (1642–1727) provided a major breakthrough in the understanding of the causes of movement of objects. He explained this in 1687 by addressing the relationship between force and one of its effects, namely motion. As a result of his studies, Newton formulated three statements known as the laws of motion.

1. Law of inertia

Every body continues in a state of rest or uniform motion in a straight line except when it is compelled by external forces to change its state.

2. Law of acceleration

The rate of change of momentum of a body is proportional to the applied force and takes place in the direction in which the force acts.

3. Law of action and reaction

To every action, there is an equal and opposite reaction.

These three laws still play a major role in the study of biomechanics, and their implications will be considered and explained later in the chapter.

FORCE

DEFINITION OF FORCE

A force is not a tangible object and should therefore be considered as a concept. It can be thought of as an entity that is generated by an action, a push or a pull for example, or imparted, as in a kick. The above examples will probably result in movement, but the object to which the force is imparted may remain stationary while it deforms, as with the force imparted to a soft chair when somebody sits down.

With this in mind, a force can be defined as *an influence that changes the state of rest or motion of a body or object*. This is essentially what Newton describes in the law of inertia, stating that a body or object that is at rest will remain at rest unless some external force is applied to it, and a body or object that is moving at a constant speed in a straight line will continue to do so unless some external force is applied to it. Similarly, inertia of a body or object (or mass) can be defined as *the resistance that a body or object offers to any changes in its motion*. It is measured in kilograms (kg).

DESCRIPTION OF A FORCE

Force is a vector quantity and therefore has a magnitude and a direction. A force is represented graphically as an arrow with the following three descriptors.

1. *Magnitude*: the longer the arrow, the greater the magnitude.
2. *Direction* or *line of action*: the arrow points in the direction of the force.
3. *Point of application*: the point of the arrow is located at the point of application.

EQUATION OF FORCE

Encapsulated in Newton's second law or the law of acceleration is the equation

$$\text{force} = \text{mass} \times \text{acceleration or } F = m \times a,$$

in which mass (m) is the quantity of matter that makes up a body or object, measured in kilograms (kg), and acceleration (a) may be the acceleration due to gravity, measured in m/s^2. The unit of force is the newton (N), while the dimensions of force are $kg.m.s^{-2}$ or $kg.m/s^2$.

FORCE SYSTEMS

It is unusual for forces to act singly, because there is usually a combination of forces acting together. For convenience, these forces can be described as force systems, which are defined as two or more forces acting together. Force systems can be described as follows.

- *Colinear* (*one-dimensional*): this is when the forces are acting in the same plane and along the same line of action. They point in either the same or opposite directions. This is shown in Figure 9.3.
- *Coplanar* (*two-dimensional*): this is when the forces are acting in the same plane but not along the same line of action, as shown in Figure 9.4.

Special cases include the following.

- *Parallel forces*: the directions of the forces are parallel in the same or opposite directions (Fig. 9.5). Parallel forces in opposite directions may produce a force couple, as in a steering wheel.

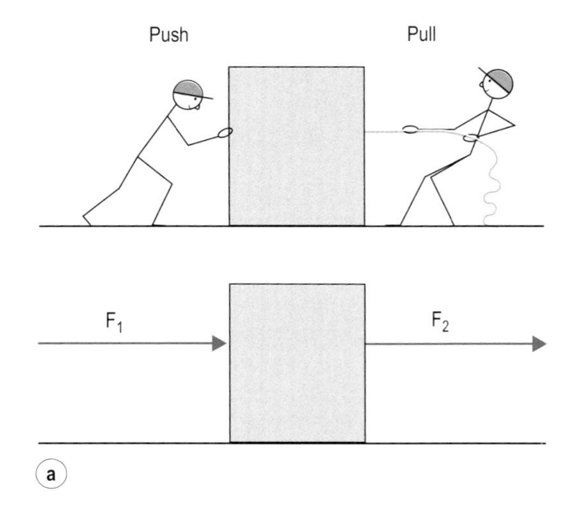

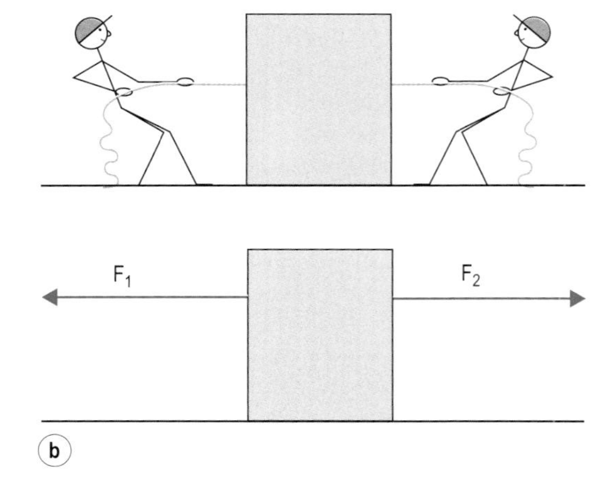

Figure 9.3 Two examples of colinear force systems: (a) two forces acting in the same direction along the same line of action, and (b) two forces acting in the opposite direction along the same line of action.

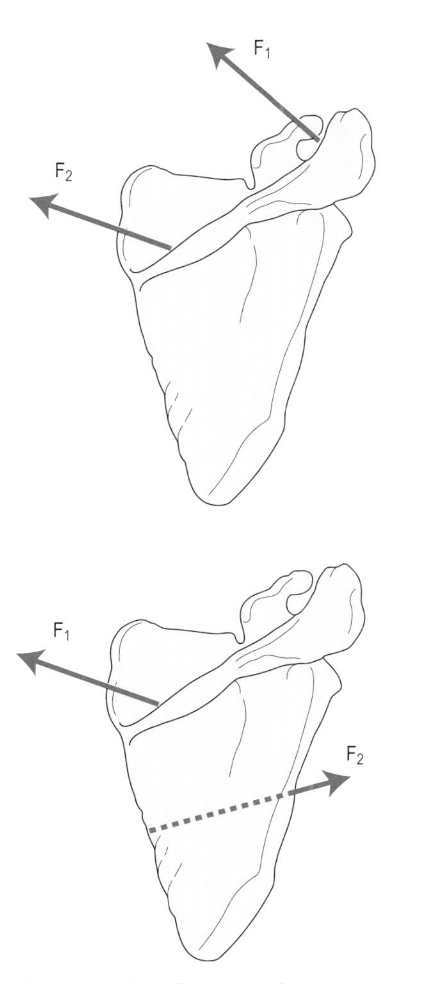

Figure 9.4 Two examples of coplanar force systems.

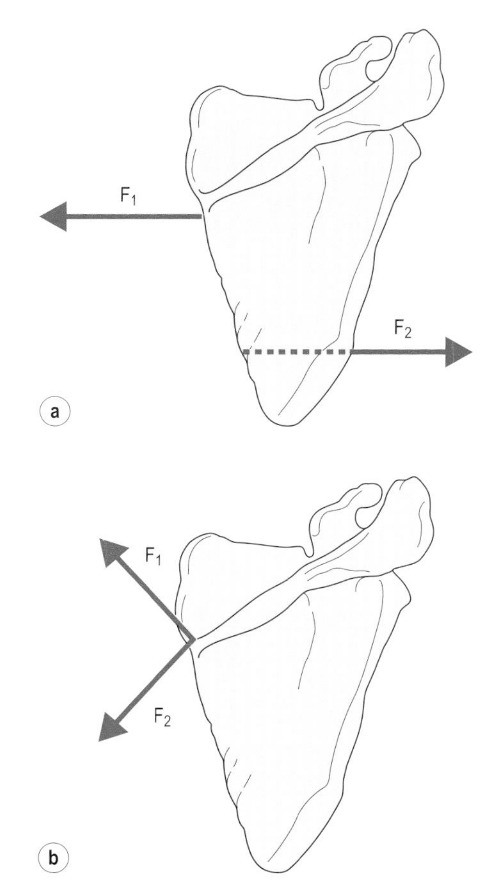

Figure 9.5 Special Cases: (a) Coplanar parallel: two forces in the same plane acting in parallel, and (b) Coplanar orthogonal: two forces acting perpendicular to each other.

- *Orthogonal forces*: the directions of the forces are perpendicular to each other.
- *Concurrent forces*: two or more forces originate from the same point of application, or their lines of action intersect at a common point (Fig. 9.6).
- *Three-dimensional force systems*: forces are acting in more than a single plane. Although this represents the situation most often encountered in everyday examples, this is more difficult to analyse. In this chapter, examples will be used of one- or two-dimensional force systems only.

FORCE ANALYSIS

Because forces are vector quantities, the analysis of force systems will involve trigonometry. Two basic techniques for manipulating vector quantities are important. These are summation and resolution of forces. Summation involves adding the force vectors to find the resultant force or the force that could replace the combined effect of all the forces acting on the body. Splitting a force into its components to establish its effects in two or three principal directions is called resolution of forces. Simple force systems can be analysed using a graphical method. In a vector diagram, each force vector is represented by an arrow drawn to scale. The resultant of two forces is determined by completing a parallelogram and joining the diagonally opposing corners (Fig. 9.7). This diagonal is then measured and the magnitude determined by using the scale.

In the case of more complex force systems, the more useful method of force analysis is by means of trigonometry. Each force is first split into its orthogonal (x and y) components (in three

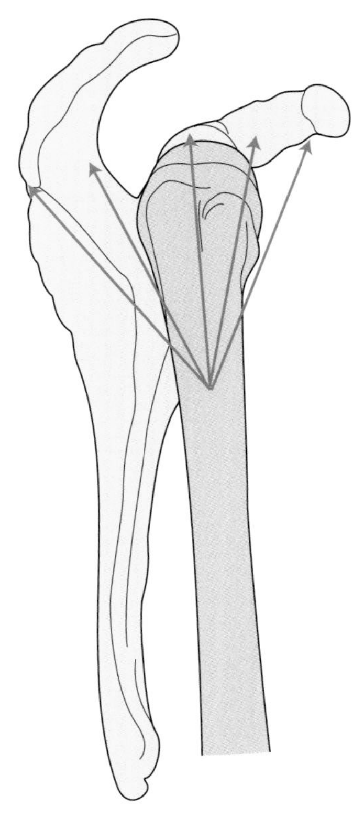

Figure 9.6 An example of a concurrent force system. The deltoid muscle is made up of multiple components that can act together.

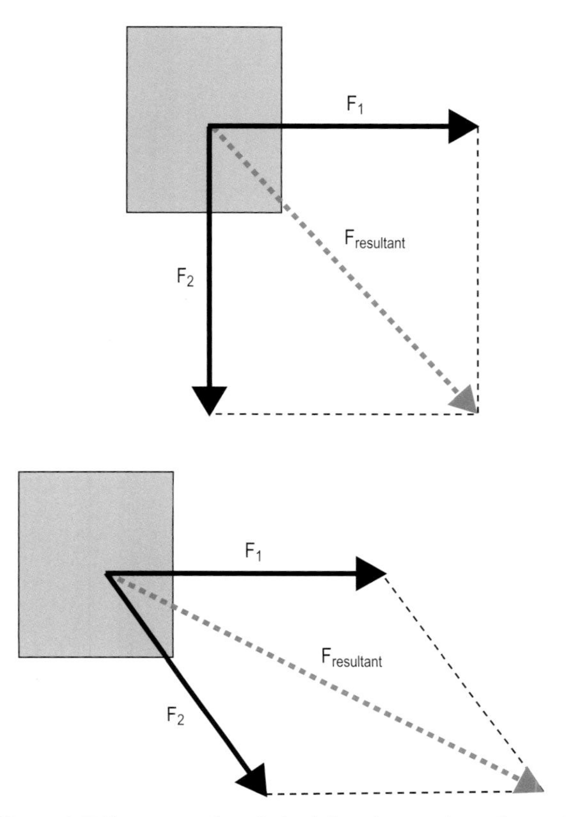

Figure 9.7 Two examples of obtaining the resultant force of a simple force system by means of the graphical method.

dimensions, there is also a z component), as illustrated in Figure 9.8:

$$F_x = \text{force} \times cos\ \phi$$

$$F_y = \text{force} \times sin\ \phi,$$

in which F_x is the force component in the x direction and F_y is the force component in the y direction.

All the x components are then added up or subtracted to obtain the resultant on that axis. The same is done with the y components. The resultant x and y are then added up by means of the following equation:

$$F_{resultant} = \sqrt{(F_x^2 + F_y^2)}.$$

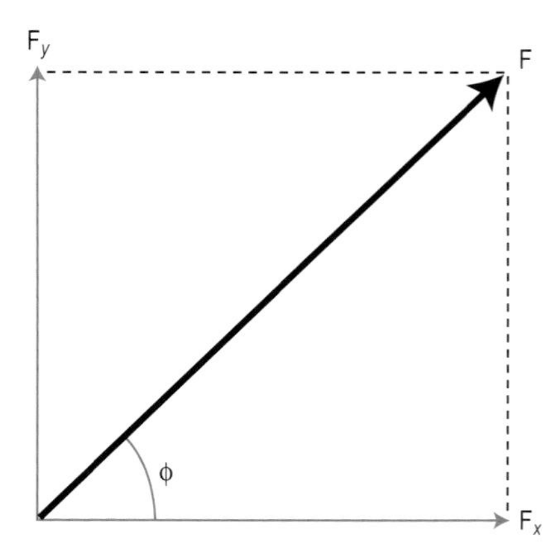

Figure 9.8 Resolution of a force: the angled force is resolved into two orthogonal components, F_x and F_y. The magnitude of the two components can be calculated by means of trigonometry.

TYPES OF FORCES

Force due to gravity

There is a force of attraction of the earth to any object on or near to its surface. The acceleration due to the gravitational pull of the earth is seen as the acceleration any mass has when it is allowed to freely fall to earth, and has the value of 9.81 m/s^2. The weight of an object is the force exerted by the earth on the mass of the object or weight = mass × 9.81 (acceleration due to gravity). Weight is expressed in newtons (N).

Ground reaction force

Newton's third law is demonstrated when considering the forces that are involved when standing on the ground. In this situation, a force is applied by the feet to the ground, which is equal to the weight of the person standing. This force is reflected back up into the feet through the same action line with the same magnitude. This is the ground reaction force (GRF) and is shown in Figure 9.9.

Centripetal force

As stated earlier, a body or object that is moving at a constant speed in a straight line will continue to do so unless some external force is applied to it. Therefore, if an object is to move in a circle, such as during a hammer throw, a force will have to be applied to make the object change direction continuously. This force is called a centripetal force. In the hammer throw, this centripetal force is produced by the thrower. The thrower will, however, perceive himself as being pulled by the hammer, which is often called the centrifugal force. In a biomechanical analysis, it is better to consider the centripetal force only. As soon as the thrower lets go of the hammer, it continues moving in a straight line because the centripetal force is no longer operating.

Frictional forces

When two objects move or tend to move over each other, they experience a resisting force. This is referred to as frictional force and occurs if the objects are solid, fluid or a combination of both. Friction can occur between two surfaces moving over each other: between a wheel and the ground (rolling resistance) or between air or water and an object (drag).

If a force is applied to an object that does not move over a surface, then there must be a frictional force (f) that is equal and opposite to the applied force (Newton's third law). As the applied force increases so will the resistance until it reaches a critical value (F_{max}). At this point, the object will move. Until the object moves, the friction is referred to as static friction, but when the object is moving the friction is referred to as dynamic or kinetic friction.

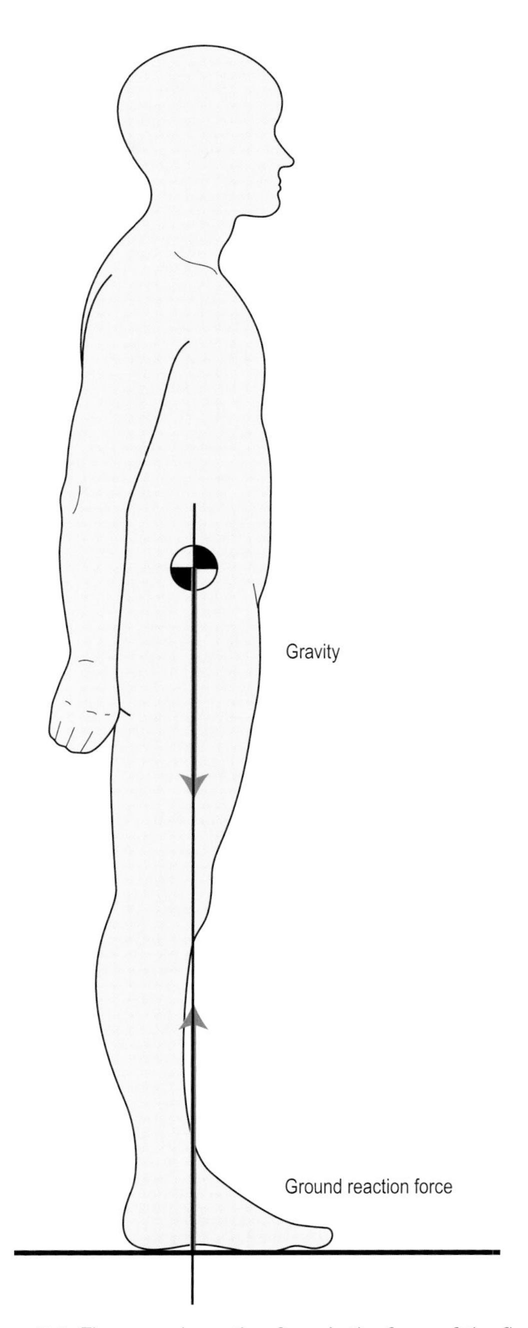

Figure 9.9 The ground reaction force is the force of the floor acting on the body, opposing the effect of gravity.

The coefficient of friction (μ, mu) is dependent on the conditions under which the friction is occurring (static or dynamic), the condition of the surfaces of the material, and the type and direction of the applying force. The frictional force can be calculated by multiplying the coefficient of friction and the force perpendicular to the contact surface of the two objects (also called normal force). In some instances, frictional forces are useful as they allow controlled movement to take place (as in walking), or they may be a hindrance as they require energy to overcome them (as in some types of exercise). Friction can produce heat and wear and tear to the surfaces that are moving over each other; this may be detrimental to the system (either machine or human body), and there may be a need for a lubricant to decrease these adverse effects.

Elastic forces

Elastic forces will be discussed in more detail in the section below about deformation of materials.

External and internal forces

Forces that act from outside the body – for example gravity, moving objects, GRF, and wind or water resistance – are called external forces. Forces can also be generated inside the body by a muscle or be transmitted between body parts by, for instance, ligaments or bone on bone contact. These are examples of internal forces. It is important to realize that muscle forces that are under voluntary control are just one aspect of the force system acting on our body. Movement of the body depends on the resultant of the force system, which includes external and internal forces, and not only on muscle forces.

PRESSURE

Pressure is the manifestation of force when the surface area over which the force is acting is taken into consideration. Pressure can be defined as the force per unit area (in m^2), or $P = F/A$. The official unit of pressure is the pascal (Pa, N/m^2), although the kilopascal (kPa, 1000 Pa) is more useful for human application.

It can be seen from the equation above that if a force remains the same and the surface area increases, then the pressure exerted by the force will be less. Conversely, if the surface area is decreased while the force remains the same, greater pressure will be felt from the force. This is important when considering the pressure felt on the human body when lying in a bed. If the force is being channelled through a small surface area, such as a bony point, then the pressure over that area will be much greater than if the force from the body were acting on a large surface area. In the latter case, pressure sores would be less likely to occur.

MOMENT OF FORCE

When application of a force produces a turning or rotary effect, the force is said to be producing a moment. The twisting effect of a force, however, is often referred to as torque. There are no real differences between rotation and twisting, so the words *moment* and *torque* cannot be considered separately. To calculate the magnitude of the moment of force, the equation of moment or torque is used:

$$\text{moment} = \text{force} \times \text{moment arm or } M = F \times d.$$

In the above equation, d stands for distance, representing the moment arm. The moment arm is the shortest distance (perpendicular) from the line of action of a force to the axis of rotation. In Figure 9.10, the moment arm is shown for a weight on a see-saw (a), and the muscle moment arm is shown for the elbow flexor (b). A line is drawn perpendicular to the force line of action and through the axis of rotation. The moment arm is the distance between the two. The unit of moment or torque is the newton.metre or N.m, while the dimensions of moment or torque are $kg.m.s^{-2}$ or $kg.m/s^2$.

Internal and external moments

When forces act from outside the body (e.g. gravity, GRF, and wind or water resistance) and produce rotatory effects, they are called external moments. When the forces are acting inside the human body, they produce internal moments. These are often muscle forces acting on the various segments of the body, but they can also be forces acting from one segment to the next by means of the ligaments.

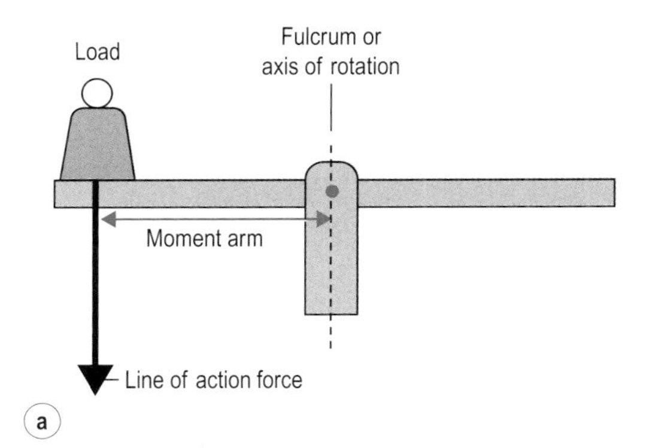

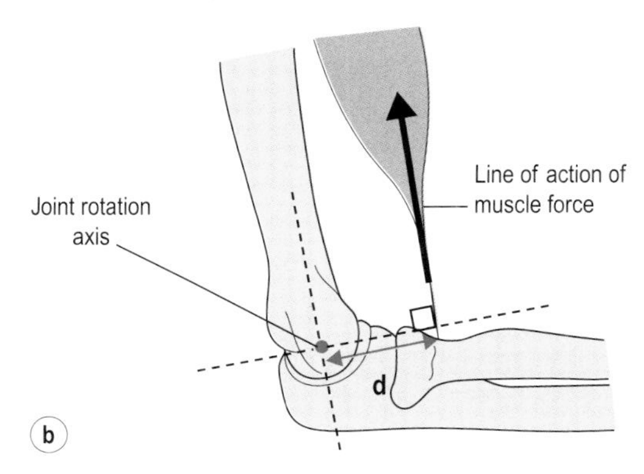

Figure 9.10 Typical levers within the physical world and the body are shown: (a) the moment arm is shown for a weight on a see-saw, and (b) the muscle moment arm is shown for the elbow flexor. A line is drawn perpendicular to the force line of action and through the axis of rotation to determine the distance between the two.

LEVERS

When levers are considered, the moment of force is an important factor. A lever is defined as *a rigid bar that rotates around a fixed point or fulcrum*. The rigid bar may be represented by a crowbar in the physical world or a bony segment inside the human body, while the fulcrum may be the point about which the crowbar turns or it may be a joint.

Figure 9.10 represents a typical lever within the physical world and the body. The resistance has to be met by a force on the other side of the lever to balance or lift the load. In Figure 9.11, the moment arms of a lever system are shown. The moment arm between the fulcrum and the load or resistance to be moved is also called the load or resistance arm ($d_{resistance}$). The moment arm between the fulcrum and the effort or force is also called the force or effort arm (d_{force}).

As can be seen from Figure 9.12, there are three distinct forms that the lever can take, the formation of which depends on the relative position of the fulcrum to the resistance and the force. The form that the lever takes decides its function. A lever is often used to make work easier (as in the case of the crowbar). This does not always happen, however. Figure 9.12a represents a lever when the fulcrum lies between the force and the resistance, as in a see-saw. This is called a first-order lever. In this example, the moment arms are equidistant. If the resistance lies between the fulcrum and the force (Fig. 9.12b), this is referred to as a second-order lever. The lever is helpful for lifting a load, as in a crowbar or wheelbarrow. The final combination can be seen in Figure 9.12c, when the force lies between the fulcrum and the resistance. This is

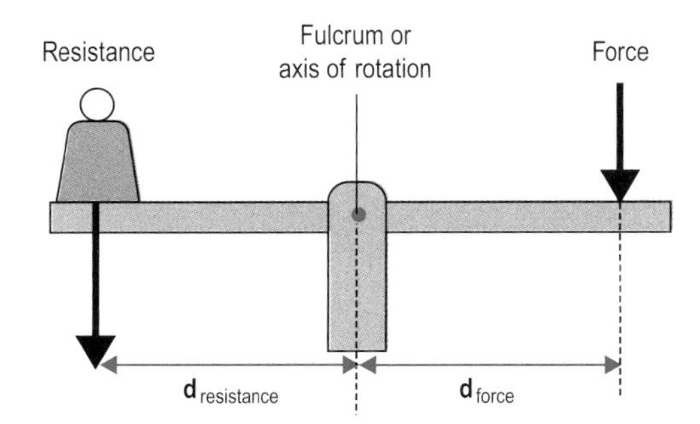

Figure 9.11 The moment arm is the distance between a force and the fulcrum. In this system, the resistance has to be met by a force on the other side of the lever to balance or lift the load. The moment arms for both are shown.

ACTIVITY 9.1

This activity concerns moment arms during strength testing. When testing the maximum muscle strength of the knee extensor, placement of the hand on the leg is important because of the effect on the moment arm of the resistance. The muscle moment arm of the knee extensor will not change during an isometric strength test. If the left knee test produces a maximum resisted force of 300 N at a distance of 25 cm distal from the knee joint axis and the right knee test produces a maximum resisted force of 250 N at a distance of 30 cm distal from the knee joint axis, can you calculate which side is stronger?

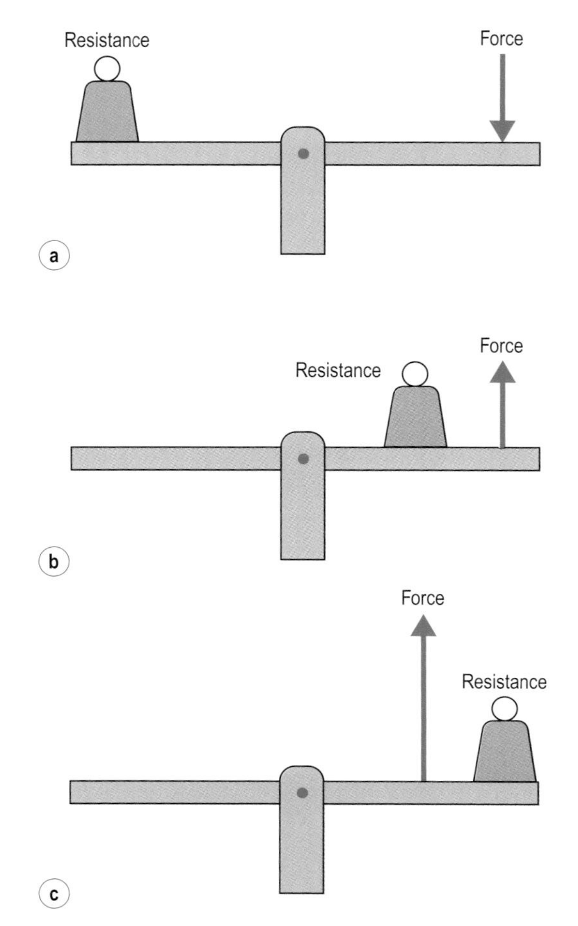

Figure 9.12 Depending on the position of the resistance, the force and the fulcrum relative to each other, levers are divided into (a) first-order, (b) second-order and (c) third-order levers.

called a third-order lever, and at first it appears quite difficult to see any advantage in this. However, this arrangement frequently occurs inside the body when the muscle insertion lies closer to the joint axis than the load. The advantage for the muscle is that the distance and velocity of shortening during contraction are smaller. The tissue loading is obviously large.

ACTIVITY 9.2

Are there situations in which a muscle in the body has an arrangement different to a third-order lever? For shoulder flexion, consider the position of the fulcrum (joint), the resistance (for instance, weight of the arm) and the force (point of application of the muscle force). What type of lever is operating and has it got a force or a speed advantage?

Mechanical advantage

The explanation of why some levers help (mechanical advantage, MA) and others do not is mathematical. The equation to calculate the MA is given as

$$\text{MA} = \text{force arm}/\text{load arm or MA} = \text{load}/\text{force},$$

in which force arm is the distance from the fulcrum to the force and the resistance arm is the distance from the fulcrum to the resistance (Fig. 9.13).

From the example of the see-saw in equilibrium, it can be seen that the MA will be 1. This means that there will be no advantage or disadvantage. If the example of the wheelbarrow is taken, in which the force arm is greater than the load arm, then the MA is always going to be greater than 1. Therefore any lever with an MA of greater than 1 will have a true advantage. This advantage is sometimes called a force advantage, which distinguishes it from a lever, in which the load arm is always greater than the force arm, giving an MA of less than 1. This last lever has a speed advantage over the other two types of lever. This is because, despite requiring a greater force to overcome the resistance, once this is achieved the load end of the lever will move with a greater velocity than the point at which the force is applied. Thus it has a speed advantage.

EFFECTS OF FORCE AND MOMENT OF FORCE

It has already been said that a force can be described by its effects, and there are two main effects, namely motion and deformation. Deformation will be described later in the chapter. Motion can be categorized as either linear or angular.

Linear motion

Linear motion is also referred to as translation. During a linear movement, all the particles of the body describe equal and parallel paths. If the trajectory of the linear movement is a straight line, it is called rectilinear movement, and if the trajectory is curved, it is called curvilinear movement.

Displacement is the shortest distance between two points, as opposed to distance, which may be travelled by a more circuitous route. This is shown in Figure 9.14.

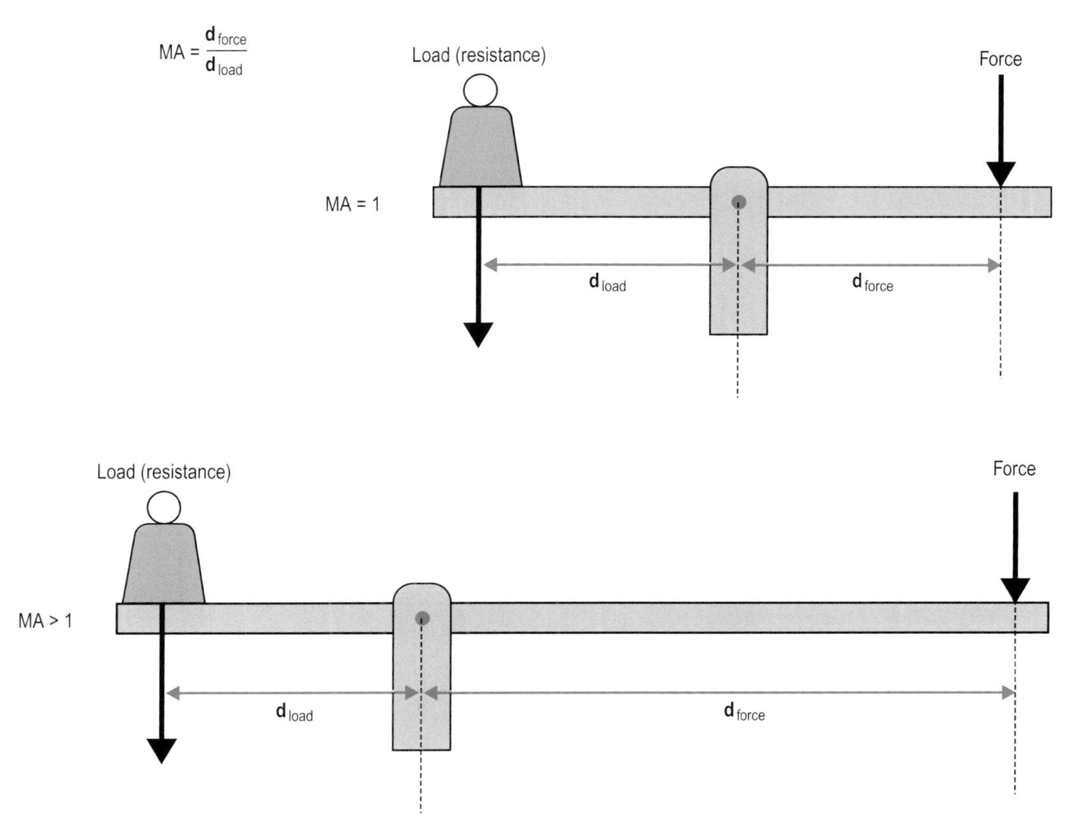

Figure 9.13 The mechanical advantage (MA) of levers can be calculated to determine whether the lever is beneficial in assisting to lift a load. A MA greater than 1 indicates that the lever makes lifting easier.

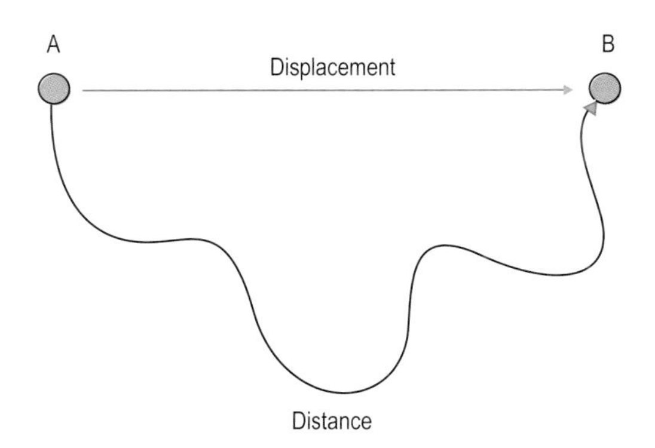

Figure 9.14 When moving from A to B, the displacement between the two positions does not have to be the same as the distance travelled.

Velocity can be defined as the rate of change of position (displacement) and is expressed in metres per second (m/s). Acceleration is defined as the rate of change of velocity and is described in metres per second squared (m/s^2). Figure 9.15 shows the linear displacement, velocity and acceleration of a person running and stopping. The velocity was between 2 and 3 m/s at first and then rapidly declined as the person decelerated (see Fig. 9.15b). Maximum deceleration was over 15 m/s^2 (see Fig. 9.15c). The velocity then gradually declined to zero. The total displacement during this period was almost 2 m (see Fig. 9.15a). It is obvious from this example that velocity and acceleration can vary from one moment to the next. The graph therefore represents the instantaneous velocity and acceleration.

Angular motion

Angular motion of a body is also known as rotation. The displacement of angular motion (*j*) is measured in degrees (one circle = 360°). In biomechanical calculations, this is often expressed in radians (1 rad = 360°/(2pi) = 57.3°).

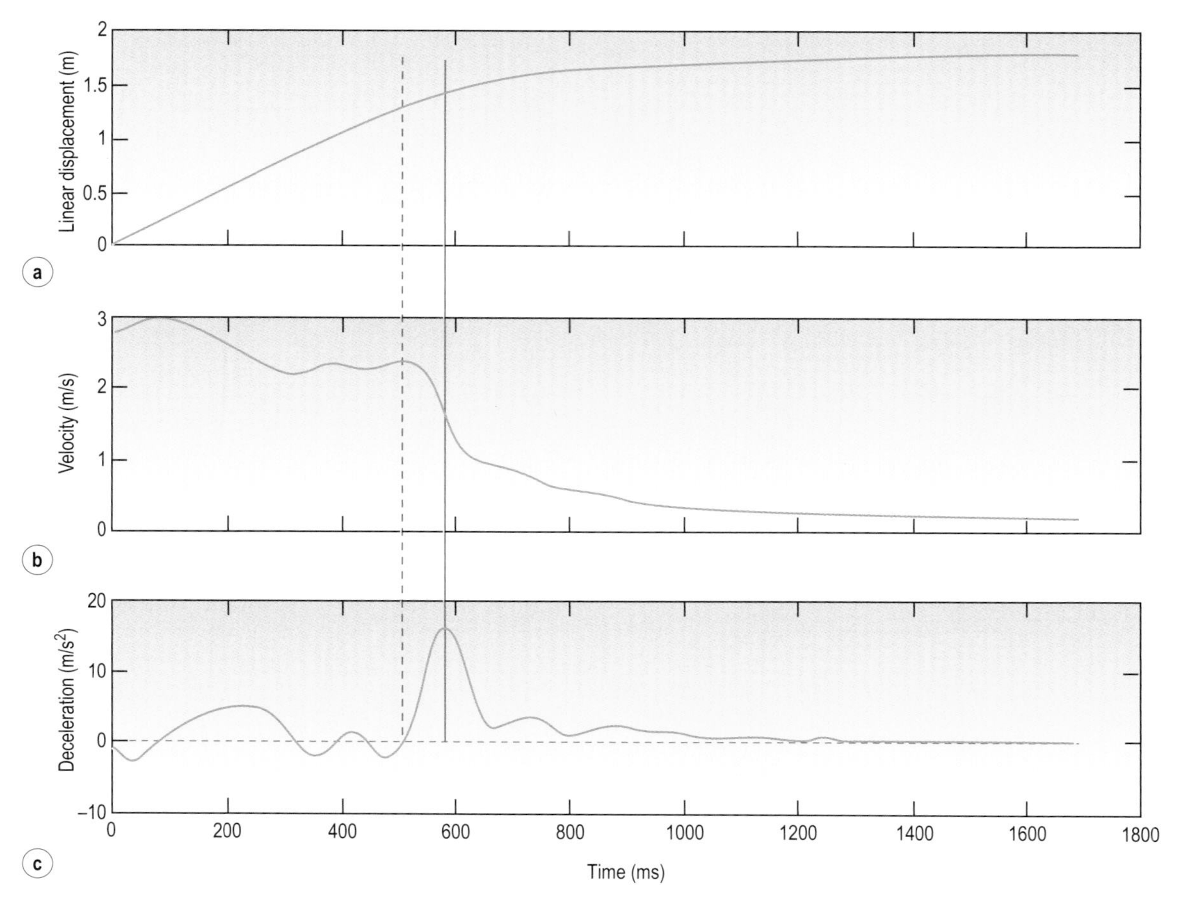

Figure 9.15 The linear displacement (a), velocity (b) and deceleration (c) of a person jogging and then stopping. In all graphs, the point at which deceleration was started is indicated by the first vertical line. The second vertical line indicates when the subject was decelerating the most, which is at the peak in the bottom graph.

Angular velocity (ω) is defined as the *rate of change of angular displacement*. It is measured in degrees per second or radians per second. Angular acceleration (α) is the rate of change of angular velocity and is expressed in degrees per second squared (°/s^2) or radians per second squared (rad/s^2). Figure 9.16 shows the angular movement of the right knee during the same activity of a person running and stopping as displayed in the previous figure. Peak knee flexion is close to 70°, peak knee angular velocity (towards knee flexion) is well over 500°/s and peak angular acceleration (towards knee flexion) is more than 10,000°/s^2.

Linear and angular motion resulting from a single force

A single force can result in a variety of effects depending on where the force is applied to the object. Once the force analysis of a system has revealed what the magnitude, direction and point of application of the resultant force are, the movement resulting from the force system can be determined. If the line of action of a (resultant) force travels through the centre of gravity (COG) of an object or body and is parallel to the direction of movement, the force will result in linear acceleration (Fig. 9.17). Obviously, if the force is applied in the direction opposite to the direction of movement, a deceleration will occur.

If the line of action of a (resultant) force does not travel through the COG of an object or body and is parallel to the direction of movement, the force will result in linear and angular acceleration (Fig. 9.18).

If the line of action of a (resultant) force does not travel through the COG of an object or body and is not parallel to the direction of movement, the force will result in linear and angular acceleration and a change of direction (Fig. 9.19).

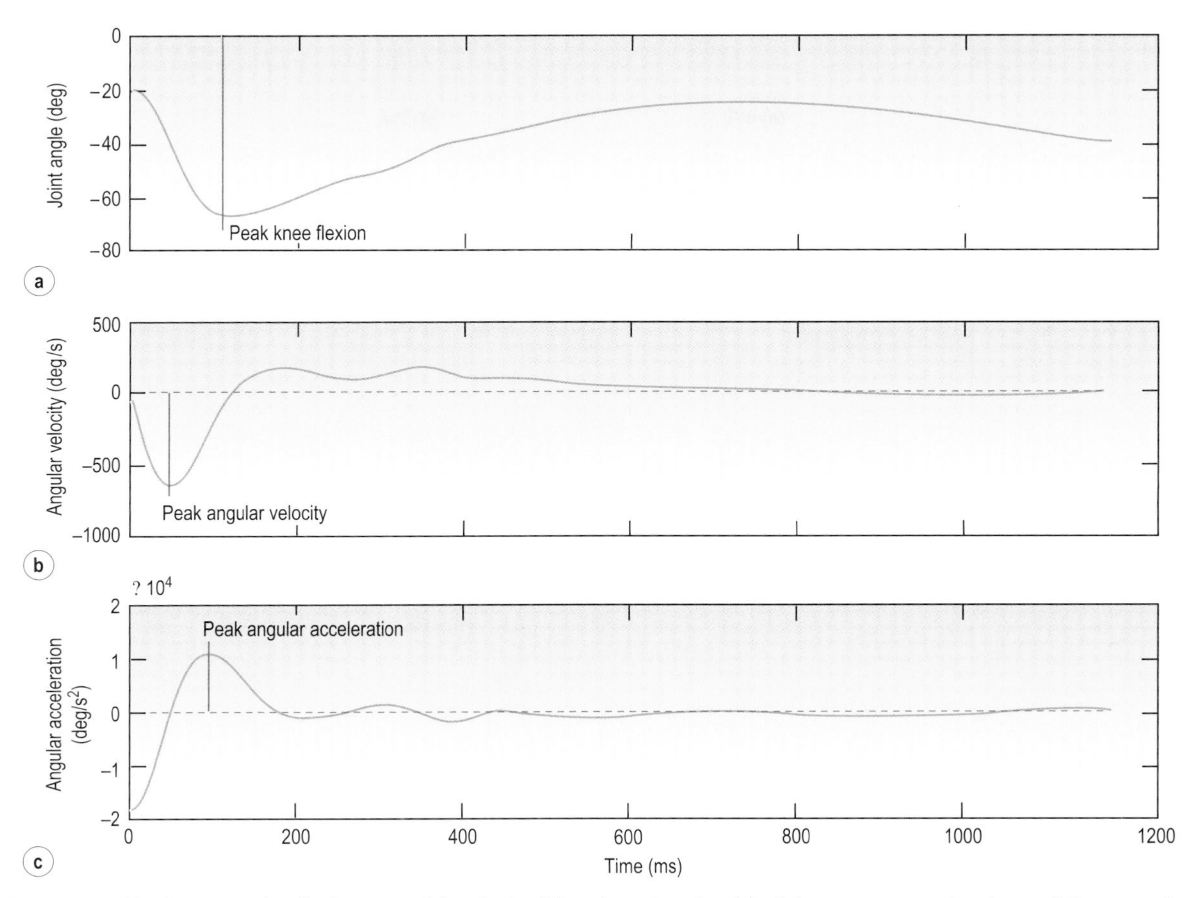

Figure 9.16 The knee angular displacement (a), velocity (b) and acceleration (c) of the same person jogging and then stopping as in Figure 9.15. A vertical line in each graph indicates the peak angular angle, velocity and acceleration. It is obvious that these peaks do not occur at the same point in time.

Figure 9.17 A force resulting in a linear effect only.

Figure 9.18 A force resulting in a linear and angular effect.

Linear and angular motion combined

Human movement often involves a combination of linear and angular movement. For instance, during gait, rotation of the lower limbs is used to achieve curvilinear movement of the trunk, represented by the dotted line in Figure 9.20. At the start of Figure 9.20 (left side), the first rotation occurs around the ankle (a), resulting in trunk linear movement (b). In the middle, the second rotation occurs around the hip, resulting in curvilinear movement of the

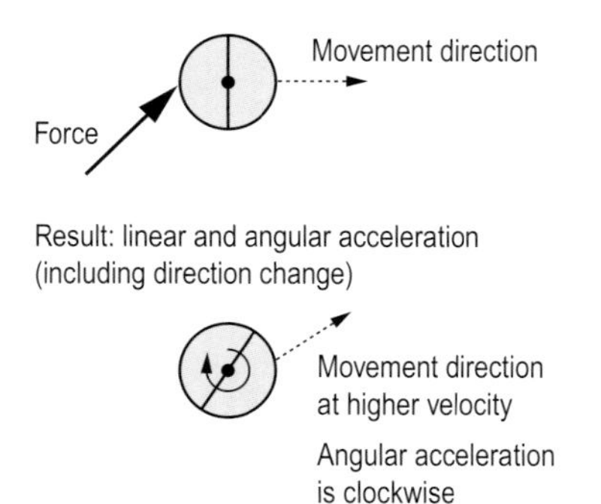

Figure 9.19 A force resulting in a combined effect.

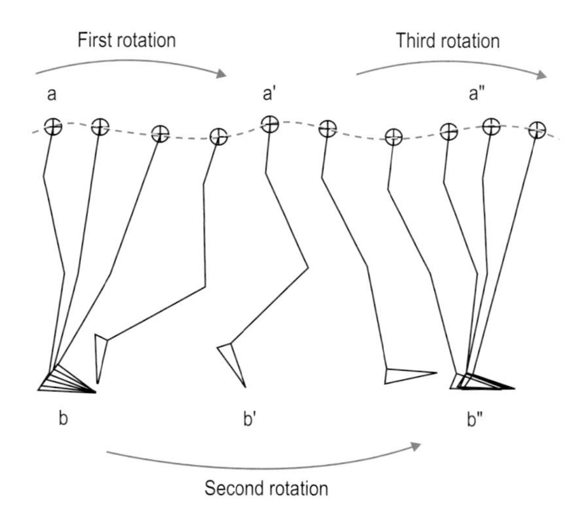

Figure 9.20 Gait involves rotation of the limbs to achieve curvilinear motion of the trunk (dotted line) through space. A single step is shown, in which a sequence of ankle (a), hip (a′) and ankle (a″) rotation provides for the forward progression of the body.

ankle (b′). The trunk continues to move forwards because of the effect of the other lower limb. At the end, rotation occurs again around the ankle (b″).

Angular motion and moment of inertia

As discussed above, objects offer resistance to any changes in motion. For linear motion, this is termed *inertia*. The tendency for an object to resist changes of angular motion is termed its *moment of inertia* or I. It can be represented by the equation

$$I = m \times r^2,$$

in which m is the mass of the object and r is the distance of the mass from the axis of rotation.

As the mass of the body increases so does its moment of inertia, but as the mass distribution moves away from the axis of rotation, the radius increases and the moment of inertia increases in proportion to the square of that increased distance. The human body has different moments of inertia depending on the direction of rotation considered. Rotation about the longitudinal axis is easier to generate than rotation about the frontal axis, because in the latter case the mass of the body will be located at greater distances from the axis of rotation.

CENTRE OF GRAVITY AND BASE OF SUPPORT

Theoretically, mass is distributed throughout the body and gravity (weight = mass × acceleration due to gravity) acts on every particle of mass. It would be almost impossible to perform any calculation if this were to be considered in calculations, so instead the centre of mass (COM) is used. It is defined as the *point about which the mass of an object is evenly distributed*. The COM is closely associated with the COG. The COG is sometimes explained as the point at which the force of gravity is said to act. The COM of an object is the geometrical centre of that object if it is symmetrical and regular (cube, cylinder or cone). The body segments do not fit exactly into these descriptions but can be approximated in this way. The percentage weight of the body segments and the position of the COM for each body segment can be seen in Tables 9.2 and 9.3.

Table 9.2 Mass of each body segment as a percentage of total body mass

SEGMENT	PERCENTAGE OF BODY MASS
Trunk	49.7
Head and neck	8.1
Arm	2.8 each
Forearm	1.6 each
Hand	0.6 each
Upper limb	5.0 each
Thigh	10.0 each
Leg	4.7 each
Foot	1.4 each
Lower limb	16.1 each

Table 9.3 Location of the centre of mass of each body segment as a percentage of segment length

SEGMENT	LOCATION
Trunk	50% between greater trochanter and glenohumeral joint
Head and neck	At the point of the ear canal
Arm	43.6% from proximal joint
Forearm	43% from proximal joint
Hand	50% from proximal joint
Thigh	43.3% from proximal joint
Leg	43.3% from proximal joint
Foot	50% between lateral malleolus and fifth metatarsophalangeal joint

(After Winter 1990, with permission.)

BODY CENTRE OF GRAVITY, LINE OF GRAVITY AND CENTRE OF PRESSURE

The COG of the body as a whole can be thought of as the point about which the mass of all body segments is evenly distributed. In the anatomical position, it is thought to be at the level of the second sacral vertebra, inside the pelvis. However, as soon as the configuration of the body differs from the anatomical position, the COG will shift and can even be located outside the body. For instance, if both arms are elevated to a horizontal position, the COG moves forwards and upwards relative to its location in the anatomical position. This change of the COG location is an important consideration when discussing balance and equilibrium and the different human postures. See Chapter 5 for a full discussion of body posture and balance.

The line of gravity (LOG) can be said to be the projection of the COG on the ground, represented by a line perpendicular to the ground through the COG. In Figure 9.21, the COG is represented by the target and the dotted line represents the LOG. Although the position of the LOG is given relative to the different joints, it should be realized that there is a wide variation in posture between different people. The LOG is not very useful for biomechanical calculations.

The centre of pressure (COP) is the point of application of the GRF. This force reflects Newton's third law, the law of action and reaction, in that the force exerted by the body on to the ground is reflected back at the COP. On average, during

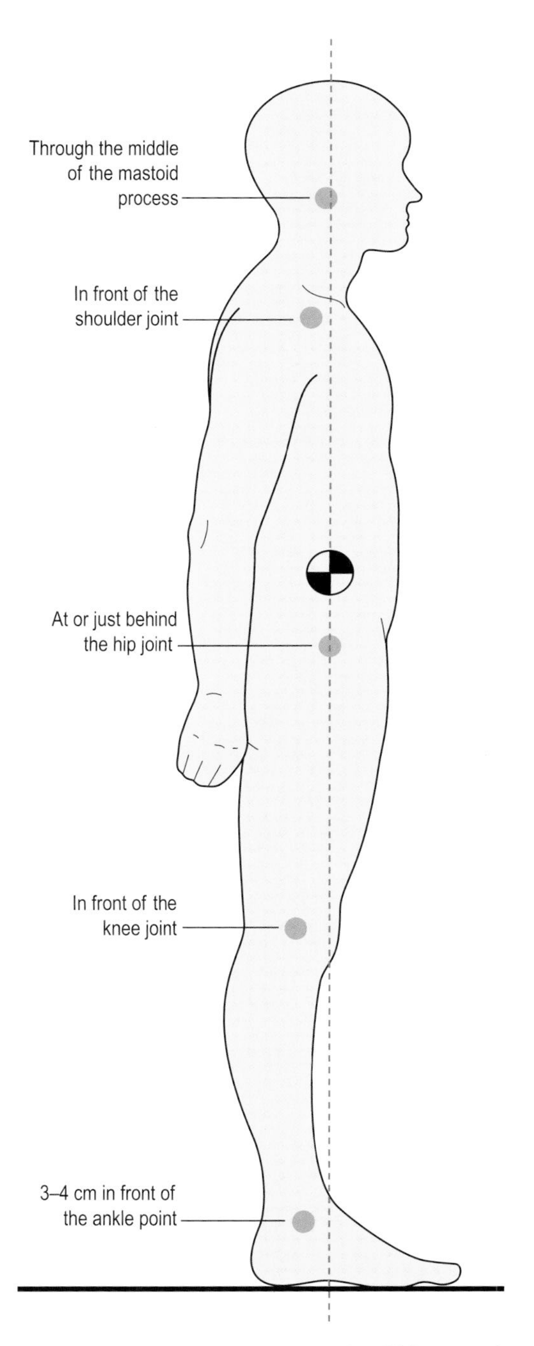

Figure 9.21 A line projected through the COG on to the floor is called the line of gravity. In the picture, this line is shown on a person standing upright. Some anatomical landmarks give a better indication of where the line is located.

quiet standing the GRF and gravity pulling on the COG will be colinear, but this is not necessarily the case during movements.

BASE OF SUPPORT

Every object, unless it is floating in space, has to rest on a supporting surface. The surface area of the part that is involved in support of the object, be it inanimate or a human body, is known as the base of support (BOS). The shape and size of the BOS depend on the posture that the body adopts (lying, sitting or standing, for example), the position of the feet and hands, and the use of extra support (crutches or a chair, for example). When standing upright unsupported, the BOS is in between and underneath the feet (see Fig. 9.22).

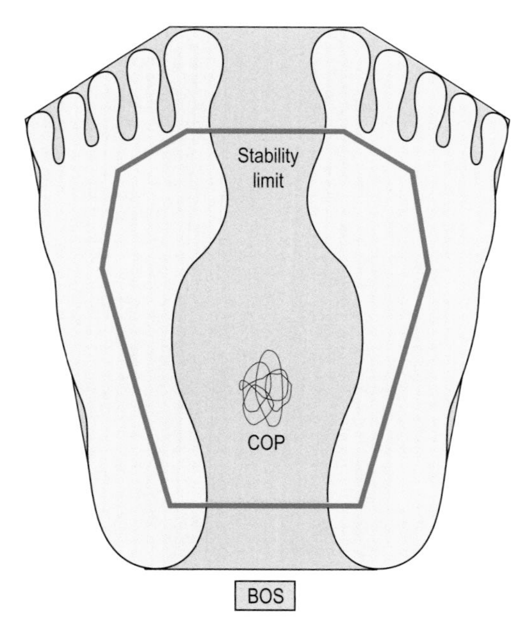

Figure 9.22 During quiet upright standing, the contact area with the floor underneath the feet and the area in between the feet is the base of support (BOS, darker area in the diagram). Because people cannot easily move their line of gravity (LOG) to the outer edges of the BOS, the stability limit has been defined as the area within which people can move their LOG without losing balance. The centre of pressure (COP) of the ground reaction force is located well within those two areas. Because a person will always sway a bit, the COP oscillates with a certain amplitude.

BALANCE, EQUILIBRIUM AND STABILITY

These terms are often used interchangeably, but they are all slightly different paradigms. If the LOG is within the BOS, then the body is said to be in balance. When all the resultant forces and moments acting on a body are equal to zero, then equilibrium is said to occur. If the body is stationary when all the forces add up to zero, then the body is said to be in static equilibrium. If, on the other hand, the body moves with a constant linear velocity, it is said to be in dynamic equilibrium. If, after a displacement by a force of short duration, the body tends to return to its original starting position, then it is said to be stable.

Types of stability

If an object tends to return to its original starting position after a force of short duration is applied, or if the object is placed such that an effort to disturb it would require its COG to be raised, then the object is said to be stable. In other words, the LOG remains well within the BOS when the object is tilted.

In the situation in which an object tends to continue its displacement under the influence of gravity after a force of short duration is applied, then it is said to be unstable. When the relationship of the LOG to the base and the height of the COG are the same after displacement (rolling of a tube, for example), then the stability is said to be neutral. The stability of a body depends on:

- the surface area of the BOS
- the location of the LOG within the BOS
- the height of the COG above the BOS
- the mass of the body.

Base of support, stability limit and centre of pressure

In Figure 9.22, quiet standing with the feet parallel and slightly apart, the BOS is the area within a line drawn around the outer edges of the feet and the area in between. In theory, a person should be able to move their LOG to the edge of the BOS and still be stable. However, this is very difficult to do, because it would require a lot of muscle force at the ankle joint. There is therefore a limited area within the BOS within which a person can move their LOG safely. This area is called the stability limit (see Fig. 9.22).

The COP was defined as the point of application of the GRF. During quiet standing, the COP lies well within the BOS and the stability limit. The

COP is usually located a few centimetres in front of the ankle joint and moves with an amplitude of approximately 1 cm, because a person sways when standing (see Fig. 9.22).

CASE STUDY 9.1

Dan is off to school. He has rugby practice and the normal full day of lessons, and after school he helps at a local swimming club for people with physical disabilities. Consequently, he has a very full and heavy large rucksack strapped to his back. He trudges off to the bus stop.

Figure 9.9 showed us the ideal LOG that Dan should have to be the most energy-efficient. Let us consider what happens to him as soon as he straps on his rucksack. His COG, which was lying perfectly happily within his pelvis at the level of S2, has been shifted up his spine, as this extra mass has been added to his trunk. We know from this chapter and Chapter 5 that when the COG is higher the body will have the tendency to become intrinsically less stable, coupled with the fact that the relationship of the LOG to the base will also alter. On their own, these alterations would probably not cause Dan too much trouble in maintaining postural equilibrium.

Bigger problems occur because the COG has been shifted posteriorly. This has a dramatic effect on the relationship of the LOG to base. Given that Dan will probably have a small base (his feet), the new LOG will fall outside his base. This will cause Dan to fall over if he does not compensate for this extra load. What do you think Dan will do?

Dan may adopt one or all of several mechanisms to counteract this tendency to fall backwards. One of the most obvious would be to return the COG to a position in which the LOG is within the stability limits (see Fig. 9.22). Leaning forwards at the hips would achieve this if the spine was kept rigid. This would necessitate an increase in the work of the hip and trunk flexors to pull the COG forwards, and an increase in the work of the trunk extensors to keep the spine rigid.

The movement described above would be pivoting around the hips and lower lumbar spine. The change in the COM would affect the distance from the fulcrum (see Fig. 9.10). From the formula $M = F \times d$, it can be seen that the moment would be increased.

The above two biomechanical scenarios alone would mean that the efficient postural control would be disrupted, leading to stress within the body and an increase in workload and energy consumption. All this happens even before Dan has started walking!

WORK, ENERGY, POWER AND MOMENTUM

WORK

When a force moves an object in a specified direction, the extent to which it is moved is the amount of work performed. The unit of work is the joule (J). Work can either be linear or rotational and can be calculated by using the following equations:

$$\text{linear work} = \text{force} \times \text{distance of linear displacement}$$

$$\text{rotational work} = \text{moment} \times \text{angular displacement}.$$

Muscle work

Muscle work can be calculated in different ways using the following equations. If the muscle force and the distance between muscle origin and insertion are known,

$$\text{work} = \text{muscle force} \times \text{muscle length change}.$$

If the net joint moment (see section on movement analysis later in this chapter) and joint rotations are known,

$$\text{work} = \text{net joint moment} \times \text{joint angular displacement}.$$

Muscle can actively shorten only by means of a contraction. During a concentric contraction, the muscle force and the muscle length change are in the same direction. The above equation will then result in a positive outcome and therefore results in positive work. During an eccentric contraction, the muscle force and the muscle length change are in opposite directions. The above equation will then result in a negative outcome and therefore results in negative work. During an isometric contraction, there is no length change of the muscle and therefore in biomechanical terms there is no work done.

ENERGY

The capacity of a force to do work is termed *energy*, also measured in joules (J). There are many forms that energy can take. Metabolic energy is the energy obtained from food by means of the metabolic process. Heat energy is the energy that a system can derive from a heat source. Mechanical energy is a measure of the state of a body at an instant in time as to its ability to do work.

Forms of mechanical energy

Potential (gravitational) energy is defined as the capacity of a body to do work due to the location of an object in a gravitational field above a certain baseline. The equation for calculating this energy is shown below:

$$E_p = \text{mass} \times \text{gravity} \times \text{height},$$

in which mass is the mass of the body or object, gravity is the acceleration due to gravity, and height is the position of the body above the baseline.

Kinetic energy is the capacity of an object to perform work due to its motion. It can be calculated using the equations below:

$$\text{linear kinetic energy } (E_k) = \tfrac{1}{2} m \times v^2,$$

in which m is mass and v is velocity, and

$$\text{rotational kinetic energy } (E_r) = \tfrac{1}{2} I \times \omega^2$$

in which I is moment of inertia and ω is angular velocity.

Elastic (strain) energy is the capacity a body has to do work after being deformed from its original shape. It is calculated by using the equation below:

$$E_s = k \times \Delta L,$$

in which k is the spring stiffness and ΔL is the change in length.

Conservation of energy

The total sum of energy is assumed to always be constant. One form of energy can be transformed into another, but the total amount of energy is never lost.

POWER

The rate at which work is performed is termed the *power* of the system. Power is the product of force and distance in a specified time (velocity). This can be calculated by using the equations below.

For linear motion,

$$\text{power} = \text{force} \times \text{distance}/\text{time}$$

or

$$\text{power} = \text{force} \times \text{velocity}.$$

For rotational motion,

$$\text{power} = \text{moment} \times \text{angular displacement}/\text{time}$$

or

$$\text{power} = \text{moment} \times \text{angular velocity}.$$

Muscle power

The rate of doing muscle work is termed *muscle power*. Muscle power can be calculated using the following equations.

If the muscle force and the distance between muscle origin and insertion are known,

$$\text{power} = \text{force} \times \text{velocity of contraction}.$$

If the net joint moment and joint rotations are known,

$$\text{power} = \text{net joint moment} \times \text{joint angular velocity}.$$

During concentric contractions power is generated (positive power), and during eccentric contractions power is absorbed (negative power). Because there is no contraction velocity of the muscle during an isometric contraction, power will be zero in biomechanical terms.

MOMENTUM AND IMPULSE

Momentum can be defined as the quantity of motion possessed by an object, measured by the product of its mass and the velocity of its COM. A linear momentum is relevant when an object moves linearly, and angular momentum is relevant when an object rotates.

The summation of a force over time is called the impulse. It is also described as the area under the force curve (Fig. 9.23) and can be interpreted as the change in momentum. Momentum and impulse have the same units, namely newton.second (N.s).

They are useful quantities when a force applied over time does not result in a change of position, as happens during gait. The GRF applied to the foot during the stance phase of gait is substantial, but the foot remains in the same position relative to the floor. Without a distance or velocity over which this force is applied, work or power generated by this GRF would be zero. The impulse of the GRF in that case is more revealing.

These quantities can also help to explain what happens during collisions (kicking a ball, or a

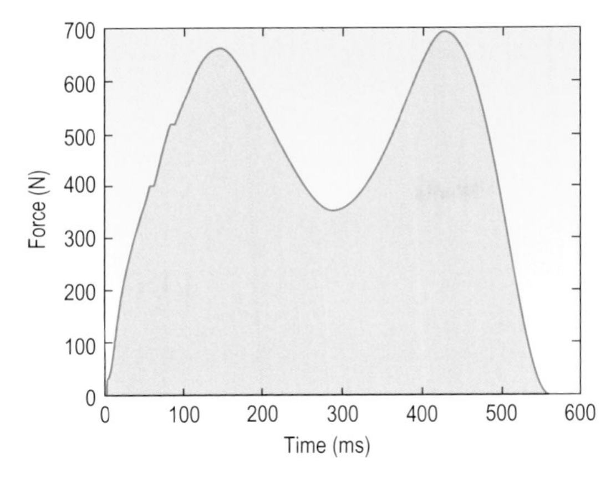

Figure 9.23 This is a typical graph of the vertical ground reaction force (GRF) under one foot during gait. The double-hump shape arises from the heel strike and push-off phases. Because the foot does not move during the stance phase, this force would not result in any work. Therefore the impulse (area under the curve) is a better indication of the impact the foot has on the ground or vice versa.

rugby tackle). The sum of the colliding objects remains constant. This is also known as conservation of momentum.

CASE STUDY 9.2

Remember the scenario from Case study 9.1. The overloaded Dan is now walking off to the bus stop. How can we use biomechanics to describe any alterations that will take place in his movement?

We have already described some of the mechanisms adopted just to maintain an upright posture. In Chapter 11, we will see that walking is a series of displacement of the COG forwards and a reforming of the BOS to regain stability. If, as was described previously, the muscle work and energy consumption have increased just to maintain an upright posture, then adding forward motion will require even greater effort and energy. As described in Chapter 11, walking can be seen as a series of accelerations in the push-off phase and decelerations in the deceleration phase. Greater muscle work (force) would be needed to overcome the inertia (in the push-off phase) that would be present because of the increased mass. Stopping this forward momentum (a product of the mass and velocity) that occurs at every step would also require a greater effort. Although this would be a minimal effect to stop the total forward momentum, the greater effort would be seen in the trunk extensor muscles to keep them upright. Can you see that the moment at the hip and lower lumbar area would be greater (as a result of the greater GRF)?

Figure 9.23 shows the GRF. The force would increase when Dan has to walk with the rucksack. Can you see the influence of Newton's laws of motion in explaining these phenomena? Look again through the section on qualitative analysis. Can you see the secondary effects on Dan's muscles, tendons and joints?

QUANTITATIVE MOVEMENT ANALYSIS

The majority of human movements are quite complex, because body segments move relative to each other and relative to the environment. This makes it difficult to predict what combination of muscle actions is required to generate a particular movement. A quantitative movement analysis can clarify which muscles should be active during a posture or movement in the context of several external forces acting on the body. For this purpose, inverse dynamics is an often-used technique. The word *inverse* indicates that the causes of movement (forces and moments) are calculated from the outcome (movement as it is measured).

The aim of such an analysis is to determine the forces and moments at the different joints. In particular, net joint forces and net joint moments are calculated. A net joint force is the resultant force of all the forces acting on the different anatomical structures at the joint interface between two body segments. In equivalent terms, a net joint moment is the resultant moment. Muscles generate the main components of a net joint moment, and therefore these net joint moments can be used as an indication of which muscles contribute to an activity. A limitation of the method is that it does not clearly predict the coactivation of muscles.

Once net joint forces and moments are known, and given certain assumptions, it is possible to estimate what loading muscle tendons, joint surfaces and passive joint structures undergo during an activity.

LINK SEGMENT MODEL AND A FREE BODY DIAGRAM

Movement analysis by means of the inverse dynamics approach involves a number of steps. First of all, a link segment model is used to represent the body in biomechanical terms. In a two-dimensional link segment model, the segments of the body can be represented by a bar or a line and the joints by a hinge. A free body diagram (FBD) is a graphical representation of the whole or part of the body. To be able to do any calculations, three different types of information are necessary. These are anthropometric information on the segments (e.g. segment mass and length), kinematic information (linear and angular segmental movement) and some information about the external forces acting on the body (e.g. GRF). Each segment is analysed separately using a FBD and the equilibrium equations of force and moment. The analysis is usually started at the distal segment. The results (net joint forces and moments) from that distal segment can then be used in the analysis of the next segment in the link. The calculation therefore progresses in steps from one segment to the next.

FINDING THE RESULTANT

To illustrate this process, we will use an example of the lower limb during gait. The phase of the gait cycle to be analysed is at the end of mid stance just before the push-off phase (Fig. 9.24). At that point, the foot is normally flat on the ground and not yet moving. To analyse the right leg, we need to start by constructing a link segment model of the right lower limb (Fig. 9.25). For each segment, we need to know the mass, the length, the location of the COM and the moment of inertia about the COM. This information is available in the literature (Winter, 1990). In our example, information about the GRF (magnitude, direction and point of application) is available from measurement.

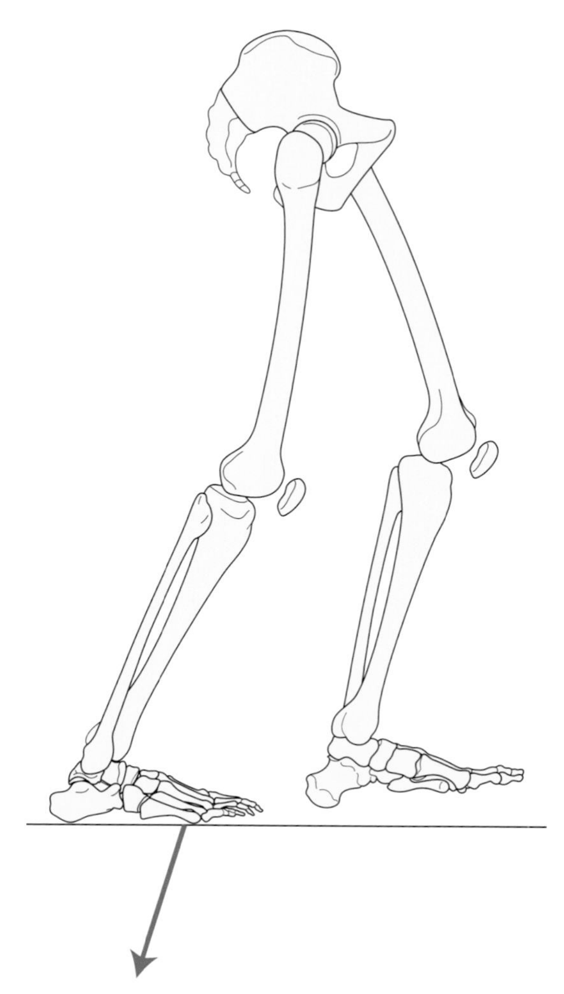

Figure 9.24 This is the phase during gait that is used in the example of movement analysis. The right foot is starting the push-off but heel-off has not yet occurred. The other limb is in terminal swing phase and heel strike will occur soon. The vector shows the force applied by the foot on the ground (magnitude, direction and point of application).

STEPS INVOLVED IN QUANTITATIVE MOVEMENT ANALYSIS

Step 1: creating a free body diagram

Calculations are started at the foot, because sufficient information about the external forces is available. Therefore a FBD of the right foot is constructed (Fig. 9.26). The free body is analysed as a separate entity. The environment and neighbouring parts of the body are represented in the FBD by arrows indicating the forces and moments that are applied by these factors to the free body.

In our example, the relevant forces and moments to be included in the FBD are the GRF, the effect of gravity on the COM of the foot (G_{foot}), the net joint force at the ankle (F_{ankle}) and the net joint moment at the ankle (M_{ankle}). The last two factors represent the net linear and rotational effects of the rest of the body on the foot acting at the ankle joint. These influences are unknown but will be solved by calculation. These forces and moments are entered in the FBD as arrows in the locations where they act. The directions of the forces and moments are not always easy to predict.

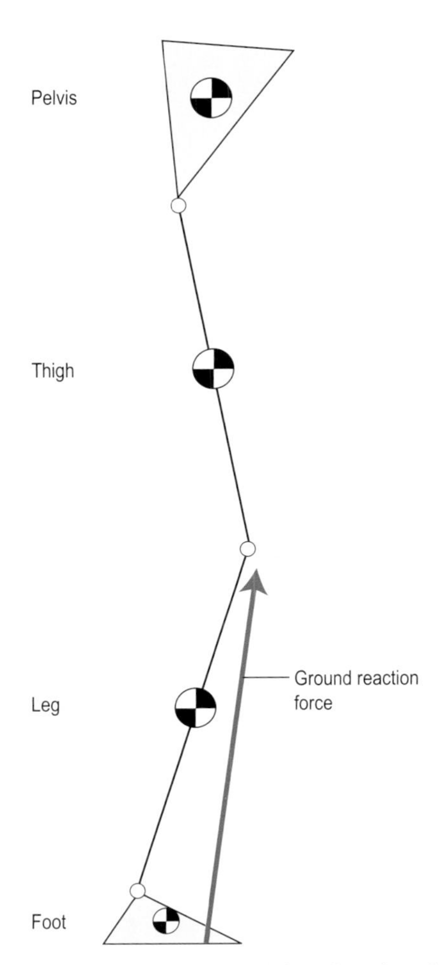

Figure 9.25 A link segment model describes the relation of the body segments relative to each other in mechanical terms, using bars to represent the segments and hinges to represent the joints. The targets indicate the positions of the centre of mass of each segment. The vector shows the magnitude, direction and point of application of the ground reaction force acting on the foot.

It is not essential to make an accurate prediction so long as the equilibrium equations are written on the basis of the FBD. The results of the calculations will then tell you whether the direction of the arrows was correct or not. A negative result indicates that the force or moment is pointing in the direction opposite to the one drawn.

Step 2: type of equilibrium

The next step is to consider whether the foot is accelerating or decelerating (dynamic equilibrium) or not (static equilibrium). In our example, the foot

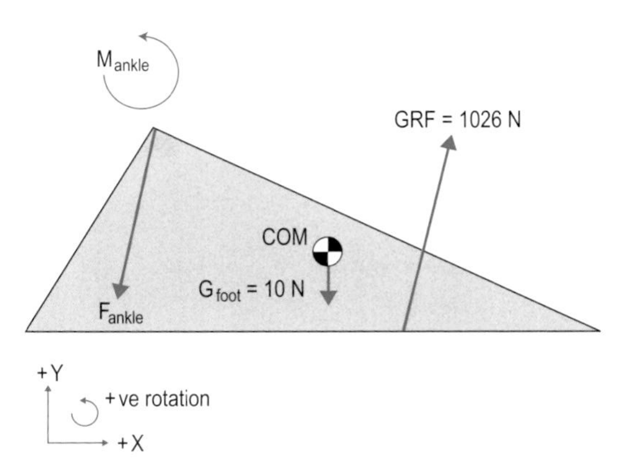

Figure 9.26 A free body diagram of the right foot. Vectors indicate forces and moments acting on the foot during the late stance phase of gait. COM, centre of mass; GRF, ground reaction force.

at that particular instant is not moving or starting to move. Therefore all the forces acting on the foot should add up to zero, because there is no linear acceleration or deceleration, and all the moments should also add up to zero, because there is no angular acceleration or deceleration (static equilibrium).

Step 3: the equilibrium equations and finding the force

With the above information, the equilibrium equations of force and moment for this FBD can be written and solved. The net joint force is solved first, followed by the net joint moment. Before the forces can be summed, they need to be resolved into their orthogonal components. For instance, the GRF can be resolved (Fig. 9.27) into a vertical component pointing upwards (positive direction) and a horizontal component pointing to the right (positive direction). The net ankle force can be resolved into a vertical component pointing down (negative direction) and a horizontal component pointing to the left (negative direction). The effect of gravity is always pointing vertically down so that it does not have to be resolved.

Step 4: the horizontal force

The horizontal forces in this FBD are the horizontal components GRF_x and $F_{\text{ankle},x}$. These should add up to zero, as stated earlier. This results in an equation that can readily be solved.

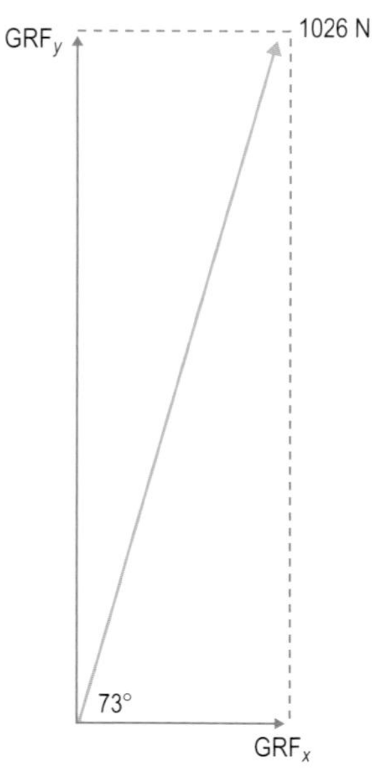

Figure 9.27 The ground reaction force (GRF) is an angled force and therefore must be resolved into its orthogonal components to be able to do further calculations.

For the horizontal forces,

$$GRF_x - F_{ankle,x} = 0$$

$$300 - F_{ankle,x} = 0$$

$$F_{ankle,x} = 300\,\text{N}.$$

The result is a positive number, therefore the direction of the arrow pointing in the negative direction in the FBD was correct.

Step 5: the vertical force

In the vertical direction, there are three forces to consider: the vertical component of the GRF_y and the $F_{ankle,y}$ and the effect of gravity on the foot (G_{foot}). They should add up to zero, so this also results in an equation that can readily be solved.

For the vertical forces,

$$GRF_y - F_{ankle,y} - G_{foot} = 0$$

$$981 - F_{ankle,y} - 10 = 0$$

$$F_{ankle,y} = 971\text{ N}.$$

The total net ankle force can now be calculated by summing the horizontal and vertical components, $F_{ankle,x}$ and $F_{ankle,y}$.

For the net ankle force,

$$F_{ankle} = \sqrt{(F_{ankle,x^2} + F_{ankle,y^2})}$$

$$F_{ankle} = \sqrt{(300^2 + 971^2)} = 1016\text{ N}.$$

Step 6: the ankle moment

Now that all the unknown forces for this FBD have been solved, the net ankle moment can also be solved. Note that the moments of force about the COM of the free body are considered and not about the ankle joint. In that case, gravity does not have a moment of force about the COM. Therefore the moment of force of the GRF and of the net ankle force and the net joint moment are to be included in the equation. The moment of force can be calculated by multiplying the force by the perpendicular distance between the axis of rotation (fulcrum) and the line of action of the force (Fig. 9.28). As mentioned before, this distance is also called the moment arm or lever arm.

Another aspect to consider is the direction in which each force would rotate the free body if acting alone. In this FBD, the GRF would result in an anticlockwise (positive) rotation around the COM if it was the only force acting on the foot. The net ankle force would result in an anticlockwise

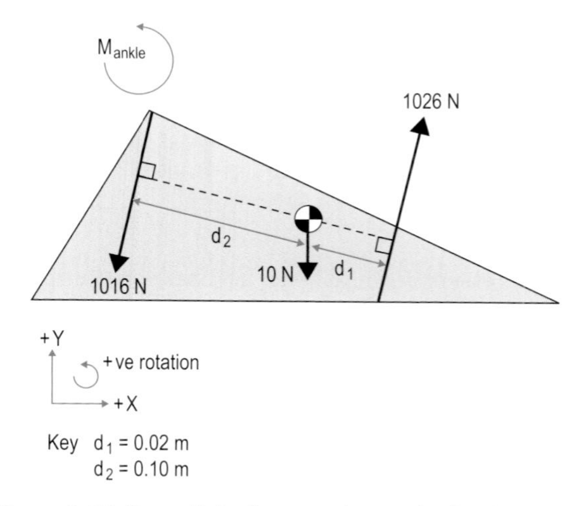

Figure 9.28 Once all the forces acting on the foot have been calculated, the rotatory effects of these forces also need to be considered. The same free body diagram used in Figure 9.26 is shown, but the moment arms of the ground reaction force (d_1) and ankle force (d_2) are entered as well.

(positive) rotation. These will therefore be entered in the moment equation as positive. The net joint moment has been drawn as an anticlockwise moment (see Fig. 9.28), so this is also entered into the same equation as positive. All these factors should add up to zero because the foot is not rotating, i.e. there is no angular acceleration.

For the net ankle moment,

$$(GRF \times d_1) + (F_{\text{ankle}} \times d_2) + M_{\text{ankle}} = 0$$

$$(1026 \times 0.02) + (1016 \times 0.10) + M_{\text{ankle}} = 0$$

$$M_{\text{ankle}} = -122 \text{ N.m}$$

Step 7: the internal moment

The result of this calculation of the net joint moment is negative. Therefore the moment should be drawn as a clockwise moment, which means that it is an internal plantar flexing moment. This internal moment is mainly generated by the plantar flexors of the ankle. The triceps surae is the major contributor because of its large physiological cross-sectional area and because of its large moment arm compared with the other flexors. Therefore it is not unreasonable to assume that this muscle almost entirely generates the ankle moment calculated in our example.

This assumption that one muscle produces the net joint moment makes it possible to calculate the force that will occur within the tendo calcaneus. A muscle moment is the product of the force within the muscle and the distance between the line of action of the muscle and the axis of rotation of the ankle joint (Fig. 9.29). This distance or moment arm for the tendo calcaneus can be estimated at 5 cm.

For the net ankle moment generated by the triceps surae,

$$M_{\text{ankle}} = force_{(\text{triceps})} \times \text{distance}_{(\text{tendo calcaneus})}$$

$$-122 \text{ N.m} = force_{(\text{triceps})} \times -0.05 \text{ m}$$

$$\frac{-122 \text{ N.m}}{-0.05 \text{ m}} = \text{force}_{(\text{triceps})}$$

$$force_{(\text{triceps})} = 2440 \text{ N}.$$

Step 8: muscles, forces and their effects

The calculated force acting on the tendo calcaneus is quite high but not out of the ordinary. In fact, the tensile loading in this example is equivalent to approximately three times the body weight of an average person. In more strenuous activities, such as jumping, this loading is expected to be much higher still.

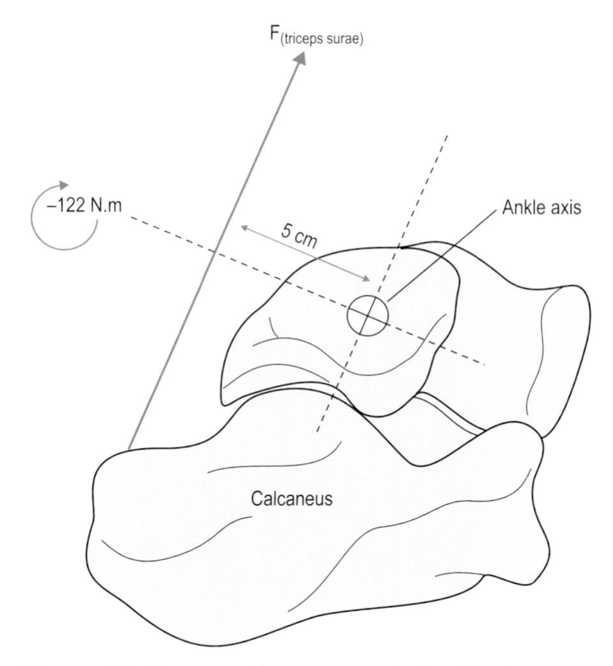

Figure 9.29 The net ankle moment calculated in the example is provided mainly by the triceps surae muscle. The moment arm of the muscle determines what muscle force is required to generate this moment. In this example, the muscle moment arm is estimated to be 5 cm.

The force of the triceps muscle is much larger than the net ankle force that was calculated earlier ($F_{\text{ankle}} = -1016$ N) and is pointing in the opposite direction. The net ankle force is the resultant force of all the different structures producing forces at the ankle joint. The muscle is just one of these structures. In this example, the other structures are therefore generating the remainder of the forces so that the resultant force equals the net ankle force that was calculated. Because all forces in this example are parallel, the force of the other structures ($F_{(\text{bone on bone})}$ in Fig. 9.30) can be easily determined.

For the bone on bone forces,

$$F_{\text{ankle}} = F_{(\text{triceps})} - F_{(\text{bone on bone})}$$

$$-1016 = 2440 - F_{(\text{bone on bone})}$$

$$F_{(\text{bone on bone})} = 2440 + 1016 = 3456 \text{ N}.$$

It is assumed that the main structure contributing to this is the force of the tibia on the talus, although the ligaments cannot be ignored.

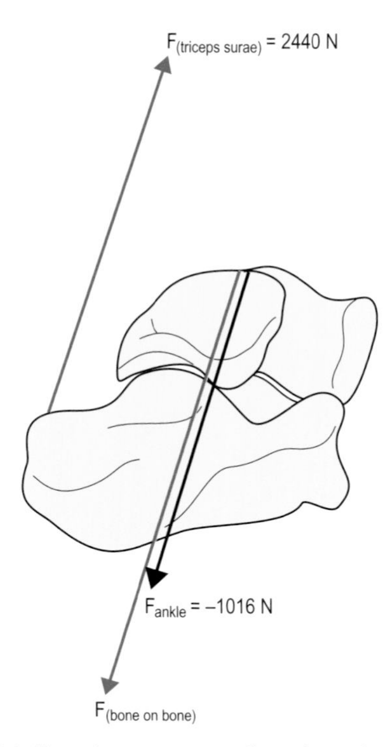

Figure 9.30 The triceps surae applies a large force in the direction opposite to the net joint force that was calculated. Therefore there needs to be a third force that balances the force system. This would have to be the force that the bones of the leg apply to the foot. The resultant of the $F_{(triceps)}$ and the $F_{(bone\ on\ bone)}$ is the net ankle force (F_{ankle}).

The bone on bone force results in a compression of the ankle cartilage. This load is the largest of all the values calculated. It is interesting to note that the activity of the triceps muscle was the major contributor to this cartilage loading. Body weight contributed to the cartilage loading to a much lesser extent. This is mainly because of the direction of pull of the muscle, which is almost parallel to the tibia, and to the small moment arm of the muscle. The tendo calcaneus has one of the largest moment arms of the muscles of the limbs. The result of this example therefore illustrates in general that joint loading is mostly due to muscle contractions.

MOVING UP THE LIMB

The calculations so far have indicated which muscle group needs to be active to provide the appropriate moment at the ankle joint and given us an idea of the magnitude of the tissue loading at the ankle joint. The same type of calculation can be carried out for the knee joint. For this purpose, an FBD of the lower leg is required. The unknowns are the net joint force and net joint moment at the knee. The net ankle force and moment have already been solved for the FBD of the foot. These can be used for the FBD of the lower leg as well, because according to Newton's third law of motion, 'to every action there is an equal and opposite reaction'. In other words, the force (or moment) applied to the foot by the leg is equal but opposite to the force (or moment) applied by the foot to the leg. Once the net knee force and moment have been solved, the analysis can be progressed to the hip joint. This will not be elaborated on in this chapter.

Note that the calculation for the example provided could be done in different ways, but the proposed method allows analysis of more complex situations as well. It is the process that is demonstrated and not so much the calculation of the example.

DEFORMATION OF MATERIALS

So far, this chapter has considered the effects of force on an object if the object moves or has the propensity to move. There are conditions when the object to which the force is being applied does not perform linear or angular motion. If sufficient force is applied, then the object will undergo deformation. The type and extent of deformation will depend on the magnitude, direction and duration of the applied force and the composition of the object itself.

STRESS AND STRAIN

The application of a force to an object is termed *loading*. The standardized measurement of loading is termed *stress*. The intensity of the applied stress is a result of the applied force divided by the surface area over which the force acts. This is expressed in $N.m^{-2}$ (pascals). The equation is

$$\sigma = \text{force}/\text{cross sectional area}.$$

Once the stress has been applied, the object will undergo a deformation, which can be defined as the change in shape or dimensions produced by the applied force. Not all materials will deform in the same way or to the same extent. Some materials will offer greater resistance to being deformed.

This resistance is known as the material's stiffness. The measure of deformation undergone by the object due to the stress that has been placed on it is termed the *strain*; it has no units and is denoted by the letter epsilon (ε) and expressed by the equation

$$\varepsilon = \Delta L / L_0$$

in which ΔL is the change in dimension and L_0 the original dimension.

Figure 9.31 gives examples of the types of stress that can act on an object. For every stress, there is a corresponding strain. The relationship between the two parameters is unique to every material.

LINEAR LOADING

There are three types of linear stress that can be applied to the object.

1. Linear tension occurs when two equal loads are applied to an object in such a way that they act along the same action line but in opposite directions. The resultant deformation is a lengthening and some narrowing. This is shown in Figure 9.32a.
2. Linear compression occurs when two equal loads are applied to an object such that they act along the same action line towards each other, i.e. squeezing. The resultant deformation will

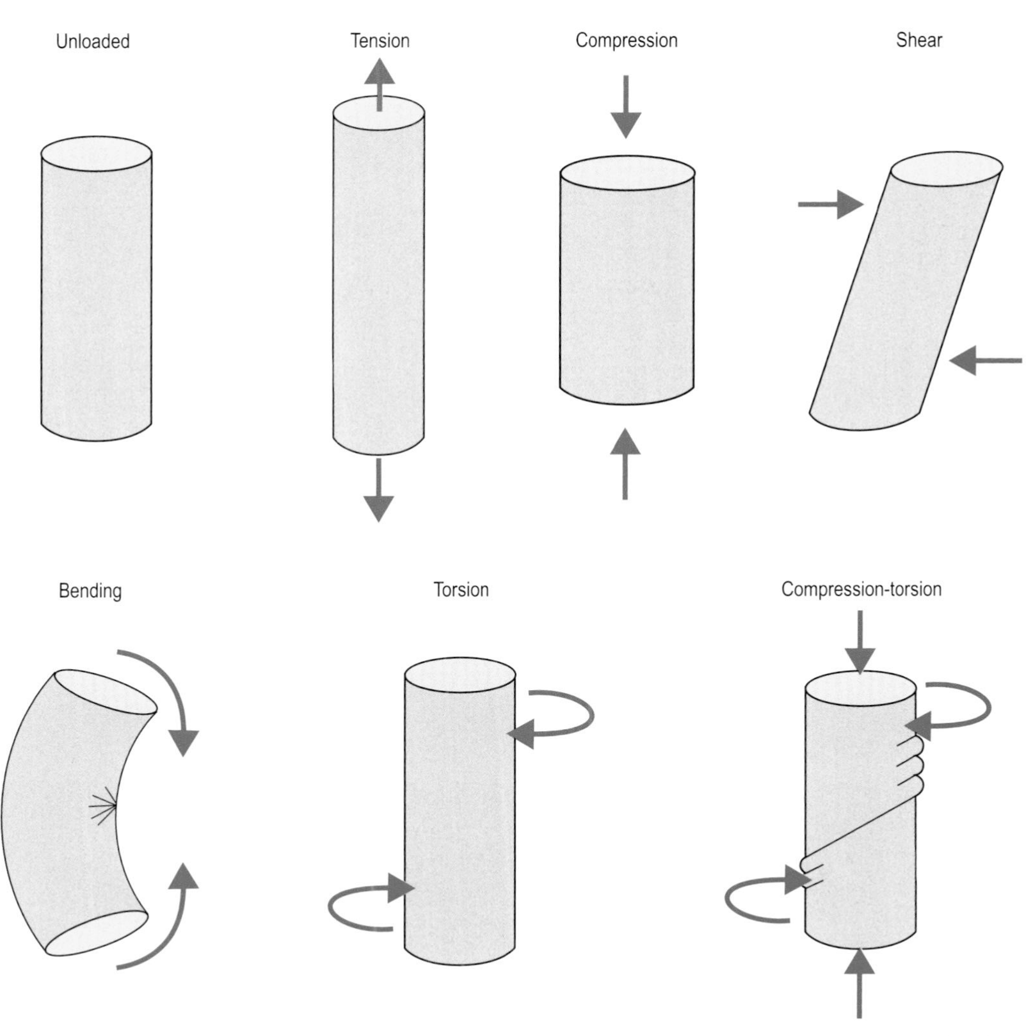

Figure 9.31 These are examples of the types of tissue loading that can occur. Included are linear and rotational types of loading.

therefore be shortening with some widening, as shown in Figure 9.32b.

3. Shear occurs when two parallel and equal loads are applied in opposite directions but not on the same line of action. This is shown in Figure 9.32c.

It can also be seen from Figure 9.32 that the calculation of the linear compression and tension strain is relatively straightforward. When considering the shear stress and strain, there is no change in length but there is an angular deformation. If, as illustrated in Figure 9.32c, the block is acted on by a stress and the top of the block moves a distance *d* relative to the bottom of the block, then there is an angle *Q* produced, as shown. This is the angle of shear strain and can be calculated by dividing the displacement by the height.

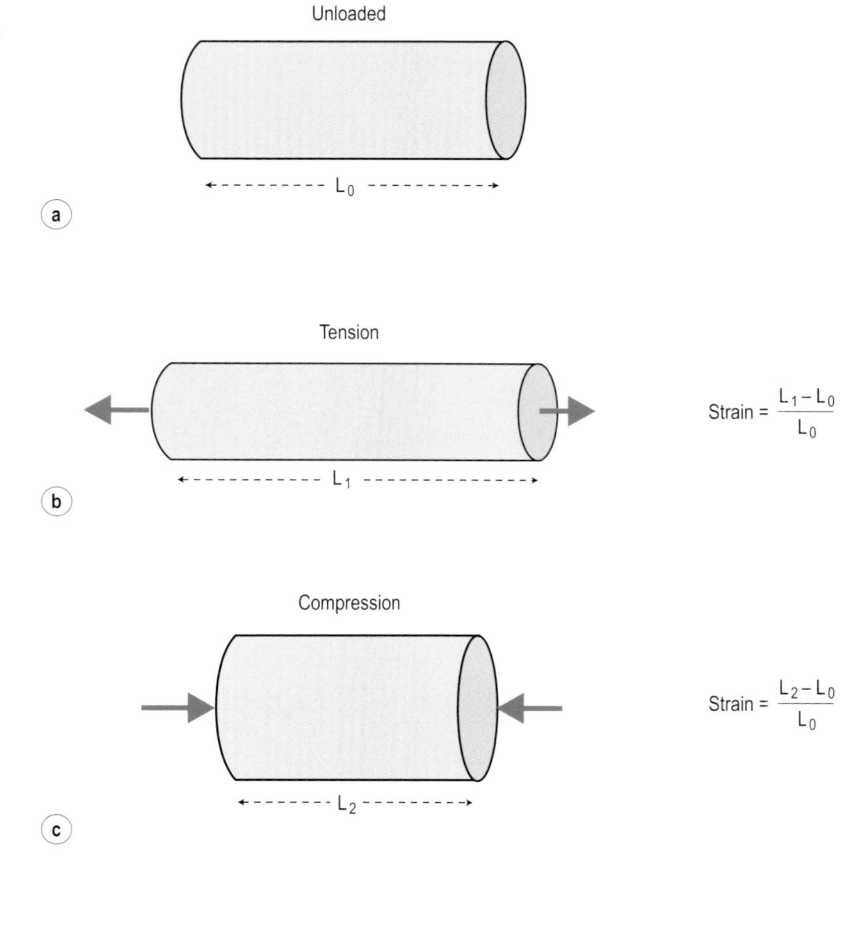

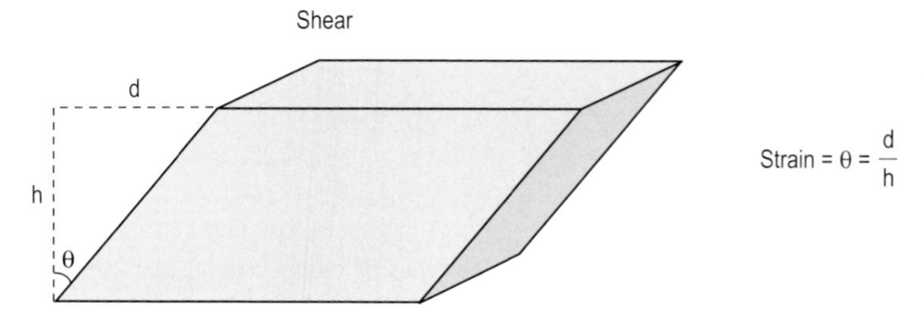

Figure 9.32 Strain occurs under the influence of stress. The calculation of strain depends on the type of stress that is active. The examples provided are for the linear stresses of tension, compression and shear.

YOUNG'S MODULUS

The relationship of the changing stress and strain results in a constant of proportionality for each individual material. This constant is called Young's modulus and is represented by the following equation:

$$\text{Young's modulus} = \Delta\sigma/\Delta\varepsilon$$

in which $\Delta\sigma$ is the change in linear stress and $\Delta\varepsilon$ is the change in linear strain.

The constant of proportionality for shear stress is the shear modulus and is represented by the equation

$$\text{shear modulus} = \Delta\sigma/\Theta$$

in which $\Delta\sigma$ is the change in shear stress and Θ is the change in shear strain.

STRESS–STRAIN CURVES

One of the more useful relationships between the stress placed on an object and the corresponding strain is seen when a graph is plotted with strain on the x-axis and stress on the y-axis. A general stress–strain curve is illustrated in Figure 9.33. The relationship found for any material is dependent only on the type of loading and the mechanical properties of the material. It is interesting to note that the relationship will stay the same for any object made from the same material irrespective of its size or shape. By experimentation, the relationships between stress and strain for different materials have been found and each displays variations of a basic pattern.

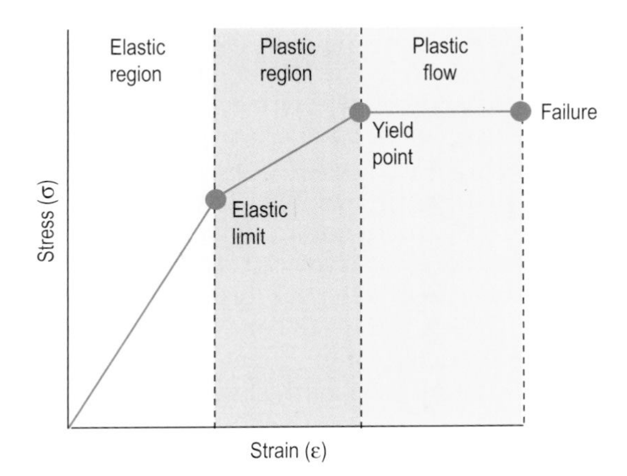

Figure 9.33 A typical stress–strain curve that applies to many types of material. Generally, materials have an elastic region and a plastic region until failure. The elastic limit and the yield point are shown in the graph, marking the transition from one region to the next.

As can be seen from the diagram, there are two distinct regions. The first is the linear section of the graph, as seen between the start and the elastic limit. This is referred to as the elastic region, in which the material under stress obeys Hooke's law. This means that for every incremental unit of stress, there is a corresponding incremental increase in strain, i.e. there is a linear relationship between the two. If the stress is released before the elastic limit is reached, the material will return to its original position. Therefore there is no permanent deformation occurring and the deformation (and material) is referred to as elastic. This section of the graph is represented mathematically by Young's modulus.

Once the stress has increased past the elastic limit, the graphical representation is seen to occur in the plastic region (up to the yield point). If the stress does not exceed the yield point, then the material will still exhibit some elastic properties but the material will not return to its original position. Some permanent deformation has taken place. The material is said to have undergone plastic change.

If the stress is maintained at the yield point, the material will undergo plastic flow, when the strain continues with no increase in stress until failure occurs and the material will fracture. The amount of stress that a material can absorb prior to failure is called its ductility and will vary for different materials.

MECHANICAL PRINCIPLES OF FLUIDS

HYDROSTATICS AND HYDRODYNAMICS

Hydrostatics is the study of the effects of force and pressure on a fluid at rest, whereas hydrodynamics is the study of fluid in motion (flow). To understand hydrostatics, we must look first at some of the physical properties of liquids. In this section, only water will be described, as it is water that is generally used in the treatment of patients. Water itself can take the form of any of the three states of matter, being solid below temperatures of 0°C and gaseous above 100°C. Hydrotherapy pools are normally heated to temperatures of between 33 and 38°C and therefore the water is always a liquid.

The structure of a liquid is such that its properties are different from those of solid objects. At a basic level, the atomic structure of a liquid can be said to have weaker cohesive bonds (attractive forces between the same type of molecules) than a solid. This means that the motion of these molecules is much greater than in solids, so much so that a liquid is unable to maintain its own shape and therefore has to take the shape of the receptacle in which it is placed. A liquid also has the property of retaining its volume, showing a molecular repulsive force when it is being compressed. It is this repulsive force that produces the almost continuous flow of the liquid when an outside force is applied to it, hence the term *fluid*.

The cohesive force is responsible for giving water its property of surface tension. This is the film-like covering on the surface of the water, which is caused by the cohesive forces of the water to water molecules, which are stronger than the adhesive forces (the attractive forces between different molecules) of the water to air molecules. The force of surface tension for a hydrotherapy pool is so weak that it can be ignored. The adhesive forces between the water and whatever is placed in it, part of the body or an oar for example, are also very weak, being greater than moving through the air but insignificant when compared with the frictional forces experienced on land.

PRESSURE

When a force is applied to a fluid in a confined container, the fluid experiences pressure. Earlier in the chapter, pressure was described as the force per unit area, and this was the pressure on a plane surface. In fluids, however, the pressure is said to be acting at a point, and this point can be thought of as a small plane surface (Bell, 1998). Even without an additional force, hydrostatic pressure is felt within the fluid in all directions as the randomly moving molecules collide into each other, the container and any object immersed in it. Pascal's law states that this force is equal in all directions and is independent of gravity. Pressure within the fluid is also the result of the weight of the fluid above a given point, and it is equal to the vertical distance from the point to the surface multiplied by the weight density of the fluid. So it can be seen that the deeper an object is in the fluid, the greater the pressure it will experience. Hydrostatic pressure is usually measured in pascals (1 Pa equals 1 $N.m^{-2}$).

DENSITY

Along with depth, pressure in a fluid will change when the density of the fluid changes. Density is defined as the mass of the fluid divided by its volume. This then gives us the fluid's mass density (r), expressed in $kg.m^{-3}$. If we multiply the fluid's mass by the acceleration due to gravity and divide it by its volume, then we have the fluid's weight density.

RELATIVE DENSITY OR SPECIFIC GRAVITY

This is the density of a material (solid or liquid) relative to that of pure water at 4°C. As the relative density of water is, by definition, 1, then the relative densities of other materials will determine whether they will float or sink in water. A relative density less than 1 will mean that the material will float, and with a relative density greater than 1 it will sink. The relative density of the human body is in the region of 0.86–0.97 (Bell, 1998). This number will vary depending on the proportion of the different body tissues and the amount of air within the lungs.

BUOYANCY

It has already been said that the pressure on a body within the water is equal on all aspects of that body, and this pressure increases with the depth of the water. The point forces will be greater on the deeper aspects of the object, therefore there will be a resultant upward force on that body. This resultant upward force is the up-thrust experienced in the water and is termed the *force of buoyancy*. The amount of force a body will experience is governed by Archimedes' principle, which states that any body that is wholly or partially immersed in a fluid will experience an upward thrust equal to the weight of fluid displaced. The force of buoyancy can be said to be acting through the centre of buoyancy, which is the COG of the displaced fluid. Therefore the centre of buoyancy does not have to coincide with the COG of the body in the water.

STABILITY

The stability of a body on dry land is subject to the relationship of the COG to the base on which it rests. In fluids, however, the relationship of the COG to the centre of buoyancy is the predominant

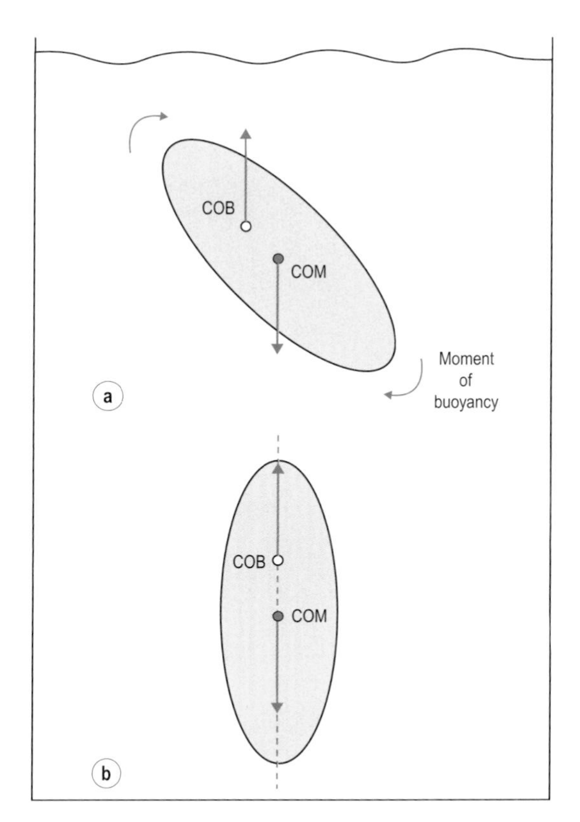

Figure 9.34 A moment of buoyancy occurs when the forces due to gravity and buoyancy are not colinear (a). The object will rotate until colinearity has been achieved (b). In this example, the centre of buoyancy (COB) and the centre of mass (COM) never coincide.

factor controlling stability. If a body is placed in water, it will rotate until the centre of buoyancy and the COG are coincident. This point is referred to as the metacentre. This is illustrated in Figure 9.34.

MOMENT OF BUOYANCY

If the centres of buoyancy and gravity do not lie in the same vertical line, then the body, as explained above, will not be in equilibrium and will rotate. The cause of this rotation is the moment of buoyancy. The moment of buoyancy is the same as any turning effect of a force on dry land, in that it causes the body to rotate until it is compelled to stop by another force or, in this example, because the forces of buoyancy and gravity are colinear.

The moment of buoyancy is calculated, as on dry land, by multiplying the force (buoyancy) by the perpendicular distance from the fulcrum. As can be seen from Figure 9.35, the nearer the body segment is to the water's surface, the greater the perpendicular distance from the knee axis of rotation and hence the greater the moment of buoyancy, thus facilitating knee extension.

TYPES OF FLOW

When a fluid flows from a point of higher to a point of lower pressure, the fluid molecules form themselves into layers or laminae. The layers at the centre of the flow move faster than those nearer the edge of the fluid, while those at the very edge may even be stationary. If these layers run smoothly along without any disturbance, then the flow is said to be laminar or streamlined.

The fluid will flow in a streamlined way until it reaches a critical velocity. At this velocity, the laminae break up and the flow is said to be turbulent. Turbulence is created because shear stress between the different laminae depends on the viscosity of the fluid and the rate of change of velocity in the direction of flow and perpendicular to the flow. Thus, with the increase in the flow rate, the laminar pattern will break up and the molecules will no longer travel in layers but take on an irregular pattern of motion. Once turbulent flow has been established, the back current or eddy currents may become exaggerated and cause areas of reduced pressure downstream (in the wake) of these eddy currents.

MOVEMENT THROUGH A FLUID

If a streamlined object moves through a fluid, then the fluid offers little resistance to its flow. Very little turbulence is caused by the movement of the object, and therefore the frictional resistance is low. If the object is not streamlined, then there will be a much greater resistance to its progress. The type of resistance it will meet will depend on the shape of the object. As can be seen from Figure 9.36a, a streamlined object produces little disturbance in the fluid. Figure 9.36b shows the formation of eddy currents and an area of reduced pressure, which together form a wake. Figure 9.36c shows the formation of eddy currents with the turbulent flow in front of the object, which provides a much greater resistance to movement. As seen from Figure 9.36d, there is also the formation of waves, one travelling out in front of the object

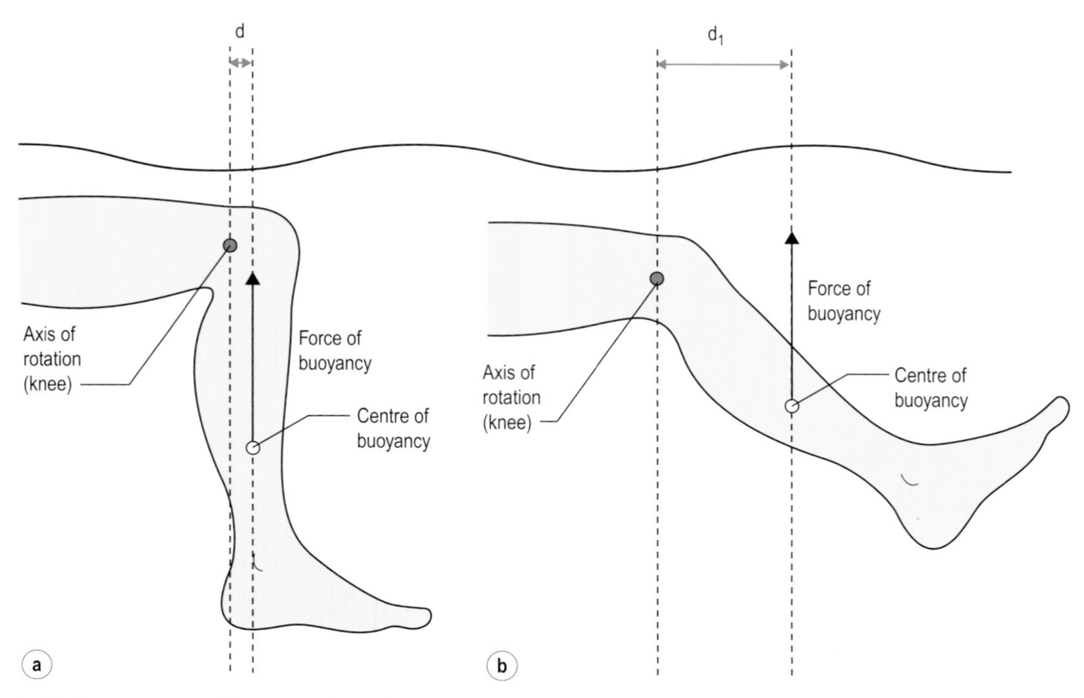

Figure 9.35 The moment of buoyancy depends on the perpendicular distance between the line of action of the force of buoyancy and the knee axis of rotation. In (a), the moment of buoyancy will be smaller than in (b) because of the greater distance (moment arm).

and one travelling out from behind. Both these waves will require energy to overcome their resistance. The resistance to movement is increased even further if the fluid in which the movement is taking place is already turbulent.

ACTIVITY 9.3

With a group of colleagues, walk around in a circle in single file in a swimming pool. After a few seconds, change the direction you are all walking (from clockwise to anticlockwise). What happens? Try this again, but this time vary the depth and speed of your walk and the frequency of your direction change. What differences do you find?

CASE STUDY 9.3

Dan has got through his day at school and is at the swimming club. He is working with Jerry, who has had his left leg amputated as a result of an accident. Jerry has also a marked weakness of his left upper limb. It is the first time that Jerry has come into the water post accident. He has not spent much time in water before his accident, but he is not frightened of the water and he is looking forward to the experience and possible benefits that the pool sessions may bring. Dan is very experienced in helping people in this situation and knows well the effects of the water.

What mechanisms will Dan have to adopt to compensate for the effects of the water on his movements? What will be Jerry's experience? What will be the differences in their experiences?

It is important that Dan will remain stable and upright in the water in order to support Jerry. Once moving, Dan will experience a greater difficulty in stopping than he would on dry land. This would be as a result of the increased inertia of the water as it continues to move as Dan stops. This force would tend to continue Dan's movement. If Dan kept his feet planted, then the rest of his body would continue to move and his COG would soon fall outside his base. This would make him unstable and of little use in supporting Jerry. One strategy would be to increase his BOS by standing with his feet wider apart than he

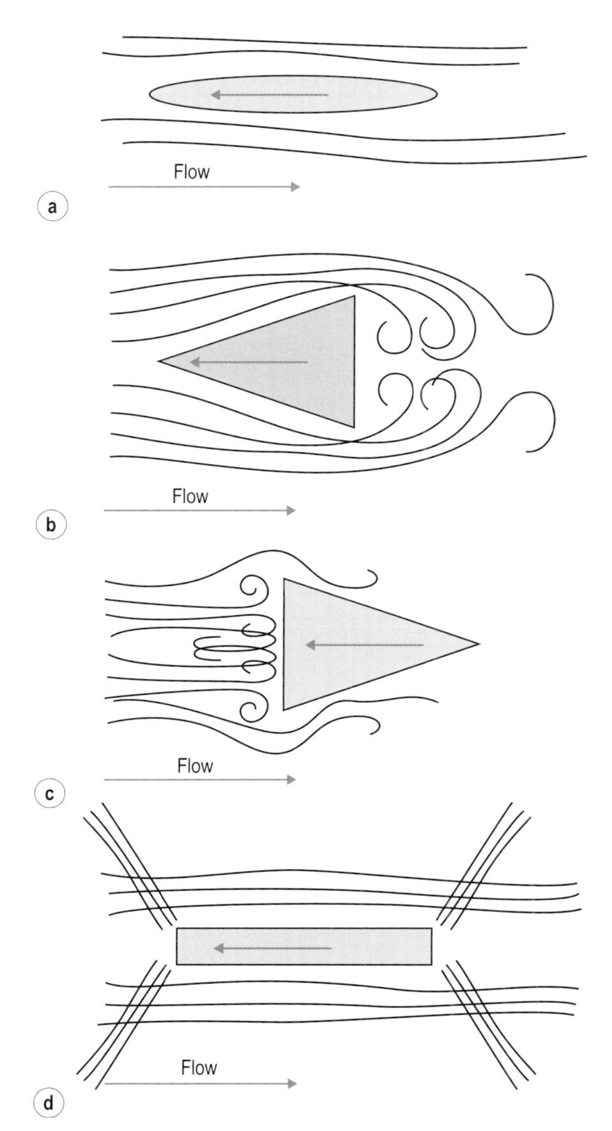

Figure 9.36 Four examples of the flow of fluid around objects of different shapes. The arrows inside the objects indicate movement of the objects from right to left. The fluid is moving in the opposite direction as indicated. In (a), there is a streamlined object that does not disturb the flow in laminae. In (b), the shape of the object results in turbulence, with an area of reduced pressure and eddy currents in its wake. In (c), there are turbulence and eddy currents in front of the object. In (d), waves form at the front and the back of the object.

normally would. This would give him more time to use his muscle to produce a force to stop the body moving in the direction of movement. Buoyancy would also be exerting an upward thrust on Dan, which again would lead to a greater muscular effort to maintain his position. The deeper that Dan was in the water, the greater these effects would be and the consequent effort Dan would have to make. There would come a point when Dan would be unable to maintain his equilibrium without great effort and therefore would have to question his ability to help Jerry. Next time you are in a pool, try to identify this point for yourself. Dan will make several other adjustments to normalize his movement because of the effects of the water, but these would be more automatic to him and best explained by considering the effects on Jerry's movement.

Left alone in the water, Jerry would probably float. This is because the up-thrust of the force of buoyancy is greater than that of the effect of the force of gravity. Usually 0.05% of the body will be above the water, but this would not really help Jerry as this 0.05% is not usually the face. This can be explained by considering the moment of buoyancy. Figure 9.34 describes this. The body will move until the COM and the centre of buoyancy are in line (the metacentre). This usually means that the body will float upright and the face turned out of the water to allow breathing. This may not happen to Jerry, as his COM will have changed because of the loss of his limb. This means that Dan will have to support Jerry until he has learned to compensate for these changes (see Ch. 6).

Dan will probably begin by supporting Jerry as he floats on his back. This may seem like an easy task, as I am sure that most of us have tried to float and found it fairly easy, especially with the support of another person. Dan needs to take care, however, as we have pointed out that Jerry has an altered COM so would begin to rotate from this position. Which way do you think Jerry would turn? Try it out next time you are in the pool by lifting your upper limb out of the water – you may be surprised at the direction of the movement. Dan knows that Jerry will rotate towards the amputated side, so he positions himself so that he can give a counter-force to that moment.

Dan knows these movements instinctively because he has experienced them many times, but they are all new to Jerry, who will have to learn to adapt his movements to the new environment.

References

Bell, F., 1998. Principles of Mechanics and Biomechanics. Stanley Thornes, Cheltenham.

Winter, D.A., 2009. Biomechanics and Motor Control of Human Movement. John Wiley & Sons, New York.

Further reading

Craik, R.L., Oatis, C.A. (Eds.), 1995. Gait Analysis: Theory and Application. Mosby-Year Book, St. Louis.

Enoka, R.M., 2008. Neuromechanics of Human Movement. Human Kinetics, Champaign, IL.

Hamil, J. & Knutzen K.M. 2006. Biomechanical basis of human movement. Lippincott Williams & Wilkins, PA.

Hollis, M. & Fletcher-Cook, P. 1999. Practical Exercise Therapy. Blackwell Scientific Publications, Oxford.

Zatsiorsky, V.M., 1998. Kinematics of Human Motion. Human Kinetics, Champaign, IL.

Zatsiorsky, V.M., 2002. Kinetics of Human Motion. Human Kinetics, Champaign, IL.

Chapter 10

Function of the upper limb

Valerie Sparkes and Mike Smith

CHAPTER CONTENTS

LEARNING OUTCOMES

At the end of this chapter, you should be able to:

1. discuss the factors that contribute to the function of the upper limb
2. identify specific factors that affect the function of the individual sections of the upper limb
3. discuss the function of the upper limb as part of the kinetic chain.

INTRODUCTION

This chapter, in outlining the function of the upper limb, focuses on specific areas that contribute to the upper limb and its function as part of the kinetic chain. It will focus on some of the anatomical and biomechanical aspects that are related to optimal function. To answer the basic question 'What is the function of the upper limb?', we need to look to the activities and demands of the upper limb during daily life, including basic activities of daily living as well as occupational and sporting demands.

When considering what anatomical and biomechanical features contribute to these functional activities, we need to appraise the role of the bones, muscles, ligaments, nerves, fascia, sensory and mechanical receptors and other soft tissues surrounding the joints, as well as the central nervous system that controls these systems (Terry et al., 1991). For full appreciation of the function of the upper limb, it needs to be considered in relation to the other parts of the kinetic chain including the trunk, pelvis and lower limb.

FACTORS AFFECTING THE FUNCTION OF THE UPPER LIMB

THE SHOULDER GIRDLE

The shoulder girdle comprises the articulation between the humerus and the glenoid fossa of the scapula (i.e. the glenohumeral joint), the articulation between the scapula and the clavicle (i.e. the acromioclavicular joint), the articulation between the clavicle and the sternum (i.e. the sternoclavicular joint) and the pseudoarticulation of the scapula lying on the thoracic wall (i.e. the scapulothoracic joint).

The primary function of the shoulder girdle is to allow freedom for the hands to function either alone or when using tools or instruments such as a hammer or a badminton racquet. The shoulder girdle is one part of the human body and can be viewed as one segment in a series of interdependent segments that are linked together. Body segments are sequentially activated to achieve functional or athletic tasks, and the activation sequence is known as the kinetic chain (Kibler et al., 2001). In assessing the function of the shoulder girdle, it is essential to view it not in isolation but as a part of total kinetic chain activity that can include trunk and lower limb activity (Kibler et al., 2001). Any disturbance in function in any of the links in the kinetic chain can have consequences in shoulder girdle function, and any disturbance in shoulder girdle function can have an impact on the function of the other components of the kinetic chain (Kibler et al., 2001). This will be further discussed at the end of the chapter.

Glenohumeral joint

The overriding feature of the glenohumeral joint is that it is a highly mobile joint but one of the least stable joints of the skeleton. However, normal function requires a balance between mobility and stability to avoid injury. Extensive range of movement and angulation is required in the upper limb to achieve the variety of tasks that daily life requires. However, because of the necessity of achieving the balance between mobility and stability the humerus and scapula are constantly adapting their positions to maintain the optimum position for function (Moseley et al., 1992). The term *dynamic stability* is often used to describe this phenomenon, in which structures such as muscles, tendons and ligaments are constantly adjusting to provide a stable base from which the limb can work and at the same time can provide the power and coordination that are required for the task. For a fully functioning upper limb, synchronized movement needs to occur in the glenohumeral, scapulothoracic, acromioclavicular and sternoclavicular joints; the elbow; the wrist and hand; and the surrounding musculature that link these structures.

CASE STUDY 10.1

Chris is a big fan of all things sporting. He is playing a game of badminton against his friend Steve down at the local leisure centre. Steve is competitive, but Chris is even more so.

Liz, on the other hand, is having a lot less fun and is returning from doing the weekly shop at the supermarket. She is carrying some very heavy bags full of food for her active son. When she gets to the front door, she uses her key to unlock it.

Try to picture these two functional activities. We will revisit them throughout this chapter in order to illustrate different aspects of upper limb function.

As described above, the glenohumeral joint is a highly mobile joint, capable of three degrees of freedom (Strandring, 2005), but one that is capable of being stable during a wide degree of activities, for example supporting the weight of the limb during painting a wall, reaching above the head and repetitive keyboard work. It has to be mobile and sufficiently powerful to support a wide variety of loads in varying positions, such as carrying bags of shopping and large boxes. The joint has to have the capacity to perform repetitive tasks of varying speeds and load, such as packing boxes on a production line or playing volleyball, tennis and badminton. The extensive range of movement is achieved through a large humeral head moving on a relatively shallow glenoid fossa (Hess, 2000). Mobility is facilitated by a large and lax capsule enveloping the joint. In normal functional movements, the humerus must roll, spin and glide to keep in contact with the relatively small glenoid fossa (see Ch. 3).

Structures supporting the glenohumeral joint

As stated, the highly mobile glenohumeral joint relies on the structures surrounding it for stability. Muscles surrounding the joint play the largest role in the effort to keep this mobile joint stable. The rotator cuff – which is made up of the tendons of subscapularis, infraspinatus, teres minor and supraspinatus – is the major stabilizing force of the glenohumeral joint and is known as a dynamic stabilizer (Pink and Perry, 1996). The rotator cuff, together with the capsule and the glenohumeral ligaments, plays a pivotal role in controlling translation of the humeral head while it spins on the small glenoid surface (Pratt, 1994). Ligaments, although contributing to the stability of the joint, function only as restraints at the extremes of movement. The rotator cuff can be viewed as the structure that steers the head of the humerus during upper limb movements, providing dynamic stability. Translation of the humeral head during elevation will also be reduced if the muscles surrounding the shoulder function in a coordinated manner, which requires preprogramming of muscle activation. In centring the head of the humerus, the rotator cuff maintains joint stability during function, especially within midrange, when the ligaments and the capsule are lax (Doukas and Speer, 2001). The rotator cuff is attached close to the axis of rotation and adds to dynamic joint compression of the humerus into the socket, enhancing stability.

All muscles generate forces that are translated to the bones they attach to. The muscle system is constantly striving to maintain a balance of forces between muscle groups to produce optimum function. Dynamic stability is achieved through certain muscles and tendons controlling shear and translational forces generated by the prime movers of the shoulder, which include deltoid, triceps, biceps and latissimus dorsi, while still allowing movement to occur. This is essential for the appropriate spinning and gliding of the humerus within the glenoid to occur. The deltoid muscle makes up a large percentage of the scapulohumeral muscle mass and, as well as being a prime mover of the glenohumeral joint, it contributes to its stability (Michiels and Bodem, 1992). Being multipennate, it is fatigue-resistant and its mechanical advantage is enhanced by its distal insertion on the humerus. During the initial phase of elevation in the glenohumeral joint, deltoid is active and produces a large degree of torque within the joint, with the effect of producing a minimal superior glide of the head of the humerus. This is counteracted by the rotator cuff, which provides a firm fulcrum on which the deltoid is able to work (Jobe et al., 1996). The force exerted by deltoid is at its maximum at 60° of elevation (Sarrafian, 1983). Early increases in tension of supraspinatus during abduction have been noted that appear to facilitate centring the head of the humerus (Poppen and Walker, 1976). Subscapularis, as a major depressor of the head of humerus, counteracts the upward shear of the deltoid muscle during upper limb function (Kabada et al., 1992).

Although the precise role in terms of stability of the long head of biceps is not fully understood, through its tendinous attachments to the supraglenoid tubercle the long head of biceps may influence the position of the head of the humerus within the glenoid by preventing anterior translation of the humerus (Kumar et al., 1989; Doukas and Speer, 2001). Both the long head and the short head appear to reinforce the glenohumeral joint anteriorly (Itoi et al., 1993).

The kinematics of the joint will determine the extent to which muscle force results in a degree of stability (Herzog et al., 1991). Rotator cuff function, and hence stability of the glenohumeral joint, is influenced by the angle of the glenoid, which in turn is dictated by the function of the muscles that control the scapula (van der Helm, 1994). Terry et al. (1991) found that the retroversion and posterior tilt of the head of the humerus and glenoid contribute to the joint stability.

Stability is further enhanced by the labrum, a fibrocartilaginous rim around the glenoid cavity (Strandring, 2005). The labrum deepens the cavity and facilitates the centring of the head of the humerus (Pink and Perry, 1996). It may protect the bone and assist in lubrication. The negative intrarticular pressure within the glenohumeral joint may have a minimal contribution to the stability of the joint (Pink and Perry, 1996). The negative intrarticular pressure increases when a downward load is applied to the dependent arm. This increase may prevent further downward glide of the humeral head (Kumar and Balasubramaniam, 1985). When the arm is abducted, the negative pressure may provide inferior stability to the humeral head (Weber and Caspari, 1989).

Muscles acting on the pectoral girdle

Many of the muscles that provide stability to the glenohumeral joint also act as movers of the joint. There are many muscles that have an influence on the scapula and the glenohumeral joint.

ACTIVITY 10.1

To refresh your anatomy knowledge, please complete the following activity.

- Name the muscles whose tendons make up the rotator cuff.
- Name the muscles that have attachments to the humerus and the spine.
- Name the muscles that attach on the humerus and the scapula.
- Name the muscles involved in the action of reaching forwards. Which joints would be involved?

The term *scapulohumeral rhythm* has been devised to describe the combined coordinated activity of the scapular muscles and the rotator cuff muscles as they work together with deltoid and latissimus dorsi (Poppen and Walker, 1976).

As noted in Chapter 2, muscles work both concentrically (shortening) and eccentrically (lengthening). In certain activities, muscles may need to work eccentrically to decelerate a limb to produce a controlled movement to add precision to the task, for example performing a chip shot in golf. In many throwing tasks in which the glenohumeral joint is placed in external rotation, subscapularis in working eccentrically might slow down the action, producing control and absorbing energy. Therefore this action produces less forces on the ligaments and to some extent may protect them (Speer and Garrett, 1993).

CASE STUDY 10.2

While playing badminton, Chris is performing an overhead smash shot. At the same time, Liz is reaching up to place her key in the front door.

Think about the muscles acting at or around the glenohumeral joint. What two main roles will they be playing?

The two main roles that the muscles around the glenohumeral joint will be playing are stabilizing and mobilizing.

For both Chris and Liz, the muscles of the rotator cuff will be working to centre the head of the humerus on to the glenoid and at the same time draw these two joint surfaces together. In this way, the rotator cuff muscles are performing the role of stabilizing the glenohumeral joint.

However, Chris and Liz also need to be able to move their arms so that Chris can hit his smash shot and Liz can place her key in the door. This requires muscles around the glenohumeral joint to perform the role of mobilizing the joint. In these scenarios, Chris will be performing his smash shot using anterior deltoid to bring his glenohumeral joint into a flexed position, and pectoralis major for the smashing action. Liz, on the other hand, will be using anterior deltoid for reaching to place her key in the door.

Acromioclavicular joint

Although supported by strong ligaments, the acromioclavicular joint is more lax than the sternoclavicular joint. It is supported by the conoid and trapezoid ligaments (coracoclavicular ligaments) and the acromioclavicular ligament. Scapula movements are translated to the clavicle through the coracoclavicular ligaments (Dempster, 1965). The ligaments surrounding the joint are fundamental in controlling movements at this joint (Pratt, 1994; Kumar, 2002).

Sternoclavicular joint

The sternoclavicular joint must be considered as an integral part of the upper limb, as it links the shoulder girdle to the axial skeleton (Pratt, 1994). The clavicle has been described as acting like a crankshaft to allow elevation and rotation at the acromioclavicular end (Pratt, 1994; Donatelli, 1997). The clavicle acts as a strut to keep the lateral aspect of the scapula away from the chest wall. The disc between the sternum and clavicle – together with the ligaments, muscle and capsule surrounding the joint – facilitates movement of the clavicle on the sternum, allowing it to act like a hinge and allowing greater movement (Pratt, 1994).

THE SCAPULA

The scapula functions primarily as a site for muscular attachments and forms the base from which the upper limb functions (Paine and Voight, 1993; Mottram, 1997; Hadler et al., 2000). The ability for normal positioning and control of the scapula is essential for optimal function (Mottram, 1997). The scapula's concave surface sits on the convex rib cage, and it is attached to the skeleton through the acromioclavicular joint and via the clavicle to the sternoclavicular joint. The scapula is a very mobile structure and glides over and around the rib cage. The scapula is primarily stabilized by muscles that attach to the spinous processes of the vertebrae and the ribs. These include trapezius, levator scapulae, the rhomboids group, serratus anterior and serratus posterior inferior, and superior and latissimus dorsi.

Role of the scapula

The scapula plays several roles to facilitate optimum functioning of the upper limb. Normally, it provides a stable base on which the upper limb functions. More specifically, in order to maintain dynamic stability of the glenohumeral joint the musculature attached to and surrounding it must work in a coordinated fashion, exhibiting appropriate recruitment patterns (Mottram, 1997, Voight and Thomson, 2000). This is in order to maintain the correct alignment of the glenoid, on which the humerus works (Kibler, 1998). Optimal muscular activity is also facilitated by the maintenance of the normal length–tension relationship of the rotator cuff (Kamkar et al., 1993) (see Ch. 2).

The scapula must move in a controlled manner during movements of the upper limb and must accommodate the continual changing positions and functional demands of the limb (Moseley et al., 1992). The scapula also acts as a base for muscle attachments that provide both stability and movement to the upper limb. The muscles surrounding the scapula work as force couples in synergistic cocontraction to control the position of the scapula (Moseley et al., 1992; Jobe and Pink, 1993; Kibler, 1998). The muscles responsible for controlling the scapula are all sections of trapezius, serratus anterior, rhomboids major and minor, levator scapulae and pectoralis minor. Pectoralis major and latissimus dorsi affect the scapula, as they attach to the humerus. Other muscles to affect the scapula are deltoid, teres major, the rotator cuff muscles, subscapularis, supraspinatus, infraspinatus, teres minor, coracobrachialis, long head of triceps, and long and short head of biceps.

The scapula needs to be able to protract, retract and rotate to allow the upper limb to be placed in the correct position for function. It needs to be able to retract to enable the cocking phase of throwing action, for example in the tennis serve, in throwing a cricket ball and in the swimming recovery phase. Throwing activities involve a coordinated sequence of movements of the scapula and upper limb. The phases of throwing are (adapted from Pink and Perry, 1996):

- preparation, or wind up, which begins the motion
- cocking phase, in which the shoulder is abducted and externally rotated
- acceleration phase, in which the humerus internally rotates, ending at
- release phase
- deceleration phase, the first one-third of the time from ball release to the end of the arm movement
- follow-through phase, the last two-thirds from ball release to the end of the arm movement.

Achievement of the cocking phase tensions the anterior chest musculature to prepare for the concentric phase of muscle action. The explosive phase of acceleration requires the scapula to protract in coordinated fashion.

As the acceleration phase begins, smooth protraction and anterior glide of the scapula around the chest wall is required. This movement allows the glenoid to be maintained in the appropriate alignment with the head of the humerus. This movement of the scapula is mainly achieved by the eccentric contraction of mainly rhomboids and middle trapezius. These muscles help dissipate some of the forces in the deceleration (follow-through) phase (Pink and Perry, 1996).

In overhead activities such as throwing, the scapula has to rotate laterally to allow full abduction, which is vital for full elevation. In most throwing activities, the arm works at an angle between 85 and 100° of abduction, therefore the muscles surrounding the scapula must draw the acromion away from the cuff to avoid compression (Kamkar et al., 1993; Kibler, 1993, 1998; Fleisig et al., 1994). In all elevation movements, it is necessary to clear the greater tubercle from the coracoacromial arch for the same reason.

CASE STUDY 10.3

If any of the muscles controlling the scapula do not function optimally, there is the potential for compression of the structures underneath the acromion, including the subacromial bursa and the supraspinatus tendon. Other factors that can contribute to compression of these structures include dysfunction of the rotator cuff, in which the compressive force on the head of the humerus is lost, which results in an upward migration of the head of the humerus. This could happen to Chris as a result of injury following trauma or through overuse.

Optimum scapula function is achieved not only by the muscles moving and controlling it but by the movements at the acromioclavicular and sternoclavicular joint. As the shoulder abducts, the clavicle has to elevate, which requires a coordinated movement at the acromioclavicular and sternoclavicular joint.

Scapulothoracic joint

Although not a classic bony articulation, the scapulothoracic joint is an important physiological joint that is vital to optimum function of the upper limb. The overall ratio of scapulothoracic to glenohumeral movement of 1:2 is achieved by movement of these two structures combined with rotation of the clavicle (Kumar, 2002). Movement of the scapulothoracic joint is between two fascial planes and is controlled by the muscles surrounding it (Pratt, 1994; Mottram, 1997). The scapulothoracic joint and structures surrounding it are constantly seeking a position of stability in relation to the humerus and to attain the optimum position of the glenoid.

Subacromial space

The space between the acromion and the head of the humerus is critical for optimum function of the upper limb. This narrow space between the head of the humerus and the arch formed by the acromion, coracoid process and coracoacromial ligament facilitates general shoulder mobility but in particular flexion and abduction. In this small space are the subacromial bursa, supraspinatus muscle and its tendon, superior aspect of the joint capsule, and tendon of the long head of biceps. In function, these structures are able to constantly move against each other without producing symptoms.

CASE STUDY 10.4

While playing badminton, Chris is performing an overhead smash shot. Think about the scapula and clavicle; what position will they be in to allow Chris to achieve this overhead position? What will be the role of the trapezius and serratus anterior muscles during this dynamic task?

The scapula will be externally rotated to allow for elevation of the glenoid (for the glenohumeral joint to achieve end of range) and optimize clearance of the humerus from the subacromial contents.

The clavicle will be posteriorly rotated to again allow for end of range glenohumeral joint positioning. These muscles have the effect of stabilizing the scapula and optimizing its position relative to the humerus. This will allow for efficient, pain-free function of the upper limb during this dynamic task. These will be providing a stable, proximal base for the upper limb, therefore allowing the distal segment to perform the task.

CASE STUDY 10.5

Liz is using her front door key to unlock the family home. What role will the scapulothoracic structures and the muscles around the scapula be playing as she performs this task? In her other hand, Liz is holding on to her heavy shopping bags. What role will the scapulothoracic structures and the muscles around the scapula be playing as she holds the heavy bags?

As for the previous case study response, trapezius and serratus anterior will be providing a stable, proximal base for the upper limb, therefore allowing the distal segment to perform the task.

Chris will be using an extended elbow, which serves to lengthen the lever arm of his smash arm. This increase in lever arm serves to increase the rotational moment (see Ch. 9, *Biomechanics of Human Movement*). In addition, the racket will be moving faster, allowing Chris to strike the shot at a higher speed.

For Liz, her elbow carrying the heavy bags will be extended. This minimizes the energy required to carry the heavy bags. In addition, it will pull the elbow into the anatomical carrying angle. For Liz, the limb that is placing the key in the lock will probably have a bent elbow. This enables her to shorten her limb and thus come closer to the poorly lit door in order to look more carefully at where the lock is.

PROPERTIES OF MUSCLE AFFECTING FUNCTION OF THE UPPER LIMB

The term *stiffness* applied to a muscle refers to the ability of the muscle to provide resistance to deformation and provide support to the bones and joints (see Chs 2 and 3). Muscle stiffness is determined by the intrinsic properties of muscle and neuromuscular control from the central nervous system. It is important to remember that muscles do not work in isolation. Coordinated muscular activity with appropriate timing is the key to optimum function. In the upper limb, as in other areas of the human body, a system of force couples act as the basis for motor control.

At the higher centres, there is integration of visual, auditory and kinaesthetic information together with the information regarding the demand of the task to produce a highly coordinated movement. (For further information, see Ch. 4, *Motor Control*.)

Muscle length as well as stiffness also contributes to the function of the joint. As a highly mobile joint, muscles affecting both the scapula and the glenohumeral joint must have the appropriate extensibility to allow full movement to occur. In the final phase of glenohumeral elevation, muscles including latissimus dorsi, pectoralis major and minor, teres major, teres minor infraspinatus and subscapularis must have sufficient extensibility to allow this final phase to be achieved.

Neural tissue

For the upper limb to function optimally, it is vital to consider the function and properties of the nerves in that region. Properties of nerves include the ability to conduct information and the ability to move and glide within the constraints of the surrounding anatomical structures, which include soft tissues and bone. Some branches of the cervical plexus, for example the supraclavicular nerves (C3 and C4), supply areas of the shoulder, with many of the branches running through muscles and close to joints. The brachial plexus (C4–T2) is intimately entwined in the structures of the upper limb after emerging from the cervical spine. In certain areas of the upper limb, there are tension points at which the nerves are relatively fixed or sit in a narrow channel and are potentially prone to compression, for example the ulnar nerve at the elbow sits in groove on the dorsum of the epicondyle, and the median nerve at the wrist lies between the retinaculum and tendons in the carpal tunnel.

Movement of the shoulder into functional positions can put nerves on tension. In normal circumstances, nerves and surrounding structures glide on and between each other. As the anatomy suggests, there is a connection between the cervical spine and upper limb movements.

ACTIVITY 10.2

- Try bending your head to the left side (cervical left-side flexion). What do you feel?
- Now abduct your right arm to 90°. What do you feel?
- Now, extend your wrist. What do you feel?
- Now, finally extend your fingers. What do you feel?
- Try the same with your right arm. Does it feel the same or different?

Proprioception

Sensory feedback about the limb's position and direction is vital to modify muscle activity when required to achieve optimum performance (Warner et al., 1996). Proprioceptors within joints provide the central nervous system with this information. Although proprioceptors are a feature of all joints, most of the research in the upper limb has focused on the glenohumeral joint. Proprioception refers to afferent information received by higher centres from receptors in articular, muscle and cutaneous structures and is vital for coordinated function. Specialized nerve endings that have proprioceptive mechanisms including Pacinian corpuscles, Ruffini endings and Golgi tendon organs are all found in capsules and ligaments of all joints.

Neuroafferents have been noted within the capsulotendinous junction in the glenohumeral joint that will give feedback on shoulder position (Grigg, 1993).

This input from receptors gives information about the position of the joint during all stages of functional activity and is vital for feedback during the task. Proprioceptive information is vital for programming movements and reflex muscular contractions, especially from the rotator cuff, to enhance stability of the glenohumeral joint (Warner et al., 1996).

All tasks rely on preprogramming with information prior to and during the task, as adjustments may need to be made during the performance (see Ch. 6). Any coordinated and skilled upper limb activity requires continual feedback from the sensory systems. Receptors must provide spatial and temporal information so that the task can be completed with optimum skill and efficiency.

This is particularly pertinent in the minimally restrained glenohumeral joint, where coordinated activity of all the contractile tissue is essential for efficient control. In a reaching task such as painting a picture or drawing, the neuromuscular system controlling the arm must integrate information such as limb position and force required together with information from the eye to produce a smooth and controlled action. When afferent information is diminished, because of injury for instance, the task may still be completed but often fine control and precision are lacking. In normal shoulders, there appears to be no difference in proprioception between the dominant and non-dominant hands and between males and females (Jerosch et al., 1996; Warner et al., 1996).

CASE STUDY 10.6

While playing badminton, Chris is performing an overhead smash shot. At the same time, Liz is reaching up to place her key in the front door at night, and the bulb in the light above the door is broken.

Think about the precision and timing required for these two activities. What are some of the similarities between them? What are some of the differences between them?

Both are tasks that each individual has performed before, so there is a degree of motor learning that will have occurred, leading to the performing of a refined, efficient motor task.

Both will benefit from visual feedback to adapt the task as required. This could include the need by Chris to alter where he is going to hit his smash so that he can win the point. For Liz, the reduced lighting will reduce the degree of visual feedback, making her more dependent on tactile feedback. Both will require a degree of precision and timing. However, the timing required to hit (and not miss!) the smash shot is crucial for the badminton task. The precision required for placing the key in the lock is crucial for opening the door successfully, and Liz putting the shopping down on the kitchen table.

THE ELBOW AND FOREARM

These two regions will be considered together, as their function is interlinked. The elbow functions in combination with the movements of the superior and inferior radioulnar joints. We must consider the upper limb as a series of links in a chain, with the elbow adjusting height and length of the limb and also the position of the hand.

The elbow consists of two articulations: the humeroulnar and the humeroradial. The superior radioulnar joint consists of the head of the radius articulating with the ulna and the inferior radioulnar joint, where the distal ends of the ulna and radius articulate with the carpus. The elbow is a stable joint when compared with the glenohumeral joint, because of the shape of the proximal ulna (Guerra and Timmerman, 1996), the radial head (Morrey et al., 1983), the collateral ligaments and the anconeus muscle.

Elbow function

The elbow is an integral part of the upper limb as the link in the chain that can position the hand and transfer loads to hand (Guerra and Timmerman, 1996). Elbow function is vital to the basic task of bringing the hand to the mouth for eating, when the elbow flexors and supinators of the forearm work together. Flexion and extension of the elbow are accompanied by some degree of rotation of the ulna (Williams, 1995). Pronation and supination, which occur at the radioulnar joints, place the hand in the most effective functional position. Pronation and supination allow the hand to be turned through 140–150°, and in extension this can increase to nearly 360° with rotation of the humerus and the scapula (Williams, 1995).

The elbow often functions as a hinge, producing flexion and extension during activities such as sawing wood or hitting nails with a hammer and lifting objects up and down.

ACTIVITY 10.3

Think of different tasks that:
- use the elbow as a hinge in a flexion–extension movement
- put a compressive force through the elbow.

The forearm in a supinated position can exert a greater pressure than in pronation, hence this governs many power activities such as screwing nuts (Fig. 10.1), turning handles and lifting heavy weights, when the most efficient position is one of forearm supination and elbow flexion.

Biceps is utilized in all activities involving supination, such as turning a key or screwdriver, as well as flexing the forearm to the upper arm. In demanding power activities, strength may be applied by grasp from the hand, but considerable

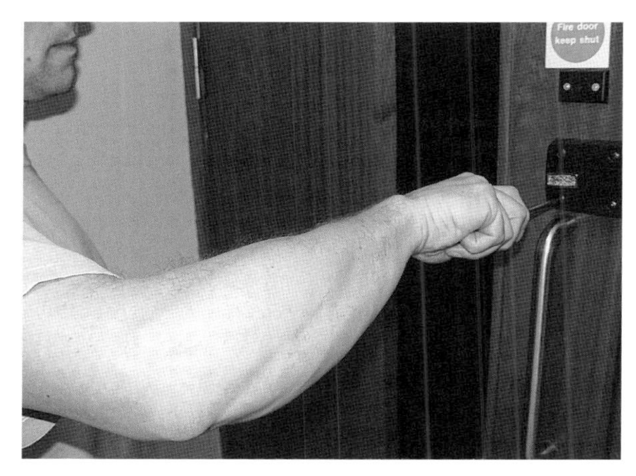

Figure 10.1 Action of using a screwdriver.

forces are applied through the forearm, the upper arm and the body. When lifting heavy loads, triceps will recruit anconeus to stabilize the joint to protect it against forces that are applied to the joint. The mid prone position is a strong functional position, as can be seen when hammering and lifting boxes. Many functional activities involve a combination of elbow flexion in the mid prone position with shoulder flexion. In a simple task such as eating, there is a complex sequence of movements and positioning to bring the food to the mouth: the scapula laterally rotates, the shoulder girdle protracts, and the glenohumeral joint flexes, adducts and medially rotates as the elbow supinates and flexes. During this limb movement, the hand must grasp the fork or spoon in an appropriate fashion to bring the food to the mouth. Muscles surrounding the elbow and forearm work in an integrated fashion, functioning as phasic antagonists as well as antagonists (see Ch. 2).

Forces through the elbow during baseball pitching have been studied, as many elbow injuries occur in this sport. Low muscle activity and low elbow joint forces have been noted during the wind-up and stride phases, which are preparatory phases for dynamic muscle activity. High muscle activity and elbow forces were noted during arm cocking, arm acceleration and deceleration phase (Fleisig and Escamilla, 1996).

HAND

The hand is a very specialized and versatile unit that has the potential for fine discrimination, coordination and dexterity as well as a powerful grip. The functional hand works as part of a coordinated unit of the upper limb. Development of hand skills is critical to the performance of activities of daily living and to function throughout life. The hand can be used alone or with tools in highly specialized and complex situations. It can be used for fine motor control skills and in situations in which gross power is needed from the whole upper limb. It can act as a lever, when the hand can be fixed with the arm while the trunk moves; it can act as a base for support and balance, and as a communicator.

ACTIVITY 10.4

Think of a task that requires:

- the hand to be used for a fine motor skill
- the hand used for a power activity
- the hand used for leverage of the body.

Manipulation of objects by the hand is a fine and complex task relying on an intact central nervous system to deliver the appropriate responses in terms of coordination and timing of activation, as in the need for the correct pressure to hold an object and the correct timing to release an object. Fine motor skills are essential to manipulate objects when there is a need for a coordinated balance between mobility and balance, for instance in writing, when the hand needs to be able to perform the pinch grip and control of paper. As well as fine control manipulation, the hand has to be capable of using tools and large objects.

The hands as a specialized sensory unit

One of the most important functions of the hand is the ability to perceive sensation (Mackel, 1996). The hand is richly innervated with sensory receptors and has a large representation area in the somatosensory cortex. The hand is an active sensory organ and is capable of detecting objects even when there is no visual input (e.g. when searching in a pocket or a bag), and it is able to perceive different sizes, shapes and textures. The ability to use the hand has developed to its highest level in the adult human because of the presence of a specialized central nervous system.

ACTIVITY 10.5

Without you seeing the items, get a friend to choose five household items of different textures that can be manipulated by the hand. Keeping your eyes closed, feel the objects and see how long it takes you to identify what they are.

Anatomy

The arrangement of the bones of the hand and the accompanying soft tissues allows for a diverse range of movements and tasks (Moran, 1989; Strickland, 1995). The skin covering the hand on the dorsum and volar aspect plays a major role in the function of the hand as a tactile and discriminatory unit. The skin of the hand, especially on the volar aspect, is richly supplied with sensory nerve endings that aid discrimination of different sensations (Moran, 1989). The skin also aids function, in particular gripping. The skin on the dorsum of the hand is mobile and aids flexion of the fingers, whereas the volar skin is less flexible but has many creases in it that follow the lines of stresses imposed on it during function (Moran, 1989; Caillet, 1994).

Bones and soft tissues

The shape of the 29 bones in the hand and the neuromuscular system that controls it give the hand the capabilities of achieving a wide range of functions. Movements of the hand are achieved by the coordinated effect of intrinsic and extrinsic muscle action as well as a balance between flexors and extensors. For example, complex movements such as writing require flexion at the metacarpophalangeal (MCP) joints and extension at the interphalangeal (IP) joints. During grip activities, the intrinsic muscles as well as the ligaments ensure joint stability while the long flexors provide most of the force needed (Palmer, 1991). Biomechanical studies have shown that the intrinsics are able to exert a strong force of more than a third of the long flexors in the gripping action (Palmer, 1991). The lumbricals contract during both MCP flexion and MCP extension. When the MCP joints extend, the lumbricals appear to be important in extension of the IP joints (Palmer, 1991). Connective tissues of the hand – for example the flexor retinaculum, palmar aponeurosis and dorsal digital expansion – protect and bind groups of muscles to allow smooth movement. To reduce friction around tension points, each flexor tendon is contained in a sheath that contains a lubricating fluid to aid smooth movement.

Some muscles that influence hand function do not directly insert to the hand, for example brachioradialis inserting on to the base of the styloid process after crossing the cubital fossa. It flexes the elbow when the forearm is in mid pronation and supination, which is essential in eating and drinking tasks.

The thumb is a vital functional unit of the hand, as the carpometacarpal joint, together with a loose capsule, allows for a wide range of motion. This joint arrangement facilitates gripping and manipulation through a wide arc of motion. The combined movements of flexion, medial rotation and adduction mean the thumb can be taken across the palm of the hand into *opposition*. In opening the hand, the thumb is extended and abducted.

Wrist and hand movements

The optimum functional position of the hand is in the mid pronation position, with the wrist in extension and the digits in a moderate degree of flexion. Most functional activities of daily living occur with the wrist in a position between 10° of flexion and 35° of extension (Blumfield and Champoux, 1984). Wrist movements are the key to the manipulations of the digits. This joint has a large influence on the long muscles of the forearm (Strickland, 1995), with maximum strength in the power grip being facilitated by extension of the wrist. In functional activities, the carpus is able to transmit forces from the hand through the forearm, for example when we get up from a chair. Gliding movements of the carpals on the radius are facilitated by capsular and ligamentous laxity, and the carpals move in the opposite direction to the movement of the hand. The ligaments also provide stability of the carpals with appropriate gliding while the wrist and fingers are mobile.

ACTIVITY 10.6

Pick some of the following day to day activities and look at the position of the wrist and forearm:

- the hand gripping a teapot
- pouring tea
- turning a tap
- holding a large glass
- using a screwdriver
- carrying a bucket or a carrier bag
- carrying a large box in front of you.

Development of gripping

For most of the first year, grasp development relates to feedback from objects via the palmar surface (McCall, 1974). The grasp reflex predominates initially. During the second year, skills that develop are associated with increasing cognitive abilities and environmental influences. Coordinated movements are developed through sensory experience and feedback through the hand. When beginning to weight bear, the hands and arms are used to support the body, to balance, to walk and to manipulate objects. As a young child, force coordination is poorly developed. At age 1, children might crush a fragile object such as an ice cream cone or a paper model, but between the ages 2 and 4 years they develop the skills to manage and handle fragile objects (Eliasson, 1995).

The crawling position, in which the hand and upper limb are used for support and propulsion, allows the child to creep and in turn strengthens the upper limb and neck musculature. Strengthening of the arms and hands also occurs as children pull on objects such as furniture to stand and support themselves (Case-Smith, 1995). The development of neck and shoulder musculature, together with increasing postural stability, is the prerequisite for the control of reach activities. The first 2 years of life show dramatic developments in the use of the hand in terms of function and skill. Basic grasps and release skills develop into mature complex coordinated movements in the early years of development (Case-Smith, 1995). At 10–12 months, the child can manipulate tools, for example a spoon to feed or a pencil to scribble.

All basic hand manipulation skills and those involving stabilization of the object in the hand were seen to be present in normal children by age 7 years (Exner, 1990). Development of all manipulation skills requires coordination of the neuromusculoskeletal system surrounding the neck, shoulder, elbow and wrist. For details of timescales for development of skills relating to feeding, self-care, dressing and washing, see Henderson (1995) and Zivani (1995).

The hook grip is when all the fingers are flexed towards the palm, for example in lifting a heavy suitcase or briefcase (Fig. 10.2). The fingers are flexed around the handle but the thumb is not involved.

The power grip is when the thumb is strongly flexed, the fingers are flexed towards the thenar eminence and the ulnar deviated. The thumb and the ring and little fingers contribute most to the power grip, and this grip is used in many activities in which power and stability are required (e.g. using a hammer or doing a bench press). In gripping a hammer, there is maximum flexion at the little finger and least at the index. In hammering, a maximum area of contact is required to provide power with precision and control. In the power grip, all the intrinsic and extrinsic muscles are used, with the hypothenar eminence muscles stabilizing the medial aspect of the hand. The wrist extensors act as a stable base on which the power grip can function. Wrist extensors increase in activity so as to direct the force on to the finger flexors so that they do not act on the wrist. Feedback from the sensory receptors in the hand is essential to exert exactly the right pressure required and coordinates the interaction of other muscles in the arm to provide power. For example, in opening a heavy door the hand grips the handle and the forearm muscles provide the power.

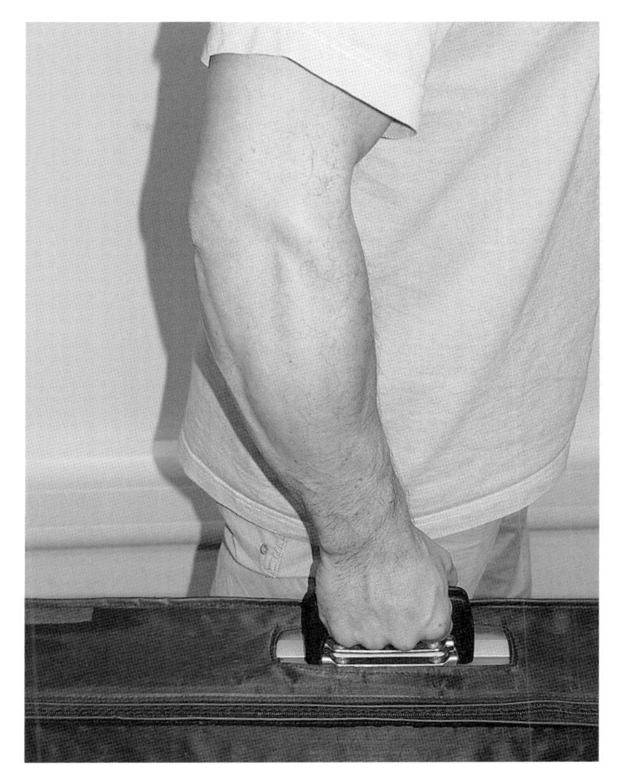

Figure 10.2 The hook grip and muscle work involved in carrying a case.

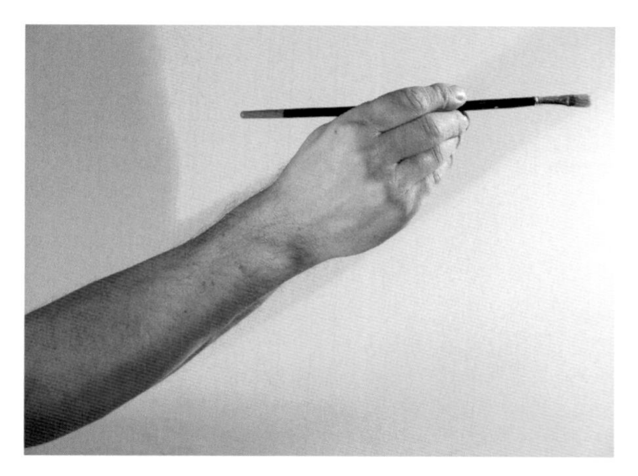

Figure 10.3 The fine control involved in using a paintbrush.

The precision or pinch grip is used for fine manipulation and requires the thumb to be placed in contact with the fingertips. To achieve this tip to tip approximation, the fingers must rotate and deviate in an ulnar direction. The precision grip utilizes the coordinated action of the lumbricals and interossei to grip the object, with the hand being positioned by the wrist and forearm. The precision grip is more advanced than the power grip, appearing at around 9 months (Fig. 10.3). In the pinch grip, the MCP and proximal IP joints of the index are flexed; the distal IP joint can be flexed or extended. The opposed thumb pad meets the pad of the digit. This pad to pad grip is used for instance when holding a pen, turning a key or sewing (Fig. 10.4).

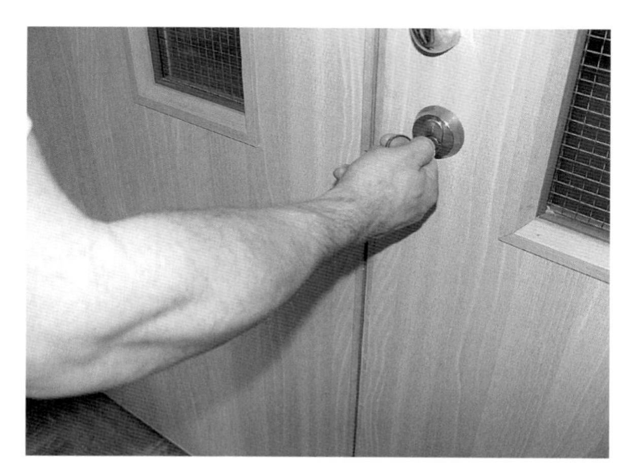

Figure 10.4 The use of the intrinsic muscles when turning a key.

In the key grip, the extended thumb is held on the radial side of the index finger.

The lumbrical grip is when the MCP joints are flexed across the palmar aspect (e.g. when holding a plate or something that needs to be kept horizontal).

ACTIVITY 10.7

Think of the following activities and decide which grip is involved:

- playing tennis
- playing chess
- digging the garden with a spade
- placing a full cup and saucer on the table
- placing a letter into an envelope
- lifting a heavy shopping bag
- using a knife and fork
- painting a picture.

Two-handed manipulative tasks are when both hands are involved in the task (e.g. playing the piano or guitar, and working at a keyboard). In some activities, one hand can provide the stability while the other hand completes the task (e.g. peeling potatoes, sawing wood and unscrewing a jar lid).

CASE STUDY 10.7

While playing badminton, Chris is performing an overhead smash shot. At the same time, Liz is placing her key in the front door (which remains poorly lit), while in her other hand she is holding on to the heavy shopping bags.

Think about the functions that the elbow, wrist and hand are performing in each of these tasks. What functions will the elbows be performing? What kind of grip will Liz and Chris each be using?

Chris will be using a power grip for maximum power on the smash shot. Liz will be using a hook grip to hold the heavy bags in the most efficient manner. In the other hand, she will be using the precision of key grip – for precise key gripping!

THE UPPER LIMB AS PART OF THE KINETIC CHAIN

The kinetic chain refers to the relationship and interplay between each segment of the body. The upper limb is part of the kinetic chain linked to the other parts of the body. As discussed earlier, its function relies on the optimum function of all the neurophysiological systems of the body, and any rehabilitation programme must consider both open and closed chain activities as well as those to enhance dynamic stability (Wilk et al., 1996). When considering the function of the upper limb, we need to consider the component parts of the trunk – including the cervical, thoracic and lumbar spine and pelvis – as the posture of the trunk can determine how effective the upper limb functions (Lannerston and Harms-Ringdahl, 1990; Crawford and Jull, 1991). Full extensibility of the soft tissues related to the trunk, for example quadratus lumborum and latissimuss dorsi, is essential for optimal upper limb function.

When power is required, for instance in a tennis serve, force generation and wind up start with positioning of the feet. Individual body segments work together to position the upper limb in the correct functional position that is required. For effective upper limb function, there needs to be a transfer of forces from the proximal segments to the upper limb, which can be achieved through the stable but active upper limb base, the scapula. Up to 54% of the total force in a tennis serve will be generated from the lower legs, hips and trunk (Kibler, 1995), with body rotation contributing up to half of the release velocity.

PROXIMAL CONTROL

The function of the upper limb is primarily dependent on a stable base on which to work. The pelvis, and more proximally the scapula, can be viewed as the two interlinking platforms on which the upper limb functions. Function is also dependent on coordination between the scapular muscles and the muscles of the humerus, forearm, wrist and hand.

Poor pelvic stability as a result of, for example, delayed muscle recruitment or atrophy of deep stability muscles (Hodges and Richardson, 1996; Hides et al., 2001), can lead to inefficient and uncoordinated activity of the upper limb. Hodges and Richardson (1996) have demonstrated that in all upper limb movements at certain speeds, transversus abdominis is activated prior to any arm movements. This preprogrammed activity of the transverse abdominal muscles prior to arm movement ensures that the pelvis is stable prior to the upper limb moving. It has also been noted that the postural muscles of the lower limb are activated before the arm moves (Cordo and Nashner, 1982), therefore adding to the stable base on which the upper limb can function. This integrated action of postural activation in the pelvis and leg muscles prior to arm movement is an important prerequisite for the production of a coordinated action of the upper limb function. The importance of postural trunk control in the development of skills in children is demonstrated in the transition from the two-handed reach to the one-handed reach (Rochat, 1992).

OPEN CHAIN, CLOSED CHAIN

The upper limb works in both closed and open chain activities. Open chain activities include daily activities such as combing your hair, dressing and reaching up, as well as sporting activities such as tennis and squash. Closed chain activities include rising from a chair and levering oneself out of bath or up from the floor after a fall. In gymnastic activities, these can include somersaults on the floor and the push off from the vault and the beam. In studying closed chain activities such as press-ups and the bench press, the highest electromyographic activity was noted in pectoralis minor in press-ups (Moseley et al., 1992).

Many activities are a combination of open and closed chain activities, such as using a walking stick, when the stick is planted on the ground, the arm is working in a closed chain pattern, and the swing phase of the arm is open chain. In loading of the upper limb, there is stimulation of all the upper limb muscles including the scapula and the proprioceptors within all the joints and ligaments. Weight-bearing activities can be used to promote stability within a notoriously mobile joint such as the glenohumeral joint by providing joint compression and muscular cocontraction, which enhances dynamic stability (Wilk and Arrigo, 1993). The upper limb in a fixed position can also aid respiration. Patients will fix the humerus on a chair or banister, for example, and pectoralis major can assist in pulling the sternum upwards and outwards to enlarge the thorax. This movement is assisted by serratus anterior and pectoralis minor.

The upper limb is used extensively to enhance the balance and stability of the body in both a closed and an open chain situation. This may incorporate one or both upper limbs.

ACTIVITY 10.8

Think of the role of the upper limb in the activities below:

- a trapeze artist walking across a tightrope
- a gymnast performing an exercise routine on the beam or a vault box
- a patient with a plaster of Paris on their leg walking with crutches
- a person with a stick who is unsteady
- the rugby player who kicks a penalty or a conversion.

FUNCTION OF THE UPPER LIMB IN ACTIVITIES OF DAILY LIVING

Any simple movement, such as brushing your hair or cleaning your teeth, involves a series of complex and coordinated movements of the upper limb segments that are dependent on the coordinated activity of many muscles, nerves, ligaments and other soft tissues. Virtually any use of the hands involves the movement of the scapula and the glenohumeral joint to position the hands in the appropriate functional position.

Many daily activities involve reaching and grasping an object and moving it, when the arm functions as one unit. The reach phase has been termed the *transportation component* and the grasp as the *manipulation component* (Jeannerod, 1984). In the phase of arm reaching, the proximal muscles and joints are used to place the limb in the appropriate position towards the object. The grasp phase starts while the limb is being moved towards the object and involves the activation of the intrinsic muscles of the hand (Jeannerod, 1981). In the grasp phase, the fingers have to anticipate the size and position of the object to attain a smooth and efficient action. This is achieved through information received through the visual senses. During any reaching task, there is a phase of acceleration and deceleration of the limb, with the muscles working in a combination of eccentric and concentric work. In activities such as eating, there is a retrieval component in which the hand is brought to the mouth.

The greater the speed of the reaching task, the less accurate the grasp activity may be. Visual feedback on the positioning of the hand is necessary for accuracy of the grasping task (Tillary et al., 1991). However, even if the reaching movement is carried out without visual feedback it can be achieved but may take more time or be less accurate. Rosblad (1995) indicates that the reaching task is probably preprogrammed prior to the task, but adjustments can be made during the task if required.

SUMMARY

This chapter has focused on the function of the upper limb, emphasizing that each segment works in a coordinated fashion to achieve optimum function. In assessing the activity of the upper limb, we need to understand the function of each component part and its relationship to the other parts of the kinetic chain.

References

Blumfield, R.H., Champoux, J., 1984. A biomechanical study of normal functional wrist motion. Clin. Orthop. Relat. Res. 187, 23–25.

Caillet, R., 1994. Hand Pain and Impairment, fourth ed. FA Davies, Philadelphia.

Case-Smith, J., 1995. Grasp release and bimanual skills in the first two years of life. In: Henderson, A., Pehoski, C. (Eds.), Hand Function in the Child. Mosby, St Louis, pp. 113–135.

Cordo, P.J., Nashner, L.M., 1982. Properties of postural adjustments associated with rapid arm movements. J. Neurophysiol. 47, 287–302.

Crawford, H., Jull, G., 1991. The influence of thoracic form and movement on ranges of shoulder flexion. In: MPPA Proceedings, 7th Biennial Conference. Australia.

Donatelli, R.A., 1997. Functional anatomy and mechanics. In: Donatelli, R.A. (Ed.), Physical Therapy of The Shoulder, third ed. Churchill Livingstone, New York, pp. 1–17.

Doukas, W.C., Speer, K.P., 2001. Anatomy, pathophysiology, and biomechanics of shoulder instability. Orthop. Clin. North Am. 32 (3), 381–391.

Eliasson, A.C., 1995. Sensorimotor integration of normal and impaired development of precision movement of the hand. In: Henderson, A., Pehoski, C. (Eds.), Hand Function in the Child. Mosby, St Louis, pp. 40–54.

Exner, C.E., 1990. In-hand manipulation skills in normal young children. A Pilot study. Occup. Ther. Pract. 1 (4), 63–72.

Fleisig, G.S., Escamilla, R.F., 1996. Biomechanics of the elbow in the throwing athlete. Oper. Tech. Sports Med. 4 (2), 62–68.
Fleisig, G.S., Dillman, C.J., Andrews, J.R., 1994. Biomechanics of the shoulder during throwing. In: Andrews, J.R., Wilk, K.E. (Eds.), The Athlete's Shoulder. Churchill Livingstone, New York, pp. 355–368.
Grigg, P., 1993. The role of capsular feedback and pattern generators in shoulder kinematics. In: Matsen, F.A., Fu, F.H., Hawkins, R.J. (Eds.), The Shoulder: A Balance of Mobility and Stability. American Academy of Orthopaedic Surgeons, Rosemont, pp. 173–183.
Guerra, J.J., Timmerman, L.A., 1996. Clinical anatomy, histology and pathomechanics of the elbow in sports. Oper. Tech. Sports Med. 4 (2), 69–76.
Hadler, A.M., Itoi, E., An, K.N., 2000. Anatomy and biomechanics of the shoulder. Orthop. Clin. North Am. 31 (2), 159–176.
Henderson, A., 1995. Self-care and hand skill. In: Henderson, A., Pehoski, C. (Eds.), Hand Function in the Child. Mosby, St Louis, pp. 164–183.
Herzog, W., Guimaraes, A.C., Anton, M.G., et al., 1991. Moment length relations of rectus femoris muscles of speed skaters/cyclists and runners. Med. Sci. Sports Exerc. 23, 1289–1296.
Hess, S.A., 2000. Functional stability of the glenohumeral joint. Man. Ther. 5 (2), 63–71.
Hides, J.A., Jull, G.A., Richardson, C.A., 2001. Long term effects of specific stabilizing exercises for first episode low back pain. Spine 26, E243–E248.
Hodges, P., Richardson, C.A., 1996. Inefficient muscular stabilisation of the lumbar spine associated with low back pain: a motor control evaluation of transversus abdominis. Spine 21, 2640–2650.
Itoi, E., Kuechle, D.K., Newman, S.R., et al., 1993. Stabilising function of the biceps in stable and unstable shoulders. J. Bone Joint Surg. (Br) 75 (4), 546–550.
Jeannerod, M., 1981. Intersegmental co-ordination during reaching at natural visual objects. In: Baddeley, L.L. (Ed.), Attention and Performance, vol. 9. Erlbaum, Mahwah, pp. 153–168.
Jeannerod, M., 1984. The timing of natural prehension movements. J. Mot. Behav. 16, 235–254.
Jerosch, J., Thorwesten, L., Steinbech, J., et al., 1996. Proprioceptive function of the shoulder girdle in healthy volunteers. Knee Surg. Sports Traumatol. Arthrosc. 3 (4), 219–225.
Jobe, F.W., Pink, M., 1993. Classification and treatment of shoulder dysfunction in the overhead athlete. J. Orthop. Sports Phys. Ther. 18, 427–432.
Jobe, C.M., Pink, M.M., Jobe, F.W., et al., 1996. Anterior shoulder instability, impingement and rotator cuff tears. In: Jobe, F.W. (Ed.), Operative Techniques in Upper Extremity Sports Injuries. Mosby, St Louis, pp. 164–176.
Kabada, M.P., Cole, M.F., Wooten, P., et al., 1992. Intramuscular wire electromyography of the subscapularis. J. Orthop. Res. 10 (3), 394–397.
Kamkar, A., Irrgang, J.J., Whitney, S.L., 1993. Nonoperative management of secondary shoulder impingement syndrome. J. Orthop. Sports Phys. Ther. 17, 212–224.
Kibler, W.B., 1993. Evaluation of sports demands as a diagnostic tool in shoulder disorders. In: Matsen, F.A., Fu, F., Hawkins, R.J. (Eds.), The Shoulder: A Balance of Mobility and Stability. American Academy of Orthopaedic Surgeons, Rosemont, pp. 379–395.
Kibler, W.B., 1995. Biomechanical analysis of the shoulder during tennis activities. Clin. Sports Med. 14 (1), 79–85.
Kibler, W.B., 1998. The role of the scapula in athletic shoulder function. Am. J. Sports Med. 26, 325–337.
Kibler, W.B., McMullen, J., Uhi, T., 2001. Shoulder rehabilitation strategies, guidelines and practice. Orthop. Clin. North Am. 32 (3), 527–538.
Kumar, V.P., 2002. Biomechanics of the shoulder. Ann. Acad. Med. 31 (5), 590–592.
Kumar, V.P., Balasubramaniam, P., 1985. The role of atmospheric pressure in stabilising the shoulder: An experimental study. J. Bone Joint Surg. 67, 719–721.
Kumar, V.P., Satku, K., Balasubramaniam, P., 1989. The role of the long head of biceps brachii in the stabilization of the head of the humerus. Clin. Orthop. 244, 172–175.
Lannerston, L., Harms-Ringdahl, K., 1990. Neck and shoulder muscle activity during work with different cash register systems. Ergonomics 33 (1), 49–65.
Mackel, R., 1996. The impact of injury and disease on the sensory function of the hand: neurophysiology and clinical implications. Bull. Soc. Sci. Med. Grand Duche Luxemb. 133 (1), 31–41.
McCall, R.B., 1974. Exploratory manipulation and play in the human infant. Monogr. Soc. Res. Child Dev. 39 (2), 1–88.
Michiels, I., Bodem, F., 1992. The deltoid muscle: an electromyographical analysis of its activity in arm abduction in various body posture. Int. Orthop. 16, 268–271.
Moran, C.A., 1989. Anatomy of the hand. Phys. Ther. 69 (2), 1007–1013.
Morrey, B.F., Tanaka, S., An, K.N., 1983. Articular and ligamentous stability of the elbow joint. Am. J. Sports Med. 11, 315–319.

Moseley, J.B., Jobe, F.W., Pink, M., et al., 1992. EMG analysis of scapular muscles during a shoulder rehabilitation programme. Am. J. Sports Med. 20 (2), 128–134.
Mottram, S.L., 1997. Dynamic stability of the scapula. Man. Ther. 2 (3), 123–131.
Paine, R.M., Voight, M.L., 1993. The role of the scapula. J. Orthop. Sports Phys. Ther. 18, 386–391.
Palmer, P., 1991. Muscle function of the hand during resisted and non resisted activity. Br. J. Occup. Ther. 54 (10), 386–390.
Pink, M., Perry, J., 1996. Biomechanics. In: Jobe, F.W. (Ed.), Operative Techniques in Upper Extremity Sports Injuries. Mosby, St Louis, pp. 109–123.
Poppen, N.K., Walker, P.S., 1976. Normal and abnormal motion of the shoulder. J. Bone Joint Surg. 58A, 195–201.
Pratt, N.E., 1994. Anatomy and biomechanics of the shoulder. J. Hand Ther. 7 (2), 65–76.
Rochat, P., 1992. Self sitting and reaching in 5 to 8 month old infants: The impact of posture and its development on hand eye coordination. J. Mot. Behav. 24 (2), 210–220.
Rosblad, B., 1995. Reaching and hand eye co-ordination. In: Henderson, A., Pehoski, C. (Eds.), Hand Function in the Child. Mosby, St Louis, pp. 81–92.
Sarrafian, S.K., 1983. Gross and functional anatomy of the shoulder. Clin. Orthop. Relat. Res. 173, 11–19.
Speer, K.P., Garrett, W.E., 1993. Muscular control of motion and stability about the pectoral girdle. In: Matsen, F.A., Fu, A.H., Hawkins, R.J. (Eds.), The Shoulder: A Balance of Mobility and Stability. American Academy of Orthopaedic Surgeons, Rosemont, pp. 159–172.
Strandring, S. (Ed.), 2005. Gray's Anatomy, 39th ed. Elsevier, Edinburgh.
Strickland, J.W., 1995. Anatomy and kinesiology of the hand. In: Henderson, A., Pehoski, C. (Eds.), Hand Function in the Child. Mosby, St Louis, pp. 16–40.
Terry, G.C., Hammon, D., France, P., et al., 1991. The stabilising function of passive shoulder restraints. Am. J. Sports Med. 19, 26–34.
Tillary, M.I., Flanders, M., Soechting, J.F., 1991. A co-ordinated system for the synthesis of visual and kinaesthetic information. J. Neurosci. 11 (3), 770–778.
van der Helm, F.C.T., 1994. Analysis of the kinematic and dynamic behaviour of the shoulder mechanism. J. Biomech. 27 (5), 527–550.
Voight, M.L., Thomson, B.C., 2000. The role of the scapula in the rehabilitation of shoulder injuries. J. Athletic Train. 35 (3), 364–372.
Warner, J.P., Lephart, S., Fu, F.H., 1996. Role of proprioception in pathoetiology of shoulder instability. Clin. Orthop. Relat. Res. 330, 35–39.
Weber, S.C., Caspari, R.B., 1989. A biomechanical evaluation of the restraints to posterior shoulder dislocation. Arthroscopy 5, 115–121.
Wilk, K.E., Arrigo, C.A., 1993. Current concepts in the rehabilitation of the athletic shoulder. J. Orthop. Sports Phys. Ther. 18 (1), 365–378.
Wilk, K.E., Arrigo, C.A., Andrews, J. R., 1996. Closed and open kinetic chain exercise for the upper extremity. J. Sport Rehabil. 5 (1), 88–102.
Williams, P.L., 1995. Gray's Anatomy, thirty eighth ed. Churchill Livingstone, New York.
Zivani, J., 1995. The development of graphomotor skills. In: Henderson, A., Pehoski, C. (Eds.), Hand Function in the Child. Mosby, St Louis, pp. 184–193.

Chapter **11**

Function of the lower limb

Tony Everett and Marion Trew

CHAPTER CONTENTS

LEARNING OUTCOMES

When you have completed this chapter, you should be able to:

1. recognize the normal patterns of movement for rising from a chair
2. discuss the importance of walking
3. identify the characteristics of normal gait
4. describe the gait cycle
5. recognize the joint movements and muscle activity that occur in normal gait
6. recognize the normal patterns of movement for ascending and descending stairs.

INTRODUCTION

Two of the main functions of the lower limbs are to support the body when standing and to enable locomotion. To be able to stand and move is a key function of an active life, and it is important to have a detailed understanding of the normal function of the lower limbs in order to plan purposeful activity or rehabilitation programmes. In this chapter, getting out of a chair, walking and stair climbing are considered in some detail, as they represent major functions of the lower limb. There are interesting similarities in the patterns of movement between these apparently different activities, and these patterns are repeated in other activities not covered by this book. Although this chapter is primarily concerned with lower limb function, it is impossible to ignore the movement patterns that occur in the rest of the body during sitting, standing and stair climbing and the motivation for performing these movements. Therefore some consideration will be given to patterns of movement in the upper limbs and spine.

MOVING FROM SITTING TO STANDING

CASE STUDY 11.1

The family have just finished their evening meal, and they are about to leave the table and carry out their varied evening activities. The chairs are standard dining chairs with no arms. Imagine how each member of the family would get out of the chair. Think of the members of your own family across the generations. Work out what abilities they need (joint range, muscle function and balance) in order to stand up in all the variety of getting up you may see. Are you able to think of any differences in the way older and younger people rise from chairs? Would the type of chair make a difference? Compare your thoughts with those given in Case study 11.2.

The ability to rise from sitting to standing is essential for the achievement of many everyday activities. It is the key that opens the door to movement, and without this ability, standing and locomotor activities are virtually unattainable. There are patterns of joint movement and muscle activity during rising from a chair that are normally stored as motor programmes in the motor cortex of the brain. Consequently, for the majority of the population, getting out of a chair is an automatic activity requiring no thought. It is only when the chair is particularly low or deep, or when the individual has a physical problem or is feeling tired or weak, that the activity requires conscious thought in order to complete. Varying the seat height or depth will change the range of lower limb joint movement and the amount of energy, but the pattern of movement remains essentially the same (Janssen et al., 2002). You may have thought of some of these while considering the activity in the case study above.

The movement of sitting to standing can be divided into two phases – a seated phase and a stance phase – both of which occur whether or not armrests are used (Fig. 11.1). In the seated phase, the subject prepares for standing by adjusting the position of the limbs and trunk to cause the centre of gravity to move forwards until it is almost over the feet. In the stance phase, weight is taken through the lower limbs as the centre of gravity is transferred forwards and upwards. There are two main subdivisions to the stance phase: first, the transfer component as the centre of gravity is transferred forwards until it is slightly in front of the ankle joint, and then the extension component when the lower limbs extend until the erect posture is achieved.

The period of time needed to complete the seated and stance phases is very variable but takes on average 1–3 s (Kerr et al., 1991; Baer and Ashburn, 1995). The seated phase consists of about 30% of the total movement time, and in the stance phase the transfer and extension components take 20% and 50%, respectively.

SEATED PHASE

Usually, people sitting in a relaxed manner lean against the back of the chair with their centre of gravity well behind their feet. This is a very stable position, as the whole chair forms the base of support. As people stand, the feet become the base of support, and for balance to be achieved, the centre of gravity must be moved horizontally until it is directly above them. The seated phase has two important components: first, that of positioning the body correctly in anticipation of weight transfer from the seat to the feet; and second, the development of momentum to assist the muscles in achieving the upright position.

The ideal preparatory position for standing up is with the knees flexed between 90 and 115° so that the feet are placed under or slightly behind the knee joints. By having the feet well back, energy is conserved because the horizontal distance that the centre of gravity has to travel is minimized. Conversely, if the feet are placed too far back the knee extensors will be in a lengthened position and may not be at their optimum length to generate force. It is suggested that the feet should not be placed more than 10 cm behind an imaginary perpendicular line dropped from the knee (Ikeda et al., 1991; Shepherd and Koh, 1996). If the person has been sitting in a chair with the knees relatively extended, then a substantial amount of knee flexion has to occur before a successful attempt at standing up can begin. This usually occurs simultaneously with the initiation of forward movement of the trunk; as the feet move backwards, the head moves forwards.

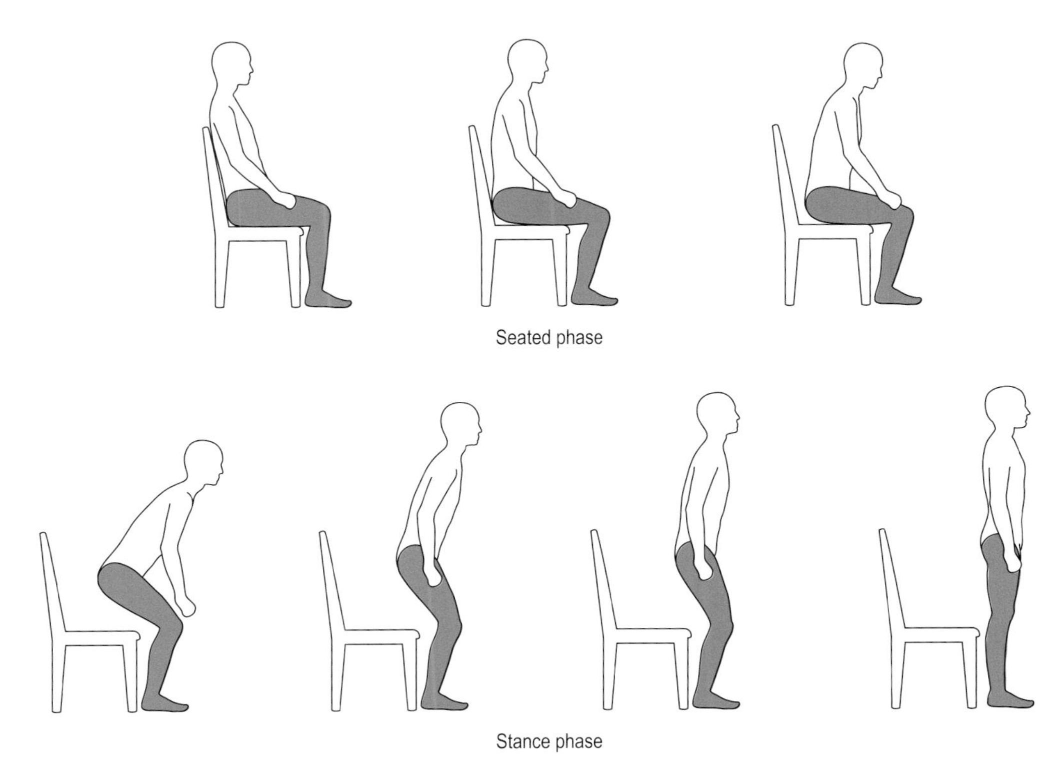

Figure 11.1 The pattern of movement when rising from a chair.

By the start of the stance phase, the centre of gravity needs to be about 2 cm anterior to the ankle joints. To achieve this position, the hips have to be flexed to about 120°, and as the hips flex, the head and trunk move anteriorly (Shepherd and Koh, 1996; Mak et al., 2003). In normal subjects, there is an initial contraction of the hip flexors in order to initiate the movement, and the anterior abdominal muscles contract isometrically to ensure that the trunk follows the hip movement. Once the hip joints have passed 90°, the flexors relax and momentum and the force of gravity cause the movement to continue. Towards the end of the hip flexion phase, there may be slight eccentric activity of the hip extensors to control the forward movement (Kelly et al., 1976; Ikeda et al., 1991). In obese people in whom the flesh of the abdomen comes into contact with the thighs, or in people who have 'stiffness' of the hip joint, the activity of the hip flexors will need to continue throughout the movement. In these cases, considerable effort may be needed to achieve the required amount of hip flexion. Some further knee flexion may occur at the end of this phase, and this will in turn cause slight passive dorsiflexion.

Very little motion is seen in the joints of the vertebral column. Schenkman et al. (1990) reported between 0 and 16° of trunk flexion in this phase, and Baer and Ashburn (1995) found negligible trunk lateral flexion and rotation. The greatest ranges of movement occur in the cervical spine, which appears to have two functions. Initially, in the seated phase, it flexes, carrying the head forwards as part of the process of developing the momentum that will translate the centre of gravity over the feet. Then, as the seated phase comes to an end and the stance phase commences, the function of the cervical spine appears to be to keep the vertex of the skull uppermost and the eyes horizontal. A pattern of movement at the cervical spine is identifiable, and the overall range of movement is, on average, about 30° (Fig. 11.2).

If the upper limbs are not being used to push up from the chair arms, they usually flex between 11 and 53° at the shoulder joint. This combination of trunk and shoulder movement serves to move the centre of gravity forwards and also provides horizontal momentum that will contribute to the transfer of body weight on to the feet and be

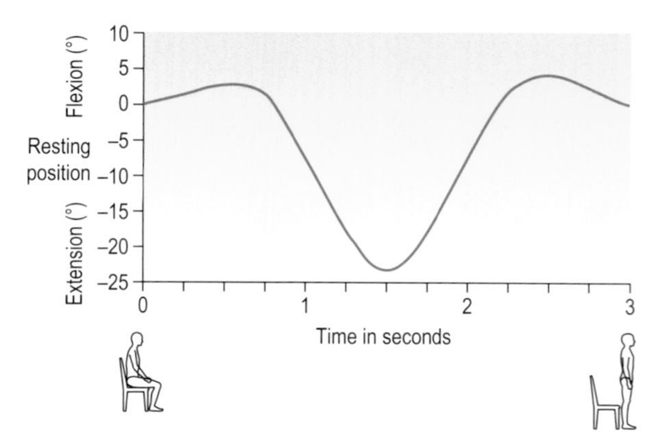

Figure 11.2 The pattern of cervical flexion and extension when rising from a chair.

translated into a vertical movement as the lower limbs extend (Riley et al., 1991).

The seated phase ends with the lift-off from the chair. There is considerable variation in the degree of flexion at the hip and shoulder joints during this phase and also in the exact position of the head. The velocity of the movement also varies between individuals and under different circumstances. The height of the chair relative to the subject's leg length, the age of the subject and the degree of joint mobility and muscle strength all have effects on the velocity of movement (Kerr et al., 1991). Despite this variability, there is an optimum speed for rising from a chair. Too fast is likely to induce loss of balance and require excessive muscle force to prevent an unwanted step or fall. If the speed is too slow, the momentum needed to assist the transfer of weight from the seat to the lower limbs will not be achieved and either the activity will fail or excessive amounts of energy will be required for successful completion. In older people, the ability to develop momentum is often lost, and when this is combined with age-related joint stiffness and muscle weakness it is quite obvious why they find rising from a chair very difficult.

ACTIVITY 11.1

Here are three small tasks that will enable you to test the importance of transferring the centre of gravity over the feet before standing up.

1. Sit comfortably in an easy chair. Move your bottom slightly towards the front of the chair, lean back and relax. Now try and stand up without leaning forwards; you may use your arms if you wish.
2. To make it easier, now sit upright towards the front of the chair; do not lean back against the chair. Your hips should be at a right angle. Concentrate hard on your hip angle and see if you can stand up without allowing that angle to decrease (without leaning forwards).
3. Finally, sit well back on the chair. In this position, look carefully at how much flexion you have at your hip joints. If you lean forwards, your centre of gravity will translate anteriorly, so you can use changes in hip angle to judge if you are leaning forwards. So stand up naturally and note, or get a friend to note, what happens to your hip angle in a normal movement.

STANCE PHASE

The stance phase can be subdivided into transfer and extension components.

Transfer

This begins on lift-off from the seat and continues until the centre of gravity is about 7 cm anterior to the ankle joints, where it will remain until the erect posture is achieved (Kelly et al., 1976; Ikeda et al., 1991). The transfer component takes about 20% of the total time needed to rise from a chair and is completed in advance of hip and knee joint extension. There is peak quadriceps and hip extensor activity at the instant of lift-off when the knees and hips extend to raise the body off the seat. Dorsiflexion of the ankle joints reaches its maximum during this phase.

The trunk and upper limbs continue the horizontal movement begun in the seated phase, and the momentum developed in the seated phase carries the centre of gravity forwards until it is substantially in front of the ankle joints. As the trunk moves forwards, the cervical spine extends through a range of about 25° in order that the head can be positioned appropriately and the eyes can look straight ahead (see Fig. 11.2).

The forward transfer of weight is completed relatively quickly, but it is vital to the whole process, as standing is not possible until the centre of gravity is positioned over the feet. It is a time of peak muscle activity in the knee and hip extensors and also a time of instability, because the body is no longer supported by the seat and the centre of gravity is moving forwards. Failure to complete the transfer phase is one of the main reasons why people are unable to stand up from a chair.

Extension

Once the centre of gravity is positioned over the feet, vertical movement replaces horizontal. Extension of the hips and knees begins in earnest, and there is also slight plantar flexion of the ankle joints. As the joints become progressively extended, the trunk also starts to extend and the cervical spine flexes to keep the vertex of the skull uppermost. The upper limbs relax and return to their normal resting position. The very last part of this activity again involves the cervical spine. Once the lower limbs and the trunk are vertical, there is a small phase of flexion and extension in the cervical spine as final head position is adjusted.

Rising from a chair is an important functional activity that requires greater joint range and muscle torque than walking and most stair climbing (Kelly et al., 1976). To be successful, it is necessary to have more than 100° of flexion at the hip and knee joints and virtually full-range dorsiflexion. It is also necessary to have good balance, because the centre of gravity is moving in relation to the base of support. If the relationship of the centre of gravity to the feet is not correct, the individual is likely to lose balance and fall. Flexibility of the cervical spine is particularly important, because balance requires correct head positioning, and if the vertebral column is stiff, then the appropriate positions may not be achieved.

The individual must also have the strength and the balance ability to complete the transfer phase, which is arguably the most crucial part of the whole process.

FORCES REQUIRED TO RISE FROM A CHAIR

The energy requirements for rising from a chair are higher than for walking, because the muscles have to raise most of the body weight vertically through a distance of about a metre. The greatest force production is found immediately on lift-off from the seat, with quadriceps torque being around 1.1 Nm/kg/m and hip extensor torque approximately 0.9 Nm/kg/m (Mak et al., 2003). Differences have been found between the forces generated by the hip flexors and the ankle dorsiflexors in younger and older adults. It appears that younger adults produce substantially more hip flexor torque in the seated phase than do elderly subjects. This may be explained by the fact that younger adults have good balance and are comfortable with generating substantial momentum when rising from a chair, as they are not in danger of falling. Older people not only have poorer balance but also have a fear of falling and may tend to avoid rapid movements when they are changing position (Pai and Rogers, 1991; Mak et al., 2003).

CASE STUDY 11.2

Having read the text and completed the activities, you should now be able to explain what is happening during this complex activity of standing up from a chair. It is essential to have an intact neurological system for both the automatic and the conscious movements. The musculoskeletal system must be functioning within a coordinated system with joint range, muscle strength and flexibility, and joint proprioception being in a good functional state. You should now be able to explain why the family would get up at different speeds with differing levels of balance and stability. You should also be able to explain why differing environmental factors would affect the ease of standing. If you have any difficulties with these questions, reread the above section after reading Chapters 2, 3, 5 and 13.

It is also worth bearing in mind that for older people a seat height of no less than 120% of leg length is recommended, although it should not be forgotten that some fit older people can rise from the squatting position.

Whether the upper limbs are involved in the process of getting out of a chair depends on the strength of the individual, the height of the chair and the presence of armrests. Under normal circumstances, the upper limbs are not essential to the activity and can be used for carrying or manipulating objects during the activity of standing up or sitting down. However, if weakness, balance problems or pain are factors, then the upper limbs will be used to assist. It is estimated that the force production by the hip and knee extensor muscles can be reduced by about 50% if armrests are used (Janssen et al., 2002).

WALKING

Human beings can perform many types of locomotion including walking, running and, less commonly, crawling, hopping, jumping and even rolling. The diversity of locomotion is even greater when you consider that each of these movements can be performed in a variety of ways and directions. Despite this variety, all these methods of locomotion have common patterns of movement, and by studying 'normal' walking in detail it becomes easier to understand the others.

Walking is a highly energy-efficient method of progression involving rhythmical, reciprocal movements of the lower and upper limbs when one foot is always in contact with the floor. People normally walk for a purpose, perhaps because they want to reach a certain place at a certain time, but they may also walk for pleasure and for health.

Although to most people walking is fully automatic and requires no thought, it actually comprises complex patterns of movement involving all body segments, particularly those of the lower limb. In addition, there is movement of the joints of the vertebral column, from lumbar to cervical spine, and when unrestricted, the upper limbs swing in a reciprocal pattern. This involvement of all the body segments requires considerable neural control, and this explains why at birth, when the nervous system is not fully developed, walking is impossible. It is not until the infant has gained control over all body parts and is able to balance that the first uncertain steps can be taken. Even then, the child is unable to walk while carrying objects and may be 7 or 8 years old before the activity becomes mature and fully automatic (see Ch. 13 for a full explanation of the development of human movement).

In the healthcare and other professions, it is quite common for the term *gait* to be used in preference to walking. Gait means the manner or way in which walking takes place and implies a detailed consideration of the kinetics and kinematics of the activity.

TERMINOLOGY OF GAIT

Walking is a complex activity, and if it is to be fully understood it needs to be broken down into phases. There are a number of ways in which gait can be described, and the most commonly used international terminology is given here.

GAIT CYCLE

This is the period of time during which a complete sequence of events takes place. While it is usual to consider the gait cycle as beginning when the heel of one foot strikes the floor and continuing until the same heel strikes the floor again, it may be measured from any moment in the gait cycle.

As can be seen from Figure 11.3, the gait cycle is subdivided into a *stance phase* and a *swing phase*, and these terms describe the periods of time when the foot is either in contact with the floor or swinging forwards in preparation for the next step.

Stance phase

The stance phase is the period of time when the limb under consideration is in contact with the floor. In walking, there is always a period of time when both feet are in contact with the floor simultaneously, and this is called *double stance* (see Fig. 11.3).

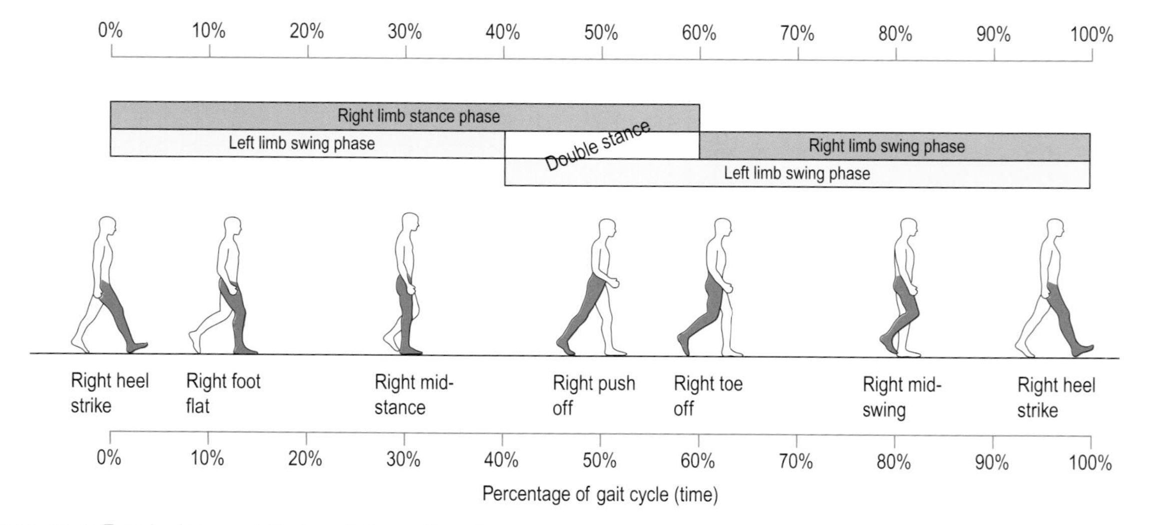

Figure 11.3 Terminology and timing of the gait cycle.

The stance phase is the most complex and arguably the most important phase of gait. During the stance phase, the lower limb has to provide a semi-rigid support for the body weight, facilitate balance and allow forward propulsion. The stance limb also has a role in compensating for uneven ground, and when positioned correctly it enables an accurate swing phase on the contralateral limb to take place. The stance phase can be subdivided into the following stages.

Heel strike

In normal walking, the leading limb initially contacts the floor by heel strike. At the moment of heel strike, the following limb is also in contact with the floor, giving a position of double stance. This is the moment when the whole body centre of gravity is at its lowest and the walker is most stable (Fig. 11.4).

Foot flat

Immediately on coming into contact with the floor, the stance limb takes the body weight. To be effective in supporting the body weight, the foot has to move rapidly to plantigrade from the position of dorsiflexion at heel strike. At foot flat, the whole foot comes into contact with the floor, allowing it to accept the weight of the body as the mid-stance phase takes place. During heel strike and foot flat, there is rapid loading of the limb, and it is important that during this phase strategies exist to absorb the sudden imposition of ground reaction forces.

Mid stance

In mid stance, the body is carried forwards over the stance limb and the opposite limb is in the swing phase. The whole body centre of gravity passes from behind to in front of the stance foot, and it is in mid stance that the centre of gravity rises to its highest position in relation to the supporting surface (Fig. 11.4). This is the position at which the walker is least stable because of the small base of support and the relatively high centre of gravity.

Heel off

This is when the heel is raised off the floor.

Toe off

This is the point just before the foot is raised from the floor.

The above two stages happen in quick succession and are designed to propel the body forwards and terminate the stance phase. Initially, the heel lifts off the ground, an event that is passive in slow gait but may require a small degree of muscle activity from the plantar flexor muscles at faster velocities. This is usually followed by a propulsive stage when the same muscles contract to plantar

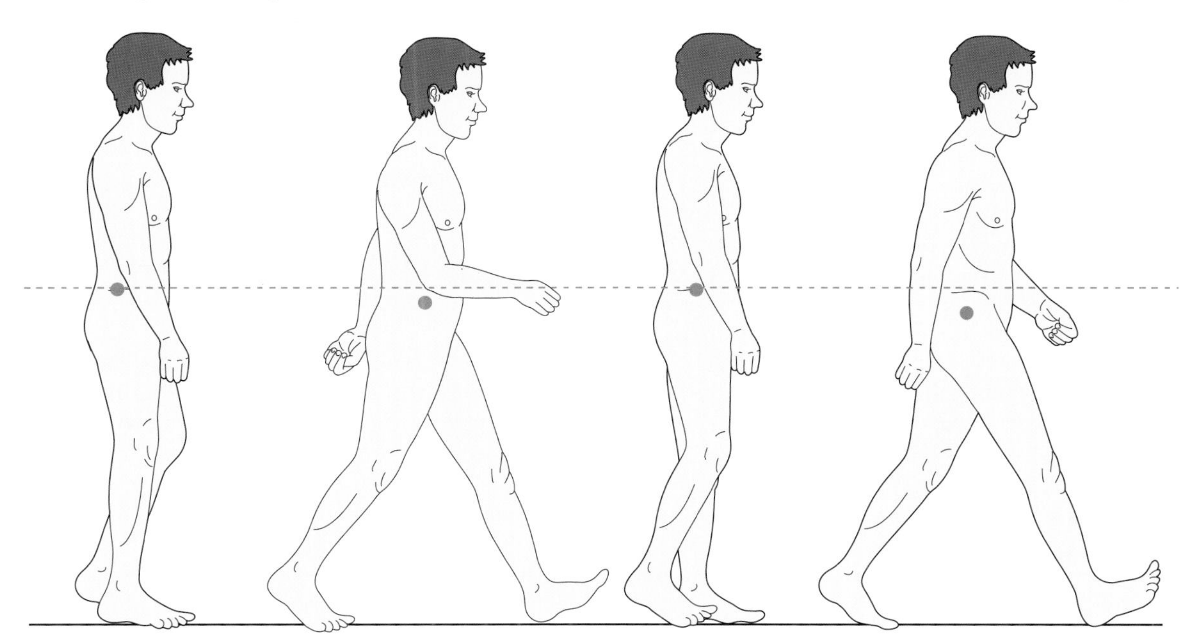

Figure 11.4 Vertical displacement of the centre of gravity when walking with an extended stride.

flex the forefoot against the floor, and this is called *push-off*. Finally, there is the moment of *toe off* when propulsion ends, the contact between the toes and the floor is lost and the swing phase for that limb starts.

In some literature, the subphases are given differing terminology. *Initial contact* correlates with heel strike, as described above. This subphase will also be described as the *deceleration phase* or *first rocker*. Flat foot may be described as the *loading phase, mid stance* or the *second rocker*. Heel off and toe off may be alternatively described as the *propulsive phase, push-off phase* or *third rocker*.

Swing phase

The swing phase is the period of time when the limb under consideration is not in contact with the floor.

The swing phase can be divided into three stages.

Acceleration

The force generated by the hip flexors, and to a lesser extent by the plantar flexors, accelerates the non–weight-bearing limb forwards.

Mid swing

This corresponds with mid stance, and it is at its shortest at the moment the swing phase limb passes the stance limb.

Deceleration

In this final stage of the swing phase, the lower limb muscles work to decelerate the swing limb in preparation for heel strike. The muscle action in this phase is usually eccentric and requires less energy than those times in the gait cycle when concentric activity is needed to accelerate a limb.

During the swing phase, the swinging limb moves in front of the stance limb so that forward progression may take place. In order to swing successfully, the limb must be shortened sufficiently to enable the foot to clear the ground, and this is normally achieved by flexion of the hip and knee joints and dorsiflexion of the ankle. Clearing the ground is key to a successful swing phase, but to conserve energy it is important that the limb is not lifted further than is necessary. In normal adult gait, the average clearance of the foot from the ground in mid-swing phase is around 2 cm, and with so little leeway for error it is quite remarkable that people do not catch their feet on the floor more often.

At an average speed, the stance phase takes about 60% of the gait cycle and the swing phase about 40% (Murray, 1967), but the relative percentages vary as the speed of walking increases or decreases. In slow walking, the stance phase can constitute more than 70% of the gait cycle, with the swing phase being less than 30%. As the velocity of walking increases, the length of time in the stance phase decreases until, on very fast walking, the stance phase may be reduced below 57% of the cycle (Smidt, 1990). The period of double stance also decreases with increasing velocity. When walking very slowly, the double stance may last as long as 46% of the total gait cycle; on very fast walking, the double-stance period may be reduced to 14%, and when walking develops into a run there is no double-stance period.

CASE STUDY 11.3

The family have met their friend Ken at the beach. Ken is recovering well from his motorcycle accident. They have met up to walk along the prom before crossing the beach to get to a café for a well-earned cup of tea. Increasingly, walking is being advocated as a safe and effective way of maintaining fitness, particularly in the later years of life (Arakawa, 1993; Hardman and Hudson, 1994; Pereira et al., 1998), and in many countries walking for pleasure and health is a popular pastime. Providing walking speeds of more than 6 km/h are achieved and hills are incorporated into the route, walking maintains reasonable ranges of lower limb joint motion and a functional level of cardiorespiratory fitness. You are an observer following the family and Ken. What do you think they will look like if you were analysing their gait? You may want to use the sequence of analysis you will find in the section on visual analysis in Chapter 14.

Before you move on from thinking about the activity posed in Case study 11.3, you may want to refresh your memory of the physical principles discussed in Chapter 9.

Basically, all people walk in the same way, with the lower limbs moving reciprocally to provide alternate support and propulsion, and if the upper limbs are unencumbered, they demonstrate a stereotyped pattern of reciprocal movement in phase with the lower limbs. The joint movements that occur in the sagittal plane (flexion and extension) are very similar in both range and direction of movement between individuals (Murray et al., 1964). The differences that set one person's gait apart from another's occur mainly in movements in the coronal and transverse planes. For example, the amount of hip rotation and therefore foot angle can vary dramatically between individuals, and noticeable variation also occurs in trunk lateral flexion. The range of trunk rotation in the transverse plane is variable, with some people having almost imperceptible rotation and others rotating through such a wide range that they appear to be swaggering (Murray et al., 1964; Smidt, 1990).

Walking is a smooth, highly coordinated, rhythmical movement by which the body moves step by step in the required direction. The forces that cause this movement are a combination of muscle activity to accelerate or decelerate the body segments and the effects of gravity and momentum.

Walking has often been described as a fall followed by a reflex recovery of balance, and to a certain extent this is true. To initiate the first step, the anterior muscles at the ankle contract to move the centre of gravity forwards in relation to the feet, and this causes a loss of balance anteriorly. Once the centre of gravity has been displaced, gravity and momentum continue the movement initiated by the muscles. In order to avoid a fall, there is a reflex stepping reaction that causes one of the lower limbs to be moved forwards so that one foot can be placed in front of the other. This alters the base so that the centre of gravity is once again above the feet. If walking is to continue, the centre of gravity must again be displaced anteriorly. However, once inertia has been overcome by muscle action on the first step, all subsequent steps benefit from the momentum accrued and only a minor amount of propulsion comes from the calf muscles. For each step, balance is disturbed so that a subsequent reflex step will occur. This continues until the purpose of walking has been achieved, and because the momentum involved in walking removes the need for much muscle activity, the whole process is remarkably energy-efficient.

Once a steady walking velocity has been reached, the actual process of walking requires low energy expenditure, and it is possible to walk for long periods of time with surprisingly little fatigue (Smidt, 1990; McArdle, et al., 2001). As would be expected, the least energy is expended when walking at a moderate speed on level ground, but an increase in velocity or a change from a firm surface to a surface such as soft sand would immediately increase energy expenditure. There are two other instances in walking on level ground when energy expenditure is relatively high. Energy expenditure is higher on the initiation of movement, when inertia has to be overcome so that the body weight can be displaced forwards, than it is during a continuous sequence of steps. Conversely, at the end of walking an increase in energy expenditure is required to stop forward movement of the limbs and trunk. The faster an individual walks, the more difficult it is to stop suddenly, and this causes a greater expenditure of energy.

Understanding the role of inertia and momentum in walking is important, as the energy-efficient nature of the activity is lost if momentum is restricted and the walker is constantly having to overcome inertia to either initiate or stop the movement. You are probably already aware of this, as you must have noticed that you feel disproportionately tired after a day wandering around the shops. When people go shopping, particularly when they are not strongly focused on what they need to buy, they constantly stop to look at goods and then move on. The repeated stopping and starting involved in shopping can be surprisingly tiring. Re-education programmes for weak people should aim to develop the rhythm of gait and the smooth swinging motion of the upper and lower limbs so that they can benefit fully from an energy-efficient gait. As they develop strength, then walking can be made more difficult by introducing stairs and slopes and by disrupting the normal rhythmical patterns.

Temporal and spatial components

It is useful to also consider gait in terms of both the temporal and spatial components. The temporal components are those periods of time during which events take place and are often measured in seconds. For example, the stance phase of walking is a temporal component and relates to the period of time that the foot is in contact with the floor. The spatial components refer to the position or distances covered by the limbs, and an example of this would be step length. When analysing gait, it is essential to consider both the temporal and spatial components because disease or trauma can affect either. The temporal components are illustrated in Figure 11.3.

The gait cycle contains a number of spatial components that are commonly measured as part of the analysis of gait (Fig. 11.5).

Stride length

This is the distance between successive foot to floor contacts with the same foot. For example, this might be the distance between the first point of contact of the right heel on the floor and the next point of contact of the right heel.

Step length

This is the distance between successive foot to floor contact with opposite feet. In this case, it could be the distance (in the line of progression) between the point of right heel strike and the point of left heel strike. There are two steps to every stride.

Step and stride length are dependent on several factors, including the length of the lower limb, the age of the subject and the velocity of walking. Short lower limb length, increasing age and decreasing velocity will all reduce the step and stride length. Restriction in the range of hip flexion and/or extension is also a common reason for reduced step and stride length.

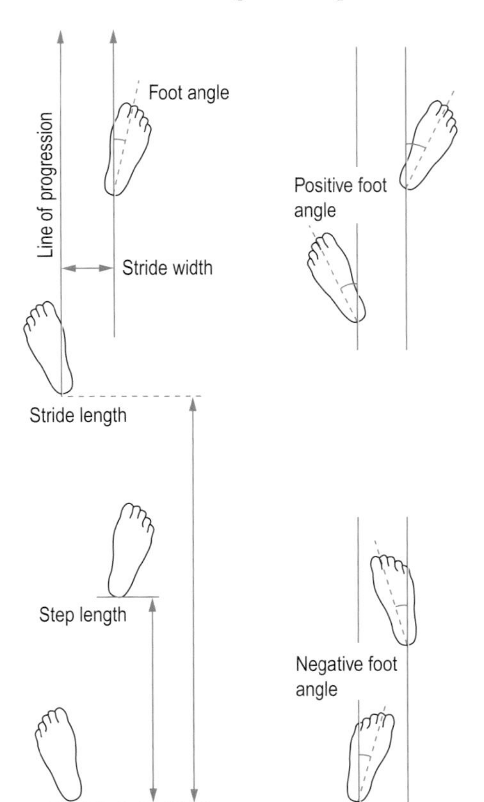

Figure 11.5 Characteristics of gait that can be measured from footprints.

Foot angle

This is the degree of in-turning or out-turning of the foot: if the foot turns in, there is said to be a negative foot angle; if the foot turns out, the angle is positive. The majority of the population walk with a positive foot angle of up to 30°. This angle is mainly associated with the degree of rotation at the hip joint and, to a lesser extent, the rotation between the tibia and the femur. In some cases, tibial or femoral torsion will influence foot angle.

Stride or step width

This is the distance between the two feet, and it is normally measured from the midpoint of the heels. This distance varies greatly between individuals and can vary between individual steps if the ground is uneven, but on average it is about 7 cm; however, when people have poor balance they tend to increase their stride width to give themselves a greater base of support. It is interesting to note that on slow walking the stride width tends to be greater than on rapid walking.

The time taken for the spatial events to occur can also be a usual objective measure when analysing gait.

Step and stride length

This is the time taken for the step and the stride to occur. It may also be possible to time the subphases of the gait cycle, such as stance phase and swing phase, although these are much easier with the use of film, as explained in Chapter 14.

Cadence

The term *cadence* is used to indicate the number of steps taken per minute. The cadence mainly depends on the velocity of walking. In slow walking, the cadence may be 40–50 steps/min, whereas moderate walking will cause an increase in cadence to around 110 steps/min, and this figure will rise with increasing velocity until running occurs. If a patient has pain, joint stiffness, muscle weakness or poor balance, the cadence will be reduced. An increase in natural cadence is often taken to indicate an improvement in a patient's walking ability, but it might be more appropriate to use the patient's ability to vary cadence as an indicator of walking skill.

Revisit Case study 11.3, adding your knowledge of temporal and special parameters. Does this make your analysis easier or more complete?

JOINT AND MUSCLE ACTIVITY IN THE STANCE PHASE

As shown in Figure 11.6, the pattern of joint movement is less complicated at the hip joint than at the knee or ankle joints. While the hip joint has only one phase of extension and one phase of flexion, the other two joints have two phases of each movement in each gait cycle. The range of movement at the hip and knee joints is more consistent between

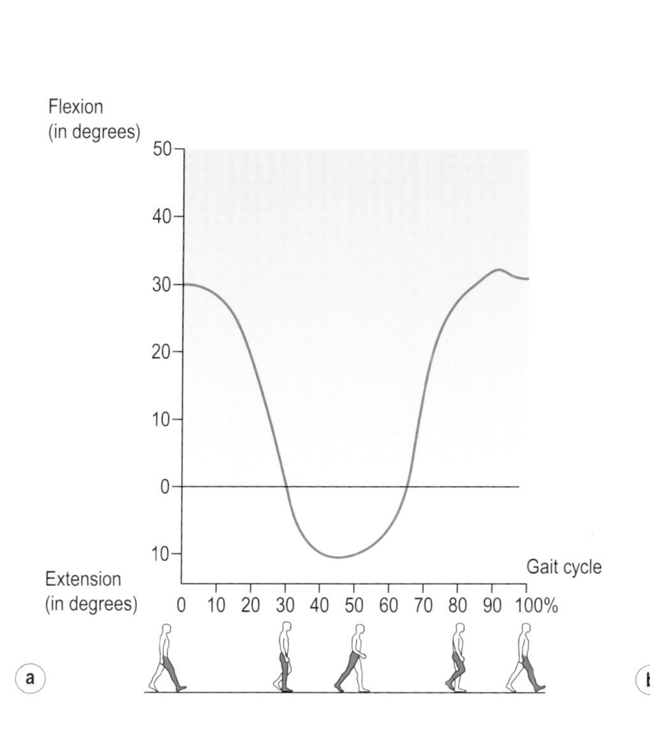

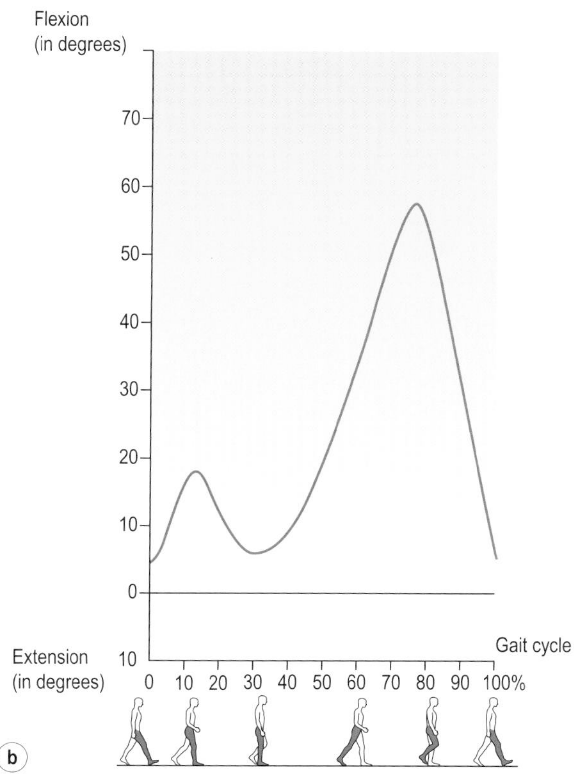

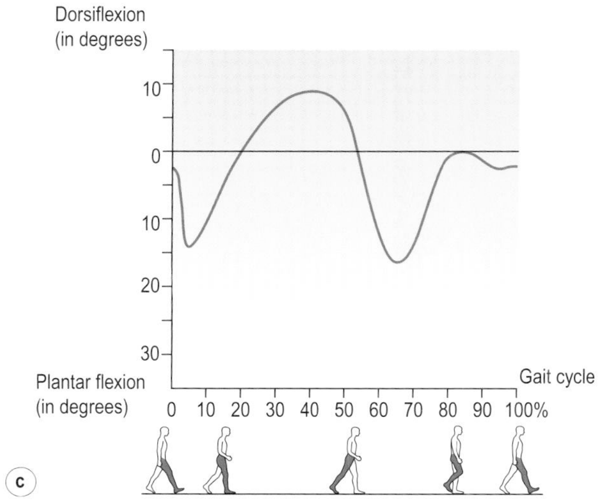

Figure 11.6 Pattern of (a) hip, (b) knee and (c) ankle joint movement during a single gait cycle (walking velocity, 3.5 km/h).

individuals, but the movement at the ankle joint can be quite variable between one person and another.

Muscle activity, as indicated by electromyography, shows variability between subjects and also when different walking velocities are chosen. A guide to the common patterns of major muscle activity is given in Figure 11.7; the data for this were gathered from subjects walking at a moderate pace. It is interesting to note that in a substantial portion of the gait cycle there is little or no muscle activity occurring in the majority of the muscle groups. This supports the theory that gait is energy-efficient.

Heel strike

At the instant of heel strike, the hip joint is partially flexed and gluteus maximus and the hamstrings contract immediately to initiate hip extension. The knee joint will be either in full extension or flexed to about 5°, and the quadriceps will be working eccentrically to control the knee flexion that follows immediately after heel strike. The ankle joint on heel strike is usually near the neutral position, although there can be a variation between individuals of up to 10° of dorsiflexion or plantar flexion. More than 10° of plantar flexion would be rare in a normal individual, as it would render heel strike difficult and leave the toes vulnerable to stubbing on the floor. This position is produced prior to heel strike by concentric action of the dorsiflexor muscles, and on heel strike there is an immediate change to eccentric activity to lower the forefoot to the floor. At the metatarsophalangeal joints, there is a similar pattern of activity with the muscles positioning the joints in extension ready for heel strike and then lowering the toes into floor contact.

Foot flat

The hip joint is beginning to move into extension by concentric action of the hip extensors, but the knee joint has flexed further in order to cushion the effect of heel strike and also to reduce the vertical displacement of the centre of gravity that would otherwise occur as the body passes over the stance limb. This knee flexion is controlled by eccentric work of the quadriceps group, and measurements of knee movement taken in our laboratory show it can often be as great as 30°. At the ankle joint, there is controlled plantar flexion to lower the foot to the floor, which is undertaken by eccentric work of the dorsiflexors. Without controlled plantar flexion, the foot would slap down uncomfortably, and this can be heard in some patients. As the foot achieves good foot to floor

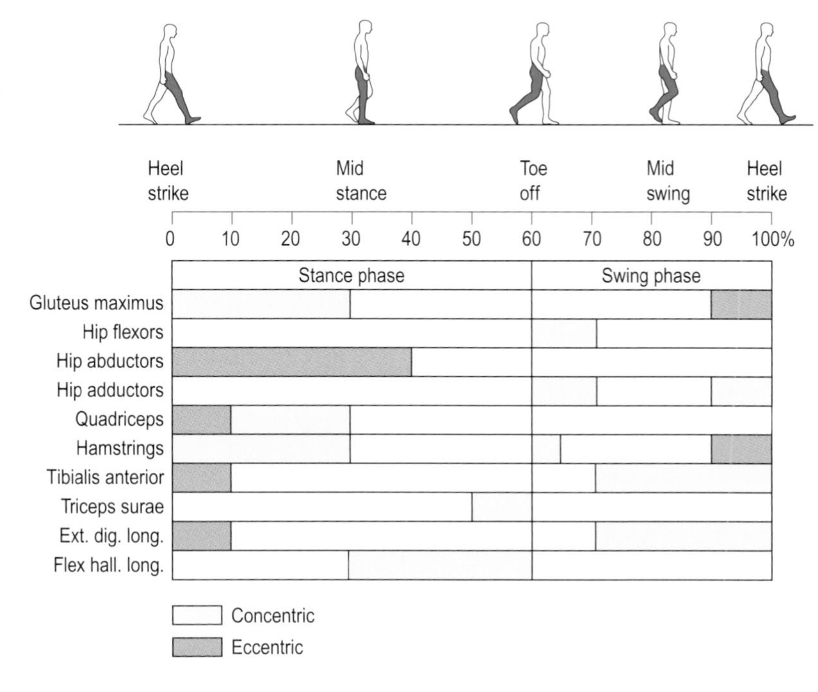

Figure 11.7 Muscle activity, as indicated by electromyography, is variable between subjects. It also varies with velocity: the faster the velocity, the more muscle input will be required. This figure shows the type and duration of muscle activity that might be expected in moderate velocity walking.

contact, there is a small amount of eversion to transfer the body weight across from the lateral border of the foot towards the great toe.

Mid stance

Hip extension continues, but by now it is being produced by momentum and the muscles are no longer active. While the contralateral limb is in the swing phase, the pelvis is unsupported on that side and the hip abductors on the stance limb initially contract to control pelvic levels and lower the pelvis towards the swing side by eccentric muscle action. The knee joint remains in slight flexion. There is sometimes a minor burst of activity in the ankle dorsiflexors to pull the tibia forwards over the foot, but once this movement has been initiated momentum and gravity will take over. Depending on the velocity of the movement, the calf muscles may need to exert a slowing influence on the tibia through eccentric muscle action.

Heel off, leading to push-off

At the beginning of this phase, the centre of gravity is in front of the stance foot, so the force of gravity will increase the range of hip extension and ankle dorsiflexion. As full-range dorsiflexion is reached, the heel will rise off the floor and the plantar flexors will contract concentrically to provide the propulsive component of push-off. In slow to moderate walking velocities, this contraction is not usually very large, as momentum is the major factor in moving the body forwards. In normal walking, the hip extensors are only slightly active at this stage, and in fact the hip and knee joints are usually starting to flex in preparation for the swing phase. At higher walking velocities, the propulsion comes increasingly from the hip extensors, with the plantar flexors playing a major role in ankle stabilization (Riley et al., 2001).

JOINT AND MUSCLE ACTIVITY IN THE SWING PHASE

Acceleration

Minor forces generated on push-off by the hip flexors and the plantar flexors accelerate the limb forwards in the swing phase, assisted by momentum and gravity. The hip and knee joints are both flexing, and there is a rapid movement towards dorsiflexion to ensure that the toes do not catch on the floor.

Mid swing

In mid-swing phase, flexion of the knee and hip joints continues to keep the foot sufficiently raised to avoid the toes catching on the floor. At this stage, the foot may be lowered into slight plantar flexion.

Deceleration

The hip continues to flex, the movement being mainly produced by momentum, and the hamstrings act eccentrically to slow down the movement at the hip joint. The knee joint moves from flexion to extension; it is interesting to note that the quadriceps play no part in this movement. The whole of the lower limb is being moved forwards by flexion of the hip, and the resulting momentum causes knee joint extension. Towards the end of the deceleration phase, knee joint extension may have to be slowed down, and this is achieved by eccentric action of the hamstrings. In preparation for heel strike, the dorsiflexors contract quite strongly to ensure that the foot is in the optimum position for heel strike.

MOVEMENT IN THE TRUNK, SHOULDER GIRDLE AND UPPER LIMBS

It is possible to walk with little movement of the trunk and no movement in the upper limbs, but in these circumstances gait is awkward and tiring (Murray et al., 1967). In the double-stance phase, when one hip is flexed and the other extended, the pelvis is rotated away from the lead limb. This causes some rotation of the lumbar spine towards the lead limb, which in turn leads to slight rotation of the thoracic and cervical spine in the opposite direction. The rotations in the spine are compensatory mechanisms designed to keep the head facing forwards. Stabilization of head position in relation to the environment is an important factor in all lower limb activities. Normally, the head is held in a fairly constant position in relation to the environment in order that visual information can be easily processed and balance maintained through unambiguous signals from the visual, vestibular and somatosensory receptors. If there is a functional need for the head to move excessively during gait, then the head stabilization mechanisms can be overridden but balance may be compromised (Mulavara et al., 2002).

Reciprocal movements of the upper and lower limbs occur in unconstrained walking; for example, on heel strike the contralateral upper limb is in front of the body. Most movement occurs at the shoulder joint, with a lesser amount at the elbow, and normally the shoulder joint starts to flex or extend slightly before the same movement is seen in the elbow joint. The range of movement varies greatly between individuals, and in the same individual it will vary according to the velocity of walking. Whereas the patterns of movement in the lower limb in the sagittal plane are very similar between individuals, they are more varied in the upper limb (Murray et al., 1967). This is not surprising, as the movement of the upper limb is not essential for the process of walking and the range of movement of upper limb joints is likely to be affected by the momentum imparted by the lower limbs and the degree of trunk rotation. There are several reasons why the upper limb moves during gait. It has been suggested that arm swing may impart momentum through the trunk to the lower limbs; subjective reports from fatigued walkers indicate that they have reduced the effort of walking by deliberately increasing their arm swing. It is also possible that arm swing acts to correct overrotation at the lumbar spine (Murray et al., 1967).

GROUND REACTION FORCES IN GAIT

Whenever the foot is in contact with the ground, there will be vertical, anterior–posterior and mediolateral forces acting between the foot and the floor. Measurement of these forces using a force plate shows a consistent intersubject pattern of vertical and anterior–posterior forces. The mediolateral forces are tiny and show much more variability between individuals (Fig. 11.8). At the beginning and end of the stance phase, the vertical ground reaction forces are normally about 25% greater than body weight but much less than they would be in running. The anterior–posterior ground reaction forces are about 25% of body weight in each direction (Craik and Oatis, 1995; Farley and Ferris, 1998). With forces passing through the body on every step, there is the potential for injury. However, the relatively small forces involved, the structural mechanisms for shock absorption and the cushioning effect of knee flexion on initial foot–floor contact mean that repetitive strain injuries caused by ground reaction forces are rare in walking.

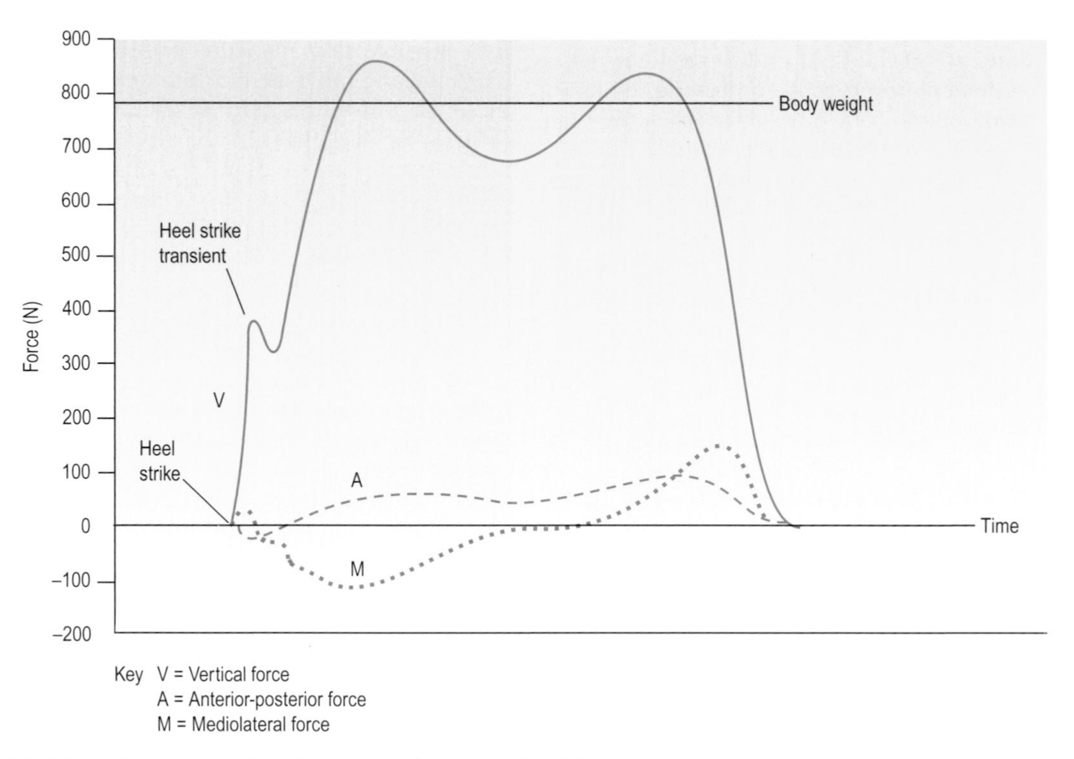

Figure 11.8 Typical ground reaction force traces for an adult walking.

ENERGY EXPENDITURE AND GAIT

The amount of energy required to walk at a comfortable pace is very small in comparison with other activities. Normal, easy walking on a firm surface requires about 0.080 kcal/min/kg. A gentle run requires 0.135 kcal/min/kg, swimming breaststroke requires 0.162 kcal/min/kg and playing squash 0.212 kcal/min/kg (McArdle et al., 2001)

Alterations in the normal structure of the human body or in the normal pattern of gait are likely to increase energy expenditure significantly. In normal gait, energy is conserved through the utilization of momentum and kinetic and potential energy. There are rhythmical, vertical fluctuations of the centre of gravity in normal gait that produce similar fluctuations in kinetic energy and gravitational potential energy. In mid stance, when the centre of gravity is at its highest, gravitational potential energy is also at its highest and kinetic energy at its lowest. This situation is reversed during double stance, and it is suggested that energy transfer mechanisms from the kinetic and potential energy contribute substantially to the overall energy requirements of gait (Farley and Ferris, 1998). Figure 11.4 illustrates the vertical movements of the centre of gravity.

RUNNING

Running differs from walking in that there is no double-stance phase and there is a period when there is no foot to floor contact at all. In general terms, running and walking are very similar, although in running the movement is much quicker and the stride is lengthened. The vertical ground reaction forces generated are of the magnitude of 2.75–3.00 times body weight, which is considerably greater than in walking, and heel strike may be replaced by toe strike (Farley and Ferris, 1998).

The trunk remains upright, although the line of gravity falls further outside the base than in walking. Movement of the upper limbs becomes essential, and the elbows are more flexed than when walking.

WALKING BACKWARDS

Although it is uncommon for anyone to walk backwards for any great distance, it is necessary in everyday life to be able to take one or two steps in that direction. The pattern of joint movement is very similar to normal gait, but the step length is reduced and heel strike is replaced by toe strike (Vilensky et al., 1987).

ACTIVITY 11.2

This task will enable you to experience normal energy-efficient gait and compare it with disrupted gait, when kinetic energy, potential energy and momentum are not used to their best effect.

Find somewhere where you can walk in a straight line for no less than 20 steps. Walk moderately quickly up and down your chosen area a few times, noting how much effort you need to complete this task. Now repeat the process, but every third step you must stop completely and then restart at a quick pace. Make a comparison between the effort needed in the second task and that in the first task. If you walk far enough, you may be able to measure differences in heart and respiratory rate between the two tasks.

WALKING UP AND DOWN STAIRS

This is a modified walking activity using similar patterns of joint movement and muscle action. There is a stance phase, a swing phase and a period of double support (Fig. 11.9). The activity is much more demanding than walking, because the range of hip and knee joint movement is greater and there is considerable vertical translation of the centre of gravity (Andriacchi et al., 1980; McFadyen and Winter, 1988). The tibiofemoral forces, patellofemoral forces and anterior–posterior shear forces are considerably higher than in walking, and therefore robust joint structures are needed if the activity is to be safely and painlessly performed (Costigan et al., 2002). Stair ascent and descent have greater potential for falls than walking has. In both these activities, there is a single-stance phase when the body is in the most vulnerable position because the base of support is at its smallest. In stair activity, substantial vertical translation of the centre of gravity also occurs in single stance, and this requires considerable muscular force. To achieve the single-stance component of stair activity requires good balance ability, and people whose balance is already challenged may be unable to go up and down stairs unless they have a banister for support.

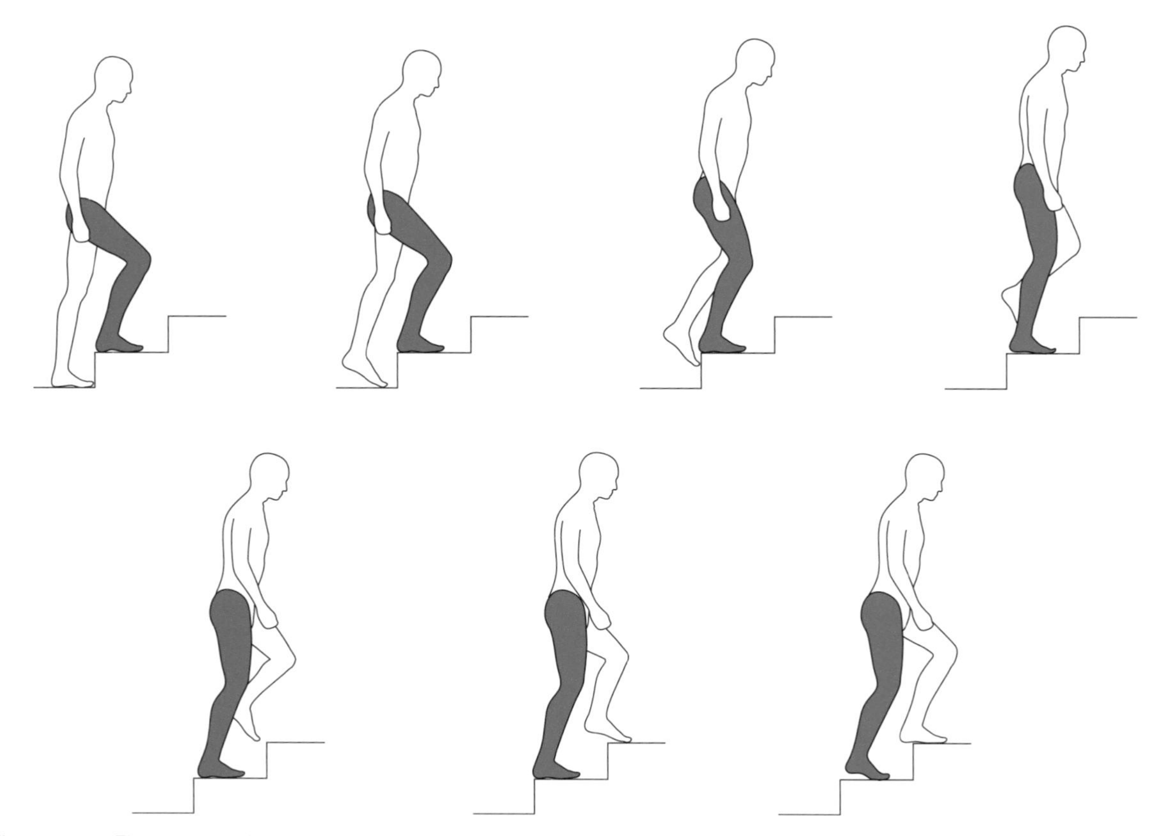

Figure 11.9 The pattern of movement in stair climbing.

When ascending stairs, the cycle is normally described as starting when the joints flex to place the foot on the step above. This is then followed by joint extension, and the muscle activity is predominantly concentric. On stair descent, the reverse pattern is seen. Most of the muscle activity is eccentric and the cycle is normally described as starting with the extended hip and knee positioning the foot on the step below, after which the joints move into flexion.

STANCE PHASE ON ASCENT

This phase is sometimes referred to as *pull-up*, and it starts from the moment of foot contact on the step above. It is normally as long as or longer than the stance phase in flat walking and has been reported to be around 60% of the cycle (Riener et al., 2002). In walking, the first instance of foot to floor contact is through the heel, but in stair ascent weight is initially taken on the anterior and middle third of the foot and then transferred to the remainder of the foot in readiness for full weight bearing.

On weight acceptance, there is strong concentric contraction of the hip and knee extensors to extend the lead limb and raise the body up to and over the step. The forces generated are substantial. The quadriceps generate force at a level of around 1 Nm/kg/m, the greatest force occurring when the knee is flexed to about 60°. The vertical force through the tibia and also the compression force between the patella and the femur are estimated to be around three times body weight and may possibly be double that amount in some subjects (Costigan et al., 2002). Gastrocnemius and soleus also work during the stance phase, moving the tibia posteriorly on the talus. As the single-support phase is entered, the hip abductors on the stance limb work strongly to prevent the pelvis dropping to the unsupported side and to pull the trunk laterally over the supporting limb.

In the latter part of the stance phase, when body weight is fully on the stance limb and the knee extended, the quadriceps work isometrically to maintain joint position while moving the centre of gravity in front of the stance foot. In some subjects, the dorsiflexors undergo a low-magnitude contraction at this stage to facilitate the movement of the centre of gravity forwards.

In the final stages of the stance phase, there is plantar flexion produced by strong contraction of gastrocnemius and soleus to accelerate the body forwards and upwards on to the new weight-bearing limb (Riener et al., 2002). At this stage, there is minimal activity in the knee and hip extensors (Andriacchi et al., 1980; McFadyen and Winter, 1988).

SWING PHASE ON ASCENT

The swing limb must swing past the intermediate step and over the top step, on which its stance will occur before the foot can be placed on that step. For this to occur, there has to be flexion of all the major lower limb joints, involving concentric work of the hip and knee flexors and the dorsiflexors. Early in the swing phase, the hip joint flexes and the hamstrings flex the knee joint to pull the leg and foot posteriorly to achieve intermediate step clearance.

By mid swing, the knee flexors are no longer contracting because the hip joint is sufficiently flexed to ensure intermediate step clearance. At this stage, there may be some eccentric work of the quadriceps to control unwanted knee flexion.

In the later swing phase, the hamstrings contract again to increase knee flexion so that the foot clears the top step, where it will eventually be placed. In order to avoid the foot catching on the top step, the amount of hip and knee flexion is quite extensive, and this results in the foot being well above the step immediately before foot to step contact. To gain step contact, the foot has to be lowered on to the step, and this is achieved by slight hip extension controlled by eccentric activity of the hip flexors.

Through most of the swing phase, tibialis anterior works isometrically to hold the ankle joint in dorsiflexion so that the toe will not stub on the steps. Immediately before foot contact, the dorsiflexors work eccentrically to lower the forefoot on to the step ready for weight acceptance on the forefoot (Andriacchi et al., 1980; McFadyen and Winter, 1988).

STANCE PHASE ON DESCENT

The patterns of movement that occur when going down stairs are illustrated in Figure 11.10, and for convenience they are subdivided into the weight acceptance phase and the lowering phase.

On weight acceptance, the initial foot contact is made with the anterior and lateral border of the foot. The ankle joint moves from the initial step contact position of plantar flexion into a neutral or dorsiflexed position controlled by eccentric work of the calf muscles. The hip joints are in very slight flexion, and the knee joint may flex up to 50° to cushion the instant of foot to step contact. This is controlled by eccentric work of the hip and knee extensors. The quadriceps then contract concentrically to extend the knee about 10° while the trunk moves horizontally to carry the centre of gravity over the stance limb. Tibialis anterior cocontracts with the calf muscles to control ankle position and to maintain weight bearing on the lateral border of the foot.

To lower the body weight (mid stance) to the next step entails controlled hip and knee joint flexion and ankle joint dorsiflexion. This mainly involves eccentric action of the quadriceps and, to a lesser degree, the calf muscles and hip extensors. The stance ankle is in maximum dorsiflexion, with the body weight tending to force the movement further. To prevent over-dorsiflexion at the joint, the plantar flexors may need to contract. Throughout this phase, the hip abductors on the stance side maintain the level of the pelvis and pull the trunk over the stance limb.

SWING PHASE ON DESCENT

In the swing phase, the limb has to be raised off the higher step and swung forwards and downwards, clearing the intermediate step until it is in position to take weight at the start of the next cycle. The hip and knee flexors work concentrically to raise the foot off the top step and pull the limb forwards. Then the limb starts to extend ready for foot placement, with eccentric work of the hip flexors controlling hip extension and the hamstrings working eccentrically to decelerate the extension of the knee joint. The ankle joint drops into plantar flexion, controlled by eccentric work of the anterior tibial muscles, which also maintain the foot in inversion in preparation for weight to be taken on the lateral border of the foot. The hip ipsilateral

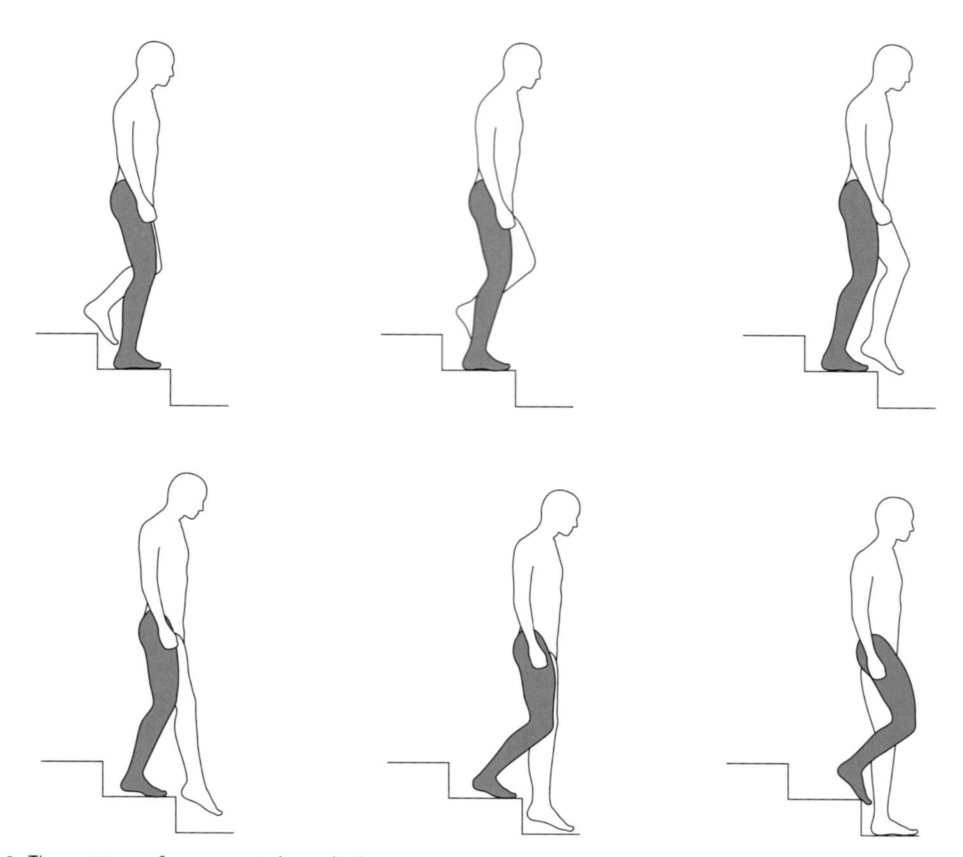

Figure 11.10 The pattern of movement in stair descent.

abductors contract just before the end of the swing phase in preparation for maintaining pelvic levels on weight acceptance (McFadyen and Winter, 1988).

The amount of joint range needed for ascent and descent of stairs depends on the depth of tread. For standard-sized tread (16.5 cm), the hip joint must be able to move between full extension and about 60° of flexion, the range required at the knee joint is 0–100° of flexion and the ankle joint needs full dorsiflexion (Tata et al., 1983; McFadyen and Winter, 1988).

When going up stairs, the period of peak muscle torque and greatest instability occurs simultaneously at the start of the swing phase. The hip and knee joints of the stance limb are in considerable flexion, and substantial effort is needed from the extensor muscles to raise the body; at the same time, effort must also be directed to the maintenance of balance (Tata et al., 1983).

When these facts are taken into consideration, it is no wonder that elderly and frail people find stair climbing difficult. To be able to use stairs safely, it is necessary to have a wide range of movement at hip, knee and ankle joints; muscles capable of generating considerable force through a wide range; and a good sense of balance. Rehabilitation programmes for patients who have difficulties with stairs should always include activities that will increase joint range and the torque-generating capacity of muscle as well as improve balance.

ACTIVITY 11.3

Think about the ankle joint. Do you need the largest range of dorsiflexion when going up stairs or down stairs? Go up and down some stairs to see whether you have sufficient dorsiflexion to complete the task easily or whether you adopt a strategy to overcome lack of range. Talk to your friends, because you may find a difference between individuals.

You will be aware from your own personal experience that going up stairs requires much more energy than walking on the flat. The main problem is caused by the need to translate the centre of gravity vertically. In walking, there are strategies to keep the vertical translation of the centre of gravity to a minimum, but on stairs there is no choice because the centre of gravity must move through a substantial vertical distance. The largest amount of energy is expended on climbing stairs, although people will often feel that going down is the most difficult. While stair descent uses less energy because the movement is in the direction of gravity and requires control only through eccentric muscle activity, it puts considerable strain on the knee joint. If people have problems with their patellofemoral joint, then going down stairs becomes problematic. In addition, going down stairs enables people to see how far the drop is, and this can make frail or injured people feel vulnerable and inhibited.

CLINICAL CONSIDERATIONS

Failure of one or more muscle groups to work can result in quite dramatic gait abnormalities. Weakness of the hip abductor muscles is quite common and will make it difficult for patients to keep their pelvis level during the stance phase. Under these circumstances, the pelvis may drop to the unsupported side, producing a *Trendelenburg gait*. Such a gait is often quite uncomfortable for patients; to avoid discomfort, they may laterally flex their trunk towards the stance side. This shifts their centre of gravity over the stance limb and the pelvis no longer drops painfully to the unsupported side. A very obvious lateral movement of the upper trunk is apparent when observing this sort of gait.

Paralysis of the dorsiflexors will lead to a high stepping gait. Dorsiflexion is normally needed to ensure that the toes will clear the floor in the swing phase, and dorsiflexion is also needed to facilitate heel strike. When patients cannot dorsiflex, they have to increase the range of hip and knee flexion in the swing phase in order to avoid dragging their toes on the ground. Under these circumstances, heel strike is not possible and the patient's toes hit the ground first, with the heel slapping down immediately afterwards. If you look at the shoes of a patient who has paralysis of the dorsiflexors, you will find that the toe area on the affected side is excessively scuffed and worn.

References

Andriacchi, T., Andersson, G., Fermier, D., Stern, D., Galante, J., 1980. A study of lower limb mechanics during stair climbing. J. Bone Joint Surg. 62A, 749–757.

Arakawa, K., 1993. Hypertension and exercise. Clin. Exp. Hypertens. 15, 1171–1179.

Baer, G.D., Ashburn, A.M., 1995. Trunk movements in older subjects during sit to stand. Arch. Phys. Med. Rehabil. 76, 844–849.

Costigan, P.A., Deluzio, K.J., Wyss, U.P., 2002. Knee and hip kinetics during normal stair climbing. Gait Posture 16 (1), 31–37.

Craik, R.L., Oatis, C.A., 1995. Gait Analysis, Theory and Application. Mosby, St Louis.

Farley, C.T., Ferris, D.P., 1998. Biomechanics of walking and running: centre of mass movements to muscle action. Exerc. Sport Sci. Rev. 26, 253–285.

Hardman, A.E., Hudson, A., 1994. Brisk walking and serum lipid and lipoprotein variables in previously sedentary women: effect of 12 weeks of regular brisk walking followed by 12 weeks of detraining. Br. J. Sports Med. 28, 261–266.

Ikeda, E., Schenkman, M., Riley, P., Hodge, W., 1991. Influence of age on dynamics of rising from a chair. Phys. Ther. 71, 473–481.

Janssen, W.G.M., Bussmann, H.B.J., Stam, H.J., 2002. Determinants of the sit-to-stand movement: a review. Phys. Ther. 82 (9), 886–879.

Kelly, D., Dainis, A., Wood, G., 1976. Mechanics and muscular dynamics of rising from a seated position. In: Komi, P. (Ed.), Biomechanics. V B International Series on Biomechanics. University Park Press, Baltimore.

Kerr, K., White, J., Mollan, R., Baird, H., 1991. Rising from a chair: a review of the literature. Physiotherapy 77, 15–19.

Mak, M.K.Y., Levin, O., Mizrahi, J., Hui-Chan, C.W.Y., 2003. Joint torques during sit-to-stand in healthy subjects and people with Parkinson's disease. Clin. Biomech. 18 (3), 197–206.

McArdle, W.D., Katch, F.I., Katch, V.L., 2001. Exercise Physiology, Energy, Nutrition and Human Performance, fifth ed. Lea and Febiger, Philadelphia.

McFadyen, B., Winter, D., 1988. An integrated biomechanical analysis of normal stair ascent and descent. J. Biomech. 21, 733–744.

Mulavara, A.P., Verstraete, M.C., Bloomberg, J.J., 2002. Modulation of head movement control in humans during treadmill walking. Gait Posture 16 (3), 271–282.

Murray, M.P., 1967. Gait as a total pattern of movement. Am. J. Phys. Med. 46, 290–333.

Murray, M.P., Drought, A.B., Kory, R.C., 1964. Walking patterns of normal men. J. Bone Joint Surg. 46A, 335–359.

Murray, M.P., Sepic, S.B., Barnard, E.J., 1967. Patterns of sagittal rotation of the upper limbs in walking. J. Am. Phys. Ther. Assoc. 47, 272–284.

Pai, Y.C., Rogers, M.W., 1991. Speed variation and resultant joint torques during sit-to-stand. Arch. Phys. Med. Rehabil. 72, 881–885.

Pereira, M.A., Kriska, A.M., Day, R.D., Cauley, J.A., LaPorte, R.E., Kuller, L.H., 1998. A randomized walking trial in postmenopausal women: effects on physical activity and health 10 years later. Arch. Intern. Med. 158, 1695–1710.

Riener, R., Rabuffetti, M., Frigo, C., 2002. Stair ascent and descent at different inclinations. Gait Posture 15 (1), 32–44.

Riley, P.O., Schenkman, M., Mann, R.W., 1991. Mechanics of a constrained chair rise. J. Biomech. 24, 77–85.

Riley, P.O., Croce, U.D., Kerrigan, D.C., 2001. Propulsive adaptation to changing gait speed. J. Biomech. 34 (2), 197–202.

Schenkman, M., Berger, R.S., Riley, P.O., Mann, R.W., Hodge, W.A., 1990. Whole body movements during rising to standing from sitting. Phys. Ther. 70 (10), 638–652.

Shepherd, R.B., Koh, H.P., 1996. Some biomechanical consequences of varying foot placement in sit-to-stand in young women. Scand. J. Rehabil. Med. 28, 79–88.

Smidt, G.L., 1990. Gait in Rehabilitation, Churchill Livingstone, New York.

Tata, J., Peat, M., Grahame, R., Quanbury, A., 1983. The normal peak of electromyographic activity of the quadriceps femoris muscle in the stair cycle. Anatomischer Anzeiger (Jena) 153, 175–188.

Vilensky, J.A., Ganiewicz, E., Gehlsen, G., 1987. A kinematic comparison of backward and forward walking in humans. J. Hum. Mov. Stud. 13, 29–50.

Chapter 12

Function of the spine

Valerie Sparkes

CHAPTER CONTENTS

LEARNING OUTCOMES

When you have finished reading this chapter, you should be able to:

1. understand how the spine supports and protects other parts of the human body
2. understand how it gives stability to the body and absorbs forces that act on the body
3. understand how it links and helps move the upper and lower limbs
4. understand how it functions to help us achieve everyday tasks.

INTRODUCTION

The spine is a complex series of joints that enable the body to move in many directions and perform complex functions. This series of joints is unlike any other in the body, as it provides a link between the upper and lower limbs. It has to be strong to absorb forces from external sources and transfer forces within the body. It has to provide stability to maintain postures but also has to be flexible to allow trunk and upper and lower limb movement in all directions. It has to protect vital structures including the spinal cord and the viscera. The spine also gives the human body its characteristic shape.

It is essential to understand the individual functional components of the spine in order to comprehend the integrated system that makes up the functioning spine that facilitates normal movement.

Although this chapter will give some detail on the anatomy of the spine, for further detailed knowledge of spinal anatomy you are advised to refer to the appropriate anatomical resources.

THE SPINAL COLUMN: A SUPPORTING LINK FOR THE HEAD, THE RIB CAGE AND THE PELVIS

The spine includes the topmost vertebrae, the atlas, and all the intervening vertebrae to the sacrum, including the coccyx. It is the key link for facilitation of normal movement, which includes movement of the arms and legs. In a seemingly simple activity such as walking, the spine has many functions. The cervical spine has to support the head at a certain angle in order that you can see where you are going, and the lumbar spine acts as a link to the pelvis in order for the lower limbs to work efficiently. When you are walking at certain speeds, your arms will swing by your sides and the thoracic spine acts as an anchor point for some of the muscles of the scapula and upper limbs to facilitate the correct positioning of the upper limbs. You can see from Figure 12.1 that the thoracic spine also acts as a connection point for the rib cage.

THE SPINE

The spine has, in total, 33 vertebrae, each separated by a disc, apart from the upper part of the cervical spine and the sacrum, with the last disc being between L5 and the sacrum. There are three main regions of the spine: the cervical, thoracic and lumbar (Fig. 12.2). In the upper cervical spine, there is no disc between the atlanto-occipital joint and the atlantoaxial joint. At the sacrum, the five vertebrae are fused together, forming a solid base on which the pelvic ring is attached (Fig. 12.1) Distally, the vertebral column has a coccygeal region at the distal end of the sacrum. This is a triangular bone consisting of four fused rudimentary vertebrae. The coccyx acts as an attachment point for muscles of the pelvis. In the thoracic region, each thoracic vertebra has ribs attached to it, forming the thorax. The top 10 are attached anteriorly via costal cartilages to the sternum. The distal two are free-floating (Fig. 12.1).

Each vertebra of the spine has a ventral body and a dorsal vertebral arch (Fig. 12.3 and Fig. 12.4). This arch has a number of bony projections, which can be considered as levers and act as points of attachment for ligaments and muscles of the spine. Centrally, there is a vertebral foramen that varies in diameter depending on the region of the spine (Fig. 12.3).

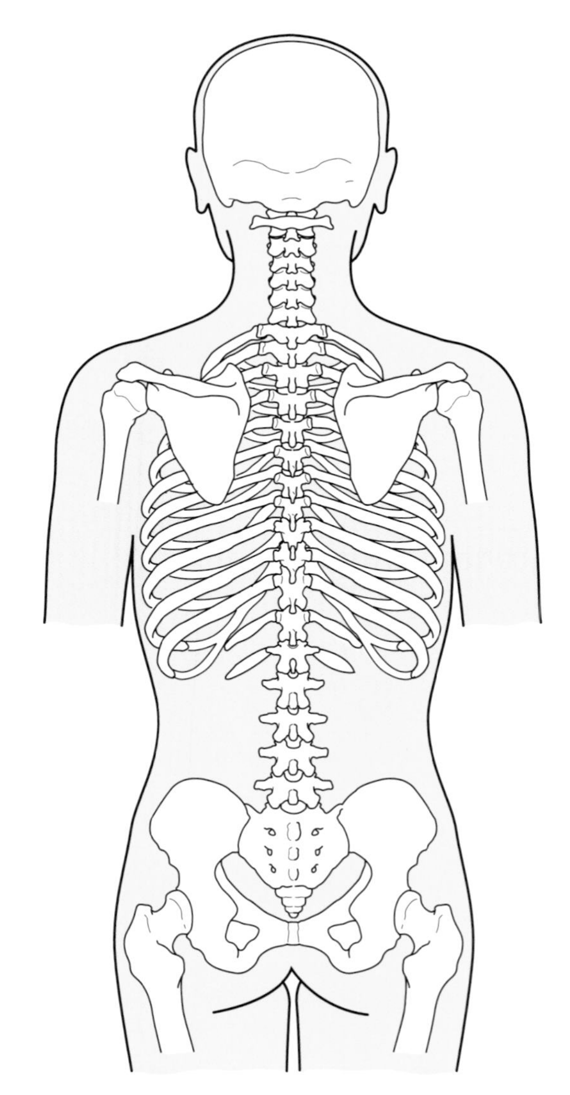

Figure 12.1 The vertebral column, posterior view. *(From Kapandji, 1974, with permission.)*

THE SPINE: PROVIDING PROTECTION OF VITAL STRUCTURES

The specialized architecture of the bones and thorax play a vital role in protecting major structures and vessels. The vertebral canal acts as a bony

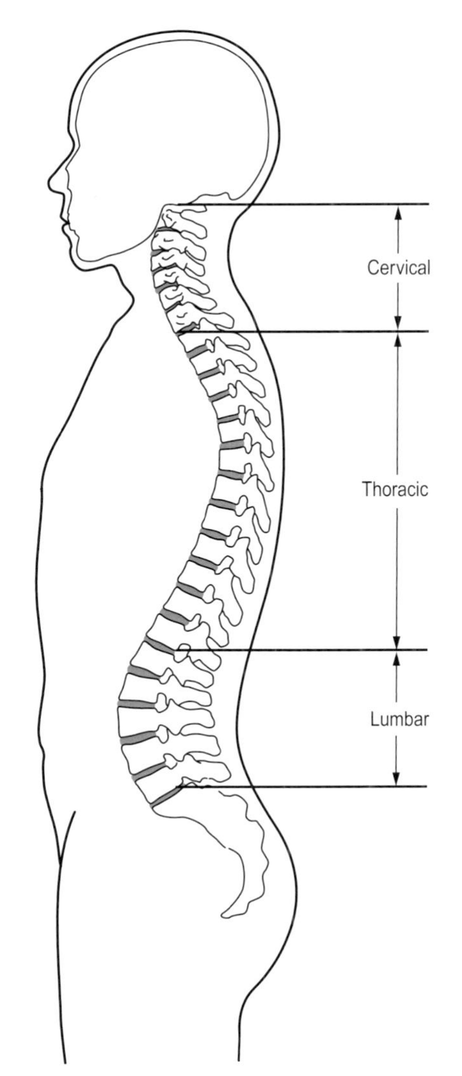

Figure 12.2 The vertebral column, lateral view indicating the spinal regions. *(From Oliver and Middleditch, 1991, with permission.)*

protective channel for the spinal cord and cauda equina together with the spinal meninges, blood vessels and lymphatic vessels. The boundaries of the canal are made of compact bone, which acts as a good protective environment for the delicate structures of the spinal cord and its meninges, as it has the ability to absorb large forces. The shape of the vertebral canal differs depending on the segment of the vertebral column (Fig. 12.4). The thoracic is smallest; at the cervical spine, it is triangular and large to allow for the relatively enlarged spinal cord near the brainstem, and the lumbar spine canal is large to accommodate the cauda equina.

Small fat pads within the canal act as cushioning for the spinal cord during rapid movements of the spine (e.g. in gymnastics, doing several somersaults during the mat exercise) or activities in which large forces go through the spine (again think of gymnastics, when landing from the vault or the beam). During movements of the spine, for example flexion, the spinal cord moves within this canal, causing an increase in the tension within the spinal cord (Butler, 2000). When lifting the leg straight, as in a straight leg raise, the sciatic nerve has to move with the tissue surrounding it by at least 12% (Beith et al., 1995).

Between the pedicles, the intervertebral foramen acts as gaps through which the nerves enter or leave the vertebral canal (Fig. 12.3). The foramen forms a space through which the spinal nerve, sinuvertebral nerve, adipose tissue, blood and lymphatic vessels can pass through. The adipose tissue helps protect the nerves within the foramen, as during movement the shape of the foramen may change as much as a 30% increase in diameter during flexion and a decrease of 25% during extension (Panjabi et al., 1983). In a non-diseased spine, the spinal and sinuvertebral nerves take up to one-third to a half of the area of the foramen, so a reduction in diameter that occurs in extension of the spine should not compromise the nerve structures.

The spine protects the blood vessels running through the cervical spine. In the cervical spine, the vertebral arteries are well protected as they run through foramen in the transverse processes (foramina transversarii). These foramen offer protection to vital blood vessels during neck movement, as these blood vessels running through the cervical spine form the circle of Willis, which supplies a large area of the brain (Fig. 12.5).

ACTIVITY 12.1

This activity concerns holes in the vertebrae.

In small groups, take a selection of vertebrae or a spine that is loosely joined together and identify the holes in the spine through which run:

- the spinal cord
- the vertebral arteries in the cervical spine
- the holes through which the spinal nerve runs. Which holes are the largest?

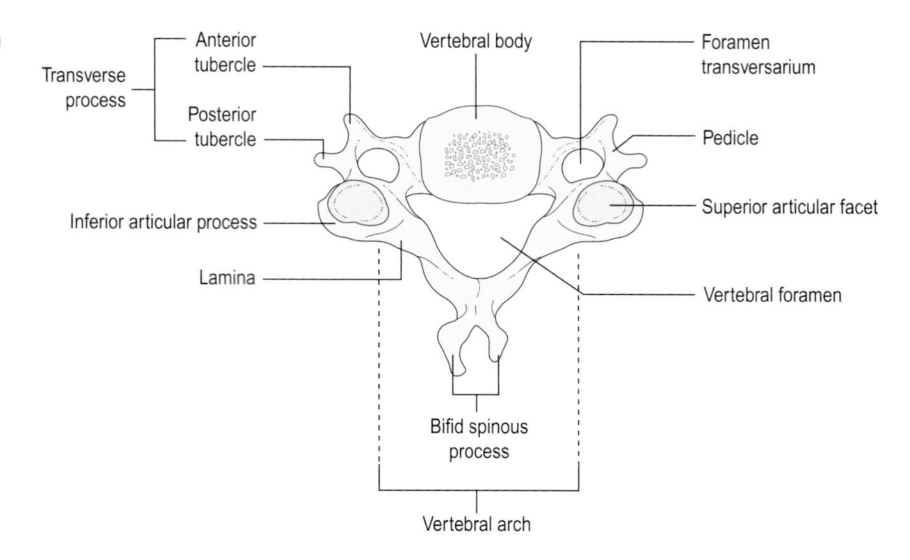

Figure 12.3 Superior aspect of a cervical vertebra showing the main bony prominences and foramen. *(From Middleditch and Oliver, 2005, with permission.)*

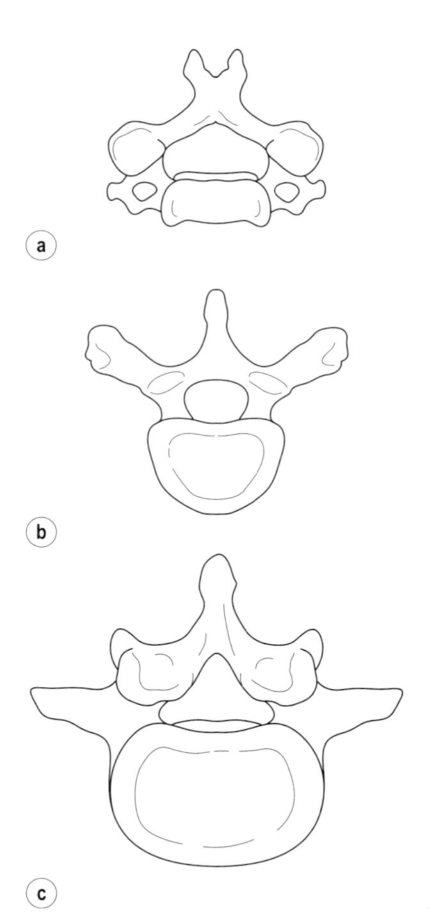

Figure 12.4 Segmental variations in the shape of the spinal canal: (a) cervical, (b) thoracic, and (c) lumbar. *(From Butler, 1991, with permission.)*

The thorax, as noted above, consists of 12 pairs of ribs. The upper 10 ribs attach posteriorly via the costotransverse and costovertebral joints, and these ribs attach anteriorly via costal cartilages to the sternum. Ribs 11 and 12 attach posteriorly via costovertebral joints only and are free-floating anteriorly. The thorax cage can be viewed as a semirigid structure that offers protection for the lungs, heart and other major vessels.

THE SPINE: ABSORBING LOAD AND FACILITATING MOVEMENT

ABSORBING LOAD

One of the functions of the spine is to absorb load. As most human movement is conducted in an upright position, the spine has to absorb weight bearing through the lower limbs as well as the influence of gravity on the body. The spine is subject to loads from many other sources, including loads carried by the upper limbs, the weight of the thorax, and when muscles attaching to the spine contract. One of the ways that the spine's ability to absorb and transmit load is enhanced is through the composition of each of the vertebrae. Each vertebra is composed of an outer layer of cortical bone (Fig. 12.6). Although this layer is strong, it is not strong enough to absorb the loads that are imposed on the spine. The composition and internal structure of the inner part of the vertebrae enhances the vertebrae's ability to withstand the axial loads imposed on the spine on a day to day basis

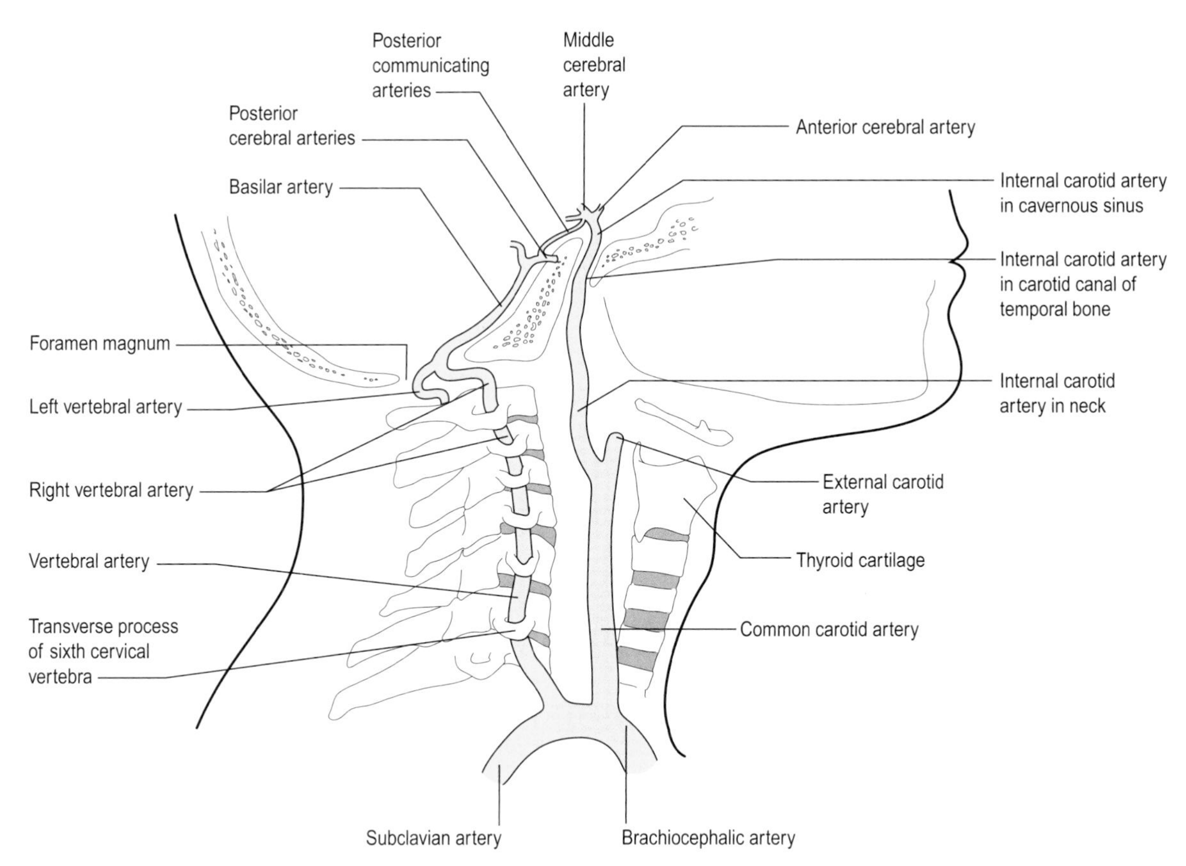

Figure 12.5 The pathway of the vertebral arteries through the cervical vertebrae to the brainstem. *(From Palastanga et al., 1994, with permission.)*

(Adams et al., 2006). This is achieved by an arrangement of a series of horizontal and vertical trabeculae or struts supporting the outer cortical shell.

The vertical trabeculae transmit compressive loads through the vertebrae, and the horizontal trabeculae reinforce the vertical ones by preventing them from buckling when subject to load (Fig. 12.6).

The vertebral bodies differ in shape according to their location in the vertebral column. The lumbar vertebrae are the largest, followed by the thoracic, with the cervical being the smallest. The lumbar vertebrae are the stoutest, as they have to absorb most of the body weight, whereas the cervical vertebrae are much more delicate, as the amount of load they have to support is much less than the lumbar spine has to. The lumbar spine has to absorb and transmit loads through the sacroiliac joints and the pelvis. The cervical spine has to support the weight of the head, which is approximately 8% of the whole body mass. The thoracic spine is not normally regarded as a weight-bearing part of the spine, but in fact it has to support the cervical spine and the head as well as the thorax and it has to absorb and transfer loads imposed during weight-bearing activities through to the lumbar spine.

ACTIVITY 12.2

This activity concerns the bones of the spine.

In small groups, take a selection of vertebrae, including cervical (including the atlas and the axis), thoracic and lumbar, and look at the differences in the construction of each of the vertebrae.

- What are the differences between the atlas and the axis?
- What are the differences between the remaining cervical vertebrae and the thoracic vertebrae?
- What are the main differences between the thoracic and the lumbar vertebrae?

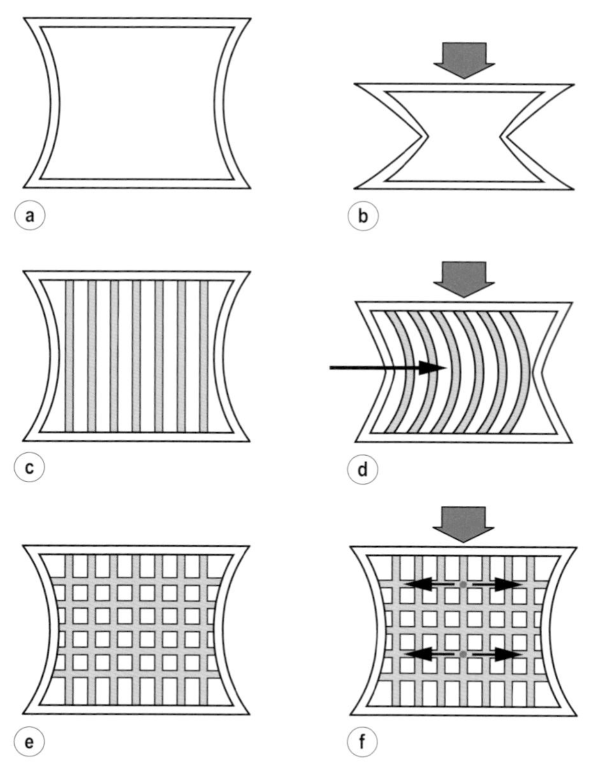

Figure 12.6 Reconstruction of the internal architecture of the vertebral body. With just a shell of cortical bone (a), a vertebral body is like a box and collapses when a load is applied (b). Internal vertical struts (c) brace the box (d). Transverse connections (e) prevent the vertical struts from bowing and increase the load-bearing capacity of the box. Loads are resisted by tension in the transverse connections (f). *(From Bogduk, 1997, with permission.)*

FACILITATING MOVEMENT

As well as absorbing loads, the vertebrae have to be flexible to permit movement. Movement occurs between the vertebrae or *interbody joints*, which are secondary cartilaginous joints. Each interbody joint is separated by the intervertebral disc. This structure allows movements such as bending and twisting to take place. The term *motion segment* is often used to describe two vertebrae and the intervening disc. The spine also acts to separate the thorax from the pelvis, which facilitates movement. Each lumbar vertebra together with each disc gives height to the region between the thorax and the pelvis. The taller the vertebral bodies and discs, the greater the possibilities of range of movement between the thorax and pelvis (Adams et al., 2006).

The intervertebral disc

The intervertebral discs are situated in between each vertebra and are the major anchoring component of the vertebrae, the first disc being between the second and third cervical vertebrae and the last being between the fifth lumbar vertebra and the sacrum. There can be exceptions to this, as some people have extra vertebrae, most commonly in the lumbar spine. The discs create a space between the vertebrae, and the total length of the discs makes up approximately one-fifth of the total length of the spine. This space helps movement to occur in the spine. The discs have a dual role: as well as transferring loads between the vertebrae, for which the discs have to be strong and resistant, they have to be pliable enough to allow movement.

As is the case with the vertebrae, the discs vary in shape depending on which region of the spine they are located in. The upper thoracic discs are the thinnest, and the lumbar discs are the thickest. One of the reasons that the cervical spine has a large range of motion is that the cervical discs are thicker in proportion to the vertebral bodies of the cervical spine.

The discs adhere by both elastic and collagen fibres (Sharpey's fibres) to the compact bone and hyaline cartilage plate on the upper and lower surfaces of the vertebrae (Johnson et al., 1982). They are also connected to the anterior and posterior longitudinal ligaments (Fig. 12.7). In the thoracic spine, they are anchored by intra-articular ligaments to the head of the ribs that articulate with adjacent vertebrae (Williams et al., 1995).

The discs are composed of an outer annulus fibrosus and an inner nucleus pulposus. The layer of hyaline cartilage that is between the disc and the vertebral body is known as the cartilage endplate or vertebral endplate (Fig. 12.5). The hyaline cartilage is permeable, and the disc and the vertebrae rely on this permeability for transport of water and nutrients. Although the endplates are only approximately 1 mm thick, they do act as a barrier in order to prevent the nucleus bulging into the vertebrae (Giles and Singer, 1998; Moore, 2000). The cartilage endplates cover almost all of the vertebral body, with a narrow ring being left uncovered that forms the hard bony rim of the vertebral endplate known as the epiphyseal ring apophysis, which is wider anteriorly (Fig. 12.8).

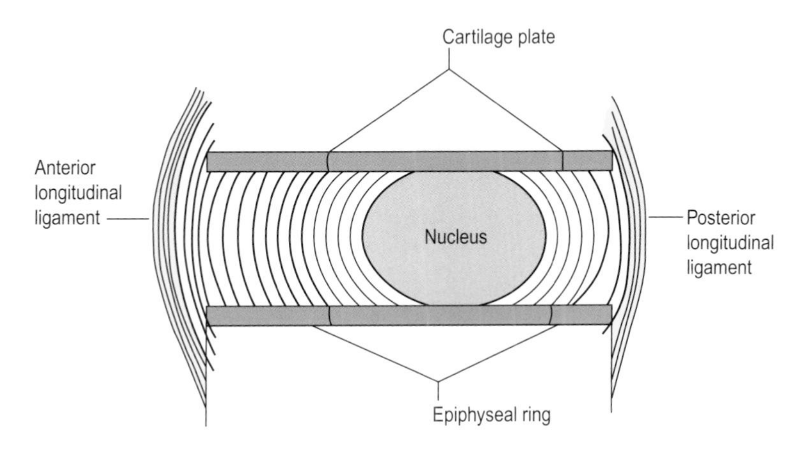

Figure 12.7 The cartilage endplate lies between the intervertebral disc and the vertebral body. The nucleus lies adjacent to the endplate. The anterior and posterior fibres of the annulus are attached to the anterior and posterior longitudinal ligaments, respectively. *(From Macnab, 1977, with permission.)*

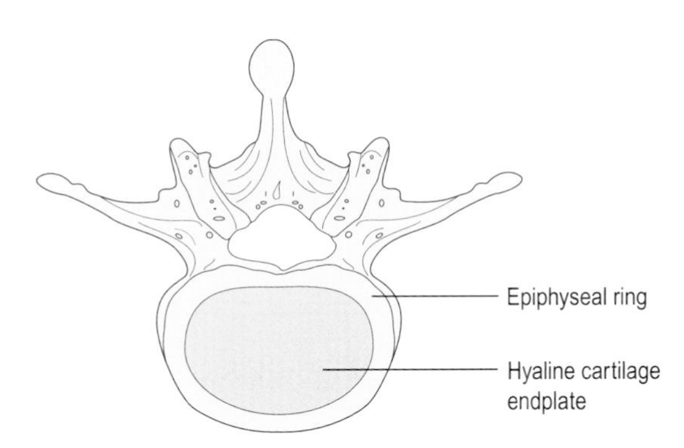

Figure 12.8 Horizontal section showing cartilage endplate and epiphyseal ring. *(From Middleditch and Oliver, 2005, with permission.)*

Annulus fibrosus

The outer part of the disc is comprised of mainly type I and type II collagen. Type I collagen is mainly nearer the periphery of the disc, where the greatest tensile strength is required (Adams et al., 2006); type II collagen is known to be able to withstand compressive loads (Bogduk, 1997). The outermost fibres of the disc contribute the most to counteracting movements and are attached to the ring apophysis; they are sometimes known as the ligamentous portion of the annulus.

The annulus has 10–20 sheets of collagen known as lamellae. These sheets of collagen are packed closely together and are arranged in parallel to each other and obliquely at 40–70° to the horizontal, but the direction of the lamellae are oriented in opposite directions. This arrangement is critical to the intactness of the disc, as it prevents leakage of any material from the nucleus. Also, this alternating direction of collagen fibres allows the annulus to resist stresses in many directions. All of the fibres will resist separation of the vertebrae. This arrangement of bundles of collagen gives the disc the ability to withstand compression, which is occurring during many hours of the day during weight-bearing activities, even in sitting. As well as absorbing and transferring load, the collagenous make-up allows movement to occur and the disc can be deformed.

Nucleus pulposus

The nucleus pulposus, the inner part of the disc, is a semifluid gel containing between 70 and 90% water. It is composed of type II collagens, which can withstand compression, and a proteoglycan gel that has the ability to imbibe and retain water. Although the disc is subject to high compressive loads during everyday activities, water content is maintained to a greater degree in a healthy disc because of the capacity of the proteoglycan gel to retain and absorb water and the discs' high osmotic pressure.

When the nucleus is compressed, it will spread and exert pressure on the annulus. In turn, this increase in pressure from the nucleus pulposus will stop the annulus from buckling inwards (Adams et al., 2006). Thus the annulus fibrosus and nucleus pulposus work together. If for whatever reason the nucleus is unable to retain water, it will no longer be able to support the annulus fibrosis and the disc's ability to withstand compressive loads will be lost.

ACTIVITY 12.3

Are you taller in the morning?

Get the people in your house and measure the height of each other just before you go to bed then immediately when you get out of bed. Are there any differences?

During normal daily activities, water is lost from the discs, so effectively the disc and you lose height. Fortunately, a healthy disc has the ability to take on water, and so after a night's sleep the water content is fully restored.

The hydrostatic pressure within the disc varies as a result of various postures adopted and any loads carried. Studies in the past have measured the hydrostatic pressure in the disc during various tasks and postures (Nachemson, 1976) (Fig. 12.9). These investigations are important, as they show in which postures or exercises the loads are greatest in the disc. Of interest is the fact that upright sitting appears to have the greatest effect in terms of load on the discs compared with other postures or exercises tested.

The benefits of movement for the disc

Discs are avascular except for the periphery, which is supplied by adjacent blood vessels. The disc has to rely on the nutrition through the permeable vertebral endplate and also from small blood vessels at the periphery. Fluid exchange is improved if the spine is moved, particularly in the sagittal plane (Adams and Hutton, 1986), but is reduced if the spine is subject to static loads, particularly at end of range flexion and extension (Adams and Hutton, 1985).

During forward bending, the anterior part of the disc is compressed slightly, whereas the posterior parts of the disc are stretched (see Fig. 12.10). During extension and side flexion, similar pattern happens in the opposite direction and in the coronal plane, respectively. In twisting movements, there is a small degree of disc movement, but this is counteracted by the arrangement of the oblique fibres of the annulus.

Most of the research undertaken on discs has been conducted on the lumbar spine. The cervical and thoracic discs are similar in many respects. However, the cervical discs take less axial load and they have a lower proteogylcan and water content (Giles and Singer, 1998). Although they take reduced loads, they are subject to large ranges of movement, which may have an effect of reducing the water and proteoglycan content of the disc in later life (Milne, 1991). The thoracic discs are the thinnest, and this contributes to the relative stiffness of thoracic spine. Although thin, the thoracic discs still bear axial load and contribute to the trunk's characteristic thoracic kyphosis (see Ch. 5 on posture).

The vertebral arch

The posterior aspect of the vertebrae is known as the vertebral arch (Fig. 12.3) and consists of processes that form attachment points for various ligaments and muscles. As such, they control the position of each individual vertebra and the spine as a whole. They also provide stability to the vertebral column and provide a resistive component to torsion, as the annulus fibrosis alone is not strong enough to withstand excessive torsion (Adams et al., 2006). Generally, the vertebral arch has two pedicles, which support the rest of the posterior elements and transfer forces between the vertebrae. From the pedicles, two laminae arise and fuse, forming the spinous process. The pedicles and the laminae, together with the posterior aspect of the vertebral body, form the space called the vertebral foramen (Fig. 12.4). Processes that arise from the laminae are the transverse processes and the inferior and superior articular processes.

The articular processes each have an articular facet, so the superior facet of one articular process articulates with the inferior articular facet of the vertebra above; this forms what is known as the zygapophyseal (apophyseal or facet) joint. These are synovial joints, covered with articular cartilage, with synovial fluid and a synovial membrane, and surrounded by a relatively lax fibrous capsule, all of which facilitate movement. Small intra-articular meniscoids, composed of fatty cartilaginous and synovial tissue, are found within the cervical and lumbar spines, which act to spread the load within these joints. The zygapophyseal joints enable but also restrict certain movements depending on their orientation. They have been described as providing a locking mechanism between the vertebrae, for example in the lumbar

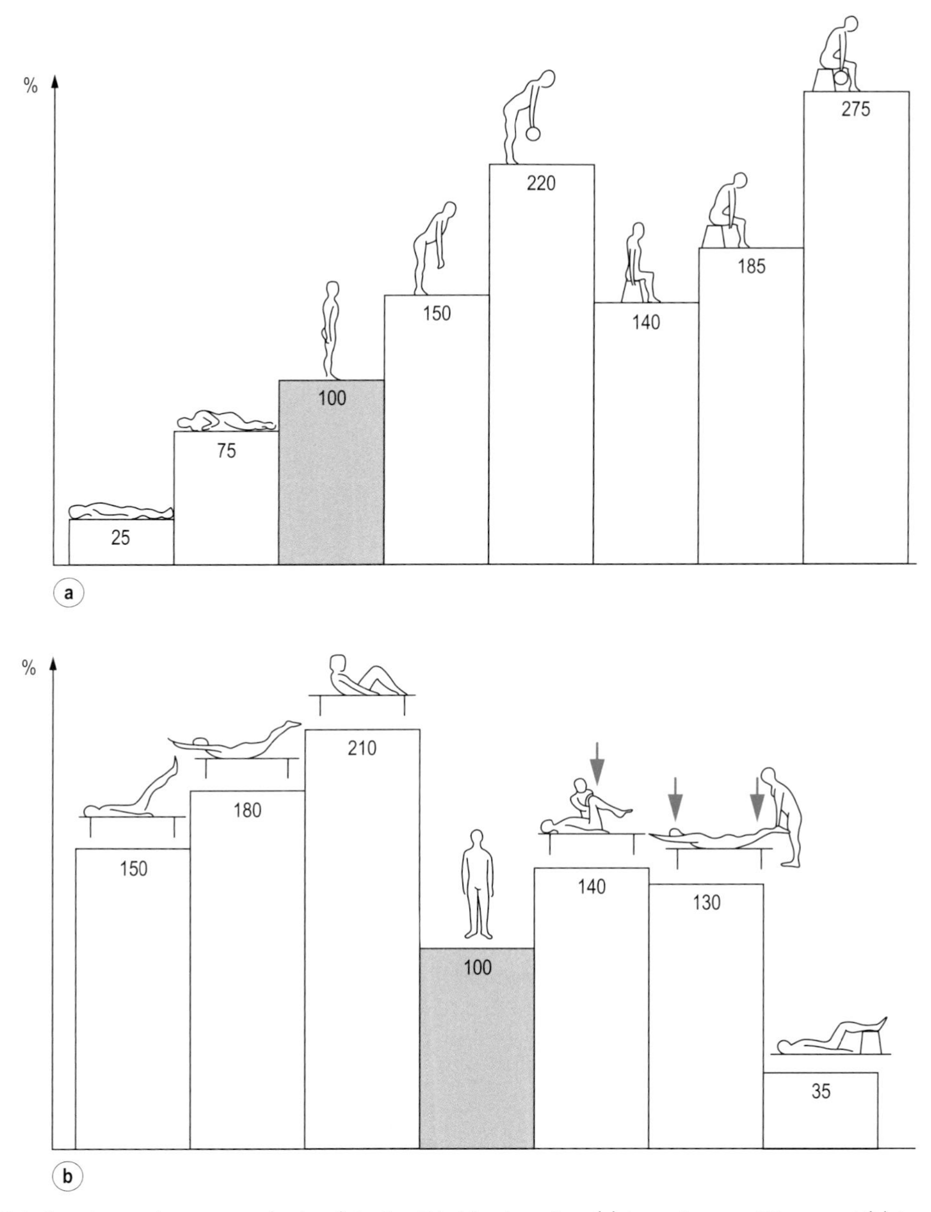

Figure 12.9 Relative change in pressure (or load) in the third lumbar disc: (a) in various positions, and (b) in various muscle-strengthening exercises. *(From Nachemson, 1976, with permission of Lippincott-Raven Publishers.)*

spine they restrict rotation and forward slide of the vertebrae (Adams et al., 2006). As a restraint to rotation, the zygapophyseal joints to some extent protect the disc from excessive torsional strains. However, the orientation of the zygapophyseal joints also serves to facilitate gliding movement between the vertebrae, for example during flexion the inferior zygapophyseal processes are relatively free to move upwards.

ACTIVITY 12.4

How does posture affect different parts of the spine?

Dan and Jenny are in their bedrooms playing on their PlayStations. Both are lying on their stomachs. Think about the structures of the lumbar spine, such as the zygapophyseal joints, discs and ligaments. Which parts of these structures will be compressed, those on the anterior or posterior aspects of the spine?

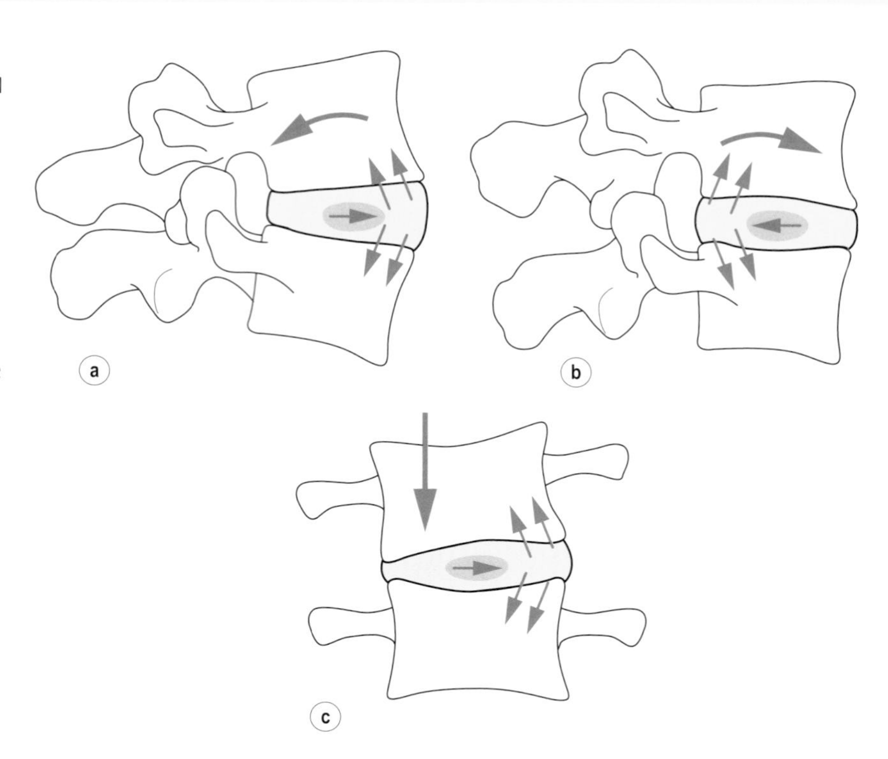

Figure 12.10 Effect of movement on deformation of the intervertebral disc. Extension (a): the upper vertebra moves posteriorly, the nucleus moves anteriorly and the annulus is tensioned anteriorly. Flexion (b): the upper vertebra moves anteriorly, the nucleus moves posteriorly and the annulus is tensioned anteriorly. Lateral flexion (c): the upper vertebra tilts towards the side of flexion and the nucleus moves in the opposite direction, where the annulus is tensioned. *(From Kapandji, 1974, with permission.)*

The moving spine: the upper cervical spine

The atlanto-occipital (CO–C1) and atlantoaxial (C1–C2) joints are atypical, as there is no disc in between each bone. The atlanto-occipital (C0–C1) joint has two articulations, which are synovial condylar joints. The occiput sits on the superior facets of C1, which are enlarged in order to support the occiput (Mercer and Bogduk, 2001). They are elongated and cup-shaped, and they face medially and lie 45° to the sagittal plane in the anterior–posterior direction. The shape and direction of the facets facilitate a large range of flexion–extension at this level; a small degree of lateral flexion and rotation is available. When flexing the head, the condyles of the occiput roll on the lateral aspects of C1 and translate forwards. The atlas translates backwards, tilts upwards and backwards. In extension of the head the reverse occurs. Movement will be restricted by tension in the capsules of the joints and the alar ligaments.

At the atlantoaxial joint (C1–C2), C2 bears the axial load of the occiput and transfers it to the cervical spine. Rotation, lateral flexion, flexion and extension occur at this level; however, rotation has the largest range by far. Because of the orientation of the facets, rotation is usually accompanied by some degree of lateral flexion.

C3–T2

Flexion, extension, rotation and to some degree lateral flexion all occur at these levels. During flexion, the vertebrae undergo anterior sagittal rotation and anterior translation, with the reverse happening during extension (Fig. 12.10). In flexion, the intervertebral canal will increase, with a decrease during extension. In these regions of the cervical spine, the inferior facets face downwards and forwards at approximately 45°, with the superior ones of the vertebrae below facing upwards and backwards.

ACTIVITY 12.5

How far does your head move?

Ask several people of different ages, as well as a mixture of males and females, to move their heads in different directions, for example rotating left and right, and flexion and extension. Do you find differences:

- between genders?
- between ages?

If so, why might this be?

CASE STUDY 12.1

This case study concerns a family activity: driving.

The family are going to a friend's house for lunch and are driving there. John is driving and puts his seatbelt on. In order to reach the seatbelt, as the shoulder is extended and taken into external rotation the cervical spine rotates and side flexes to the right. Also, the thoracic spine extends to some degree. When John sets off from the drive or when approaching any junction or roundabout, the cervical spine rotates to increase the field of vision to check for oncoming cars. Jenny is chatting to her mum, Liz, who is sitting in the front passenger seat, but Chris will not stop singing his latest favourite tune, which is making it difficult for her to hear her mum. Jenny leans forwards to speak to her mum; here, the lumbar, thoracic and cervical spine work together to position the body in the best posture for Jenny to hear her mum speak. The spine sometimes has to bend in awkward positions in order to achieve the task.

Now make a list of occupations in which the spine has to adopt awkward postures in order for people to do their jobs, as an example a plumber fixing water pipes under the sink has to flex the lumbar and thoracic spine in order to work underneath the sink but has to extend the neck in order to see the pipe work.

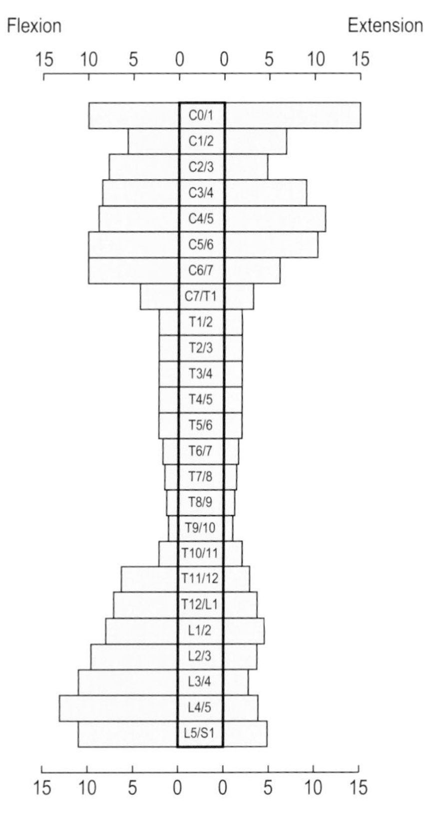

Figure 12.11 Average ranges of spinal segmental movement (flexion and extension). *(From Oliver and Middleditch, 1991, with permission.)*

Thoracic T3–T12

This area of the spine has the least range of movement because of the articulations with the ribs. Lateral flexion of the whole of the thoracic spine and flexion and extension are particularly limited at T6–T10. Rotation is always accompanied by lateral flexion because of the orientation of the zygapophyseal joints, which in the inferior facets face forwards and medially, with the superior ones of the vertebrae below facing backwards and laterally.

Lumbar spine

There is a large degree of flexion at the lumbar spine, particularly at L4–L5. The next largest movement is lateral side flexion, then extension, with the least amount being rotation. As in the cervical spine, lateral flexion is always accompanied by rotation. In the upper lumbar spine, lateral flexion to the left is accompanied by rotation to the right; at the lower two levels, the lateral flexion to the left is accompanied by rotation to the left (Percy and Tibrewal, 1984). In the lumbar region, the zygapophyseal joints of the inferior processes face laterally and forwards, with the superior ones facing medially and backwards (Fig. 12.3). The average ranges of spinal movement can be seen in Figure 12.11.

ACTIVITY 12.6

How far can you bend forwards?

In small groups, in turn (do not ask anyone to do this if they have back pain), each of you bend as far forwards as you can with your fingertips pointing towards the floor. Look at the difference between how individuals move and how far they move. As they reach their limit of movement, ask the individuals in turn what limits how far they can bend. Is everyone the same or are there differences? Make a list of all the structures that may limit flexion of the lumbar spine.

THE SPINE: PROVIDING STABILITY – LIGAMENTS AND THE THORACIC CAGE

The spine can be regarded as a series of individual bones. In order for the spine to function as a cohesive unit to produce movement and transfer loads, ligaments act to provide stability to this essentially unstable structure. One of the major supporting ligaments is the ligamentum flavum, made of elastin, which bridges the space in between the laminae of the vertebrae. This ligament is very flexible and will stretch when the spine flexes forwards, and when returning to upright from flexion the ligament will recoil and shorten without buckling and getting trapped, thus aiding the work of the back extensor muscles (Adams et al., 2006). Although allowing flexion, it also acts as one of the limiting factors to flexion (Middleditch and Oliver, 2005).

Other ligaments connect various bony points, but these are much thinner and often blend with the muscles of the spine. Intertransverse ligaments, which are thin sheets of collagen, connect the transverse processes. Interspinous and supraspinous ligaments connect the spinous processes, but the supraspinous ligament often blends with the erector spinae muscles. To provide further stability to the spine, the vertebral bodies are connected to each other. On the posterior aspect of the vertebral bodies is the posterior longitudinal ligament, which also connects to the intervertebral discs. The anterior longitudinal ligament covers the anterior aspect of the vertebral bodies and the discs.

THE THORAX

The thorax – consisting of a series of ribs, costal cartilages and sternum – provides a degree of stability to the spine as a whole. The ribs need strong ligaments to attach them to the vertebrae, as they have flat surfaces. Ligaments such as the radiate ligaments fan into three bands to attach the rib head on to the vertebrae. The intra-articular ligament links the thoracic zygapophyseal joint to the disc (Williams et al., 1995). The thorax is primarily involved in respiration, which involves movement of the ribs at the costotransverse and costovertebral joints posteriorly and the costochondral joints and sternocostal joints anteriorly. The thorax also acts as an attachment site for many muscles and fascia (see the section on muscle and movement). It also acts as a link between the cervical and lumbar spines and can influence the movement in these regions. Being a semistable cage-like structure, it is well suited to its role in protecting the lungs, viscera, heart, and other major vessels. One of the advantages of this stiffness is its ability to absorb load (Andriacchi et al., 1974). However, this protective role comes at the expense of mobility in this region (Fig. 12.12).

PELVIC GIRDLE

Distally, the spine forms an attachment to the pelvis and the lower limbs via the sacrum. The sacrum is a large flat bone that supports the weight of the vertebral column and to some extent the whole of the trunk. It also transfers loads through to the lower limbs as well as acting as a base on which the lower limbs work. The pelvic girdle consists of two pelvic bones and the sacrum, with the

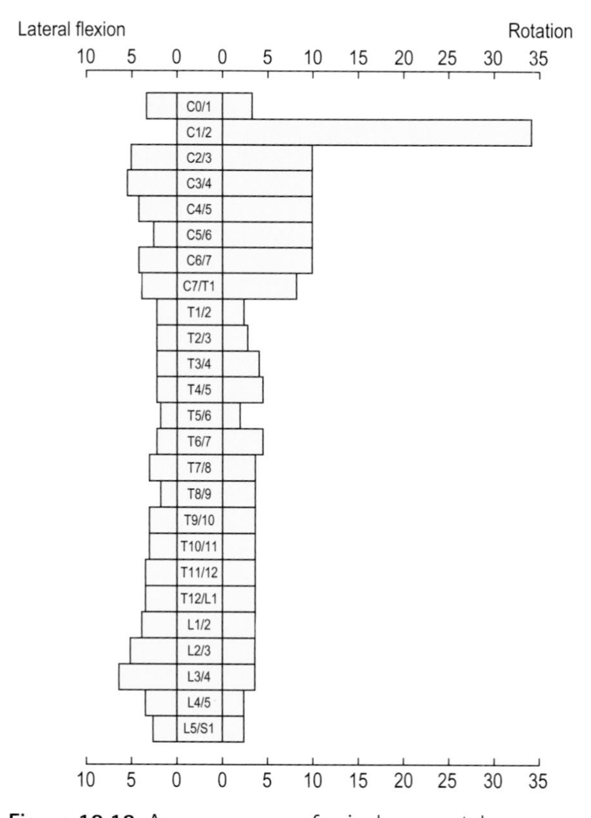

Figure 12.12 Average ranges of spinal segmental movement (values given to one side) for lateral flexion and rotation. *(From Oliver and Middleditch, 1991, with permission.)*

sacrum linking to the ilia via the sacroiliac joint. The sacroiliac joints tend to act as stress-relieving joints rather than joints that produce movement (Adams et al., 2006). Movement at the sacroiliac joint is small. Anterior rotation of the base of the sacrum, with accompanying posterior rotation of the apex, is known as *nutation*, with the opposite movement being known as *counternutation*. In lumbar spine flexion in standing, there is approximately 1° of nutation and 1° of counternutation in extension (Jacob and Kissling, 1995) (see Fig. 12.13).

Movements also occur at the sacroiliac joint and pubic symphysis during activities such as walking. Figure 12.14 shows that during left leg stance phase, shear forces arise at the symphysis pubis as well as the sacroiliac joints.

The pelvis has to absorb larger forces from the lower limbs, particularly during walking and running. Because of this, some of the strongest ligaments are found in this region to prevent unwanted movement of the bones in order to facilitate efficient movement. The iliolumbar ligament links the fifth lumbar vertebra to the ilium, stopping it from rotating and moving forwards. The interosseous sacroiliac ligament and the anterior and posterior sacroiliac ligaments all play a role in preventing separation of the sacrum from the ilia. Other ligaments that provide stability to the

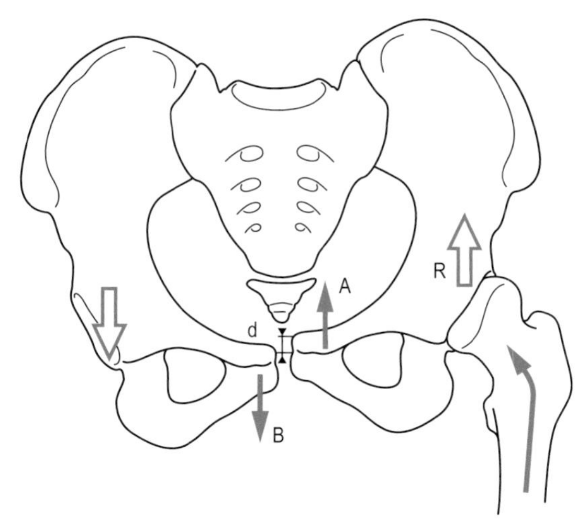

Figure 12.14 The forces around the pelvis when standing on the left leg (stance phase) and taking the right leg forwards in walking. The ground reaction force (arrow R) elevates the left hip while the right hip is pulled down by the weight of the free leg. This causes a shearing force at the pubic symphysis, tending to raise on the left (A) and lower on the right (B) (d is the distance moved because of this shear force). The forces will be in the opposite direction at the sacroiliac joints; the left ilia will tend to lower and the right ilia will tend to be raised. *(From Kapandji, 1974, with permission.)*

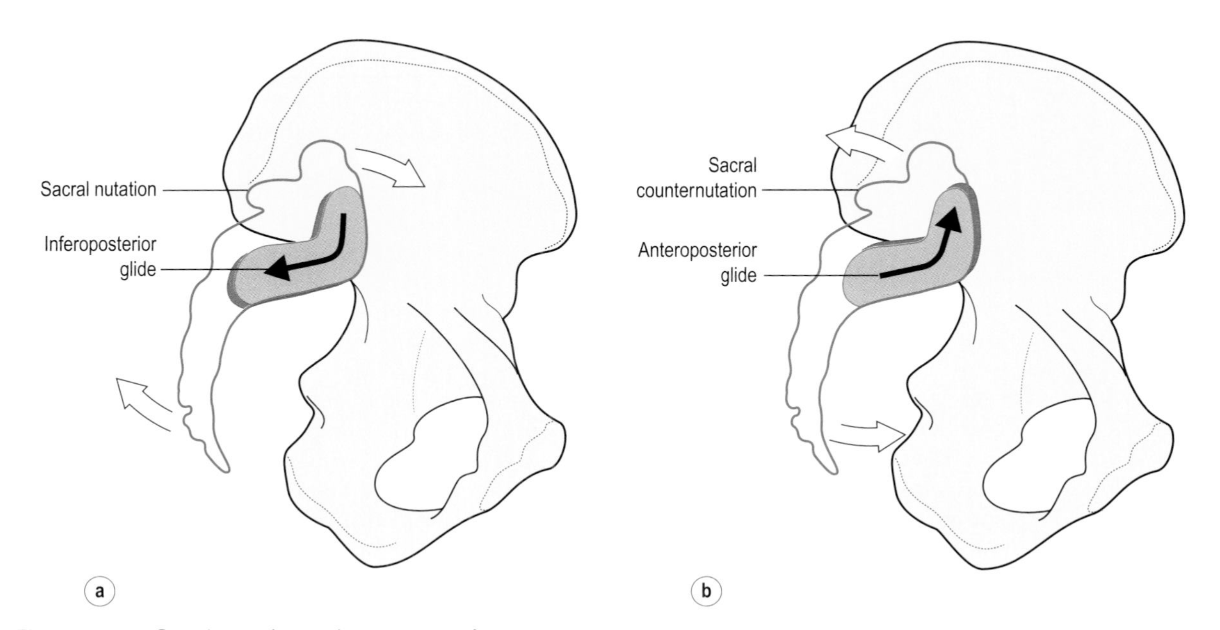

Figure 12.13 Sacral nutation and counternutation. *(From Middleditch and Oliver, 2005, with permission.)*

pelvic girdle are the sacrospinous and sacrotuberous ligaments, which prevent forward rotation of the sacrum. Anteriorly, stability of the pelvis is achieved by a fibrous union of the symphysis pubis. This stability achieved by the ligamentous connections enhances the ability of the whole body to work efficiently.

THE SPINE: ENHANCING STABILITY AND MAKING THE BODY MOVE – MUSCLES OF THE SPINE

In order for normal movement to occur, muscles have to work in a coordinated fashion (see Ch. 4 on motor control). A major function of the spine is to offer an anchor point for the attachment of muscles. Muscles with attachments on the spine have several roles. Broadly speaking, the role of the muscles can be divided into three:

1. provide dynamic stability or control of the segments of the spine
2. help to produce large movements of the trunk and body
3. help to maintain postures.

> **ACTIVITY 12.7**
>
> Bones are attachment points for muscles. Go back and look at the anatomy of any vertebrae. Find some examples of vertebrae of different parts of the spine. Count how many prominences on the vertebrae or projections of bone there are from the main body of the vertebrae that can act as an attachment point for muscles.

Muscles are known to influence the spine's stability. Even when working at low levels of activation, they can have a positive effect on the stability of the spine (Chloewicki and Magill, 1996). Muscles of the spine can produce large forces involving gross movements of the spine, such as when lifting a heavy object from the floor. It is important to remember that the spinal muscle rarely works in isolation, and gross movements are often combined with activation of other muscles, such as the hip extensors. It is important to remember that the muscles of the spine also work to maintain the posture of the spine (see Ch. 5).

Muscles consist of different proportions of type I and type II fibres (see Ch. 2). Some muscles play a predominant role in stabilizing the spine and maintaining postures, while other muscles play a predominant role in producing large movements and powerful movements. However, it is important to remember that most muscles work together in a coordinated fashion, for example helping to maintain posture while producing movement.

MUSCLES PROVIDING DYNAMIC STABILITY: SITTING STILL IS A DYNAMIC ACTIVITY

The words *dynamic stability* are used, as even when sitting the muscles providing support for the spine are constantly adjusting (Moseley and Hodges, 2006). Muscles that provide stability to the spine are in the main placed close to the vertebrae, for example multifidus. Other muscles that have an effect on the stability of the spine are further away, for example transversus abdominis, but have an effect on the spine via its attachments to the thoracolumbar fascia (Hodges, 1999). In the lumbar spine, the abdominals and lumbar spine portion of the multifidus work in a coordinated fashion to provide stability of the spine. The same is true of the cervical spine, as the muscles on the anterior part of the spine work as well as those on the extensor aspect to maintain head posture (Falla et al., 2007).

Lumbar spine

In the lumbar spine, the actions of the deeper spinal muscles such as multifidus are particularly important in providing stability to the spine when movements, for example of the upper limb, are required (Moseley et al., 2002). Lumbar multifidus has many fascicles and is arranged in deep and superficial layers, with the deepest parts providing most of the stability (Moseley et al., 2002). Some of the fascicles of multifidus span two vertebrae, with others spanning four vertebrae. The multifidus muscle arises in the sacral region and attaches to the lumbar, thoracic and lower four cervical vertebrae. The lumbar part of multifidus has the largest influence on the stability of the lumbar spine when compared with other muscles (Wilke et al., 1995). The more superficial parts of multifidus can produce sufficient torque to produce extension of the lumbar spine as well as providing a compressive force to the vertebrae (Mackintosh and

Bogduk, 1986, Bogduk et al., 1992). Other deep muscles close to the spine are rotators, interspinalis and intertransversarii. These probably have a proprioceptive role, as they contain a higher percentage of muscle spindles than other spinal muscles do. Many of the small muscles of the spine, for example multifidus, rotators and intertransversarii, are in effect a system of short, sturdy cables that prevent the individual vertebrae from separating or buckling. They play a role in stabilizing the individual segments of the spine, while the more powerful and longer muscles produce movements of the spine.

There are many other muscles that lie on the dorsal aspect of the spine. Many of these span numerous vertebrae. The erector spinae (sacrospinalis) lying over multifidus arises from the sacral region and then splits into three regions:

1. iliocostalis cervicis (lateral)
2. longissimus (intermediate)
3. spinalis (medial).

These all have portions in the lumbar, thoracic and cervical spine. In general, these superficial muscles are extensors of the spine, with the more laterally placed muscles (iliocostalis cervicis) influencing side flexion. Other muscles in the thoracic and cervical region include semispinalis thoracis, cervicis and capitis, which all aid in extension of the thoracic and cervical regions.

CASE STUDY 12.2

This case study concerns another family activity: shopping.

Jenny, Dan and Agnes are going into town. Jenny has to go to the library to get some more books for her schoolwork. Dan wants some new trainers, and Agnes wants some food from the supermarket. They first go to the library and Jenny gets her four new books, which she puts in her rucksack. They are quite heavy, and she shortens the straps on her rucksack; this means that the load is carried quite high. It has been shown that when carrying a rucksack high on the back, the trunk compensates by leaning forwards and the erector spinae muscles increase in activity when compared with during normal standing. They next go to get Dan some new trainers. He has his rucksack with long straps, like all the other boys do at school, so with his new trainers in his rucksack they set off to the supermarket. With the rucksack low on his back, the muscle activity in erector spinae will reduce to less than that in standing; however, there is an increase in psoas activity, which will increase the load on the spine.

At the supermarket, they put Agnes's shopping in the trolley and set off for the bus. In order to pull the trolley behind her, the abdominals transversus abdominis and internal oblique and external oblique, as well as the erector spinae and the deep muscle of the spine back muscles, work together to keep her balanced while slightly rotated.

Moving the body into flexion will cause the erector spinae to counteract the effect of gravity. All of the muscles on the dorsal aspect of the spine work in a synchronized fashion to counteract the effect of gravity. At 90% of flexion, a point is reached at which the erector spinae muscles stop working; this is known as the *critical point*. When returning to an upright position, the erector spinae muscles work to overcome gravity; however, once the line of gravity has been overcome erector spinae activity ceases.

CASE STUDY 12.3

This time, the family activity is sitting.

Sitting at a table, Chris is sitting in a chair in what you might call a slumped position, whereas Jenny is sitting in a better posture although still not with her back supported. We know in a slumped position the abdominals – particularly internal oblique, superficial lumbar multifidus and thoracic erector spinae – do not work as much as when you sit upright. If you sit in slumped positions for a long time, as the muscle activity is reduced this may put extra strain on the passive structures such as the ligaments of the spine. Liz is sitting in a chair using the back support; as her back is supported, this reduces the activity of the erector spinae muscles.

Psoas major and minor

Although largely considered as a muscle of the lower limb, psoas attaches to all the lumbar vertebrae. It flexes the hip with the lumbar spine providing a stable base, and working together with iliacus is a strong flexor of the trunk, for example sitting upright from a lying position. It is also suggested that through its lumbar attachments it helps to maintain the position of the lumbar spine (Toshio et al., 2002). Because of its close relationship to the vertebrae, on contraction psoas will increase intradiscal pressure.

Quadratus lumborum

Lying lateral to the vertebral column, quadratus lumborum spans from the 12th rib to the iliac crest. In side flexion, to the left for example, the right quadratus lumborum works eccentrically to control the rate of movement and it assists return from that position by working concentrically. Working bilaterally, it also aids in extension of the spine. The medial fibres, because of their close proximity to the vertebrae, may have an effect on the stability of the spine (McGill et al., 1996).

The abdominal muscles

The abdominal muscles play an important part in providing stability to the lumbar spine. Transversus abdominis plays a key role in stabilizing the spine through its link to the thoracolumbar fascia. As the thoracolumbar fascia attaches to the lumbar vertebrae, contraction of transversus abdominis has an effect on the lumbar vertebrae. It is known that in healthy subjects when moving the upper or lower limb at certain speeds and in different directions, transversus abdominis activates prior to any other abdominal muscle, thus indicating that its activation is preprogrammed (Hodges and Richardson, 1997a, 1997b). The most superficial abdominal is rectus abdominis, which assists in trunk flexion and also controls the degree of pelvic tilt. External oblique helps flex the spine by bringing the thorax closer to the pelvis, say for example when doing a sit-up. As it has an oblique orientation, it also assists in rotation of the thorax, working with internal oblique of the opposite side. Lateral flexion of the spine is assisted by internal and external oblique. Activation of internal oblique with transversus abdominis helps control the visceral contents and increases the intra-abdominal pressure, which also has the effect of stiffening the spine (Hodges et al., 2003). Internal oblique works together with external oblique as above and also assists in trunk flexion. It also works together with transversus abdominis to stabilize the spine. The horizontal orientation of the fibres of transversus abdominis assist in decreasing the laxity of the sacroiliac joints (Richardson et al., 2002).

Overall, the muscles provide support for the spine. Much work has focused on the role of specific muscles; however, muscles work in a coordinated fashion and no single muscle is dominant when assessing spinal stability (Kavic et al., 2004).

Cervical spine

In the cervical spine, the deep muscles on the posterior and anterior aspects act to provide support for the spine, which in turn supports the head. The cervical spine, in supporting the head, plays an important role in maintaining body posture and alignment. The muscles, ligaments of the cervical spine, vestibular system and eyes all play a role in providing information regarding body awareness, including posture and balance.

The deep muscles on the anterior aspect include longus colli, longus capitis, and rectus capitis anterior and lateralis. The main action of the anterior neck muscles is flexion of the head, with the more obliquely oriented parts of the longus colli and rectus capitis lateralis aiding in lateral flexion. There are a high number of muscle spindles within the muscle fibres anteriolaterally to the vertebral body. This may indicate that the neck flexors probably have a proprioceptive role (Boyd Clark et al., 2002). Also, the cervical flexors may play a role in stabilizing the cervical spine. In patients with neck pain, when an arm is moved the deep cervical flexors together with the contralateral sternocleidomastoid and anterior scalenes show delayed activation (Falla et al., 2004).

The superficial cervical muscles acting on the cervical spine that span several vertebrae include upper fibres of trapezius, sternocleidomastoid, and scalenus anterior, medius and posterior. These muscles are important in producing movement of the spine but also act as connecting points of the cervical spine to the thoracic spine (trapezius), clavicle and manubrium (sternocleidomastoid), and first rib.

CASE STUDY 12.4

John is driving home one evening when he has to suddenly brake to avoid hitting a dog that has run into the road. The following day, John's neck feels very stiff when he tries to move it forwards and backwards. If we think about what happens to the head and neck during such an accident and how it moves, this can go some way to explaining why some people might get symptoms in their necks after this type of incident. First, when the car stops suddenly the head and neck move into a hyperextended position; this movement will be limited by the headrest. The head and neck will then go forwards, together with the thorax. This movement will be limited by the seatbelt. The head and neck will then move backwards again. List all the structures that will be stretched and compressed during these movements in the neck and the thorax.

THE SPINE: GIVING SHAPE TO YOUR BODY POSTURE

The architecture of the vertebrae, combined with muscle tone, gives an individual's spine its characteristic shape. The spinal posture must be considered in both static and dynamic functional positions. In basic tasks such as eating and drinking, the spine works to maintain postures while you move the arms. For further information about posture, refer to Chapter 5.

CASE STUDY 12.5

This case study concerns posture.

Chris is complaining to his mum Liz that he has been studying hard at college, as he has exams coming up in the next few weeks and is rushing to finish his assignments. He is getting backache. As Chris is using the laptop while sitting on the bed to write his assignments, think about his posture. As he is sitting on the bed without any back support, the neck and back muscles will be working very hard in this awkward position. The cervical, thoracic and lumbar spine will be flexed, so all the muscles and ligaments, particularly on the dorsal aspect, in these regions will be working hard to support him in this position. Over time, the muscles will become fatigued, and this can lead to discomfort in his back. Now think about what is the best head posture when using any computer. For further information on this, look at Chapter 8 on ergonomics.

THE SPINE: THE LINK IN THE KINETIC CHAIN AND KEY TO FUNCTIONAL TASKS

In day to day activities, the muscles work in a coordinated fashion to achieve tasks, and the activity of all muscles is centrally programmed from the motor cortex (see Ch. 4). Muscles are continuously adjusting to the tasks that they are required to do (Kavic et al., 2004). The muscles on the dorsal aspect work together with the muscle on the anterior aspect. For example, when getting up from a chair the trunk abdominals will flex the trunk, moving the body forwards; the dorsal spinal muscles will work to control this movement.

Because of the spine's inherent flexibility but also supporting properties, the spine has the ability to allow the body to move from a relatively static position to a dynamic movement. In many instances, the head and trunk movement are the first to occur, followed by lower or upper limb movement as required, but it is often a synchronized series of movements in order to achieve normal movement. It can be something as basic as getting up from a chair and walking away, or something much more explosive, for example a swimmer diving in a sprint race from the blocks. In preparation for the dive, the spine moves forwards to allow the centre of gravity to move forwards, then the lower limb muscles work to propel the body forwards. During the dive into the water, the muscles of the spine work to maintain a streamlined posture of the head and trunk.

As noted previously, the thorax serves as an attachment point for muscles and fascia. Some of these muscles and the fascia serve to link the upper and lower limbs. Fasciae such as the thoracolumbar fascia play a vital role in linking the kinetic chain, transferring forces within the body (Vleeming et al., 1995). Although a non-contractile structure, fasciae play a vital role in acting as a link system for many muscles within the body, transferring forces throughout the body. For example, the thoracolumbar fascia helps to transfer load from the trunk to the pelvis and the lower limbs. The thoracic lumbar fascia also plays a critical role in providing some stability to the spinal column through its links to the muscles, as discussed earlier. Tension within the thoraocolumbar fascia will increase with arm and leg movements as a result of anatomical connections. For example, latissimus dorsi, although attaching on to the humerus, also has an extensive spinal attachment, but also via the thoracolumbar fascia to the sacrum. Because of this link, arm movements such as pulling down on a roller shutter door or doing front crawl or butterfly swimming stroke can influence the mechanics of the lower spinal region (McGill and Norman, 1986).

Several muscles not only span several vertebral levels but also have attachments on the limbs. These links are important when assessing movement not only of the trunk but of the lower or upper limbs. It has already been noted that psoas major attaches to the lumbar spine and the femur, and piriformis links the sacrum to the femur. Levator scapulae, latissimus dorsi, trapezius and rhomboids major and minor all have spinal attachments but are connected to the shoulder girdle. In assessing normal movement, you may need to assess the spinal component of, for example, latissimus dorsi when assessing a swimming action such as front crawl.

Also, the spine acts as a link in a chain in order to improve the reach. If you think of activities such as rock climbing or playing badminton, in order to reach for the next hand hold on a rock face or to reach for the shuttle, the spine has to side flex or extend in order for the upper limb to achieve what is desired. The large range of rotation of the head enables us to have a wide field of vision, for example if you are playing rugby and running forwards but need to look to your right to see who you can pass the ball to as a defender is approaching, turning the head and thorax can give you a large field of vision.

Sporting situations can be demanding in terms of a combination of dynamic muscle control and explosive power. Think about gymnastics, for example. In the female balance bar exercise or mat exercise, the spine has to go through many extreme ranges of movement, often in explosive bursts, but also has to maintain a sustained posture. If you think about ballet dancing, this is again a combination of explosive power, often with a large degree of control of the body posture.

SUMMARY

The spine has many roles and works in a highly coordinated fashion to achieve complex motor tasks. Ligaments, discs and muscles link it together, which provides stability but also facilitates movement in many directions. The spine also serves as a link to the upper and lower limbs, enhancing normal coordinated movement. It provides attachment points for muscles and fascia that link the spine to the limbs.

In many functional tasks, the spine needs to keep relatively stable while the upper or lower limbs complete the task. Muscles, fascia and ligaments around the spine serve to provide that stability. The spine also provides protection for major organs such as the heart and lungs.

References

Adams, M.A., Hutton, W.C., 1985. Gradual disc prolapse. Spine 10, 524–531.

Adams, M.A., Hutton, W.C., 1986. The effect of posture on diffusion into lumbar intervertebral discs. J. Anat. 147, 121–134.

Adams, M., Bogduk, N., Burton, K., Dolan, P., 2006. Biomechanics of the Lumbar Spine, second ed. Elsevier.

Andriacchi, T., Schultz, A., Beltyschko, J., Galante, J., 1974. A model for studies of mechanical interactions between the human spine and the rib cage. J. Biomech. 7, 497–507.

Beith, I.D., Robins, E.J., Richards, P.R., 1995. An assessment of the adaptive mechanisms within and surrounding the peripheral nervous system during changes in nerve bed length resulting from underlying joint movement. In: Shacklock, M.O. (Ed.), Moving in on Pain. Butterworth Heinemann, Chatswood.

Bogduk, N., 1997. Clinical Anatomy of the Lumbar Spine. Churchill Livingstone, Edinburgh.

Bogduk, N., Mackintosh, J.E., Pearcy, M.J., 1992. A universal model of the lumbar back muscles in the upright position. Spine 18 (8), 897–913.

Boyd Clark, L.C., Briggs, C.A., Galea, M.P., 2002. Muscle spindle distribution, morphology and density in longus colli and multifidus of the cervical spine. Spine 27 (7), 694–701.

Butler, D.S., 1991. Mobilisation of the Nervous System. Churchill Livingstone, Edinburgh.

Butler, D.S., 2000. The Sensitive Nervous System. NOI Group, Adelaide.

Chloewicki, J., Magill, S.M., 1996. Mechanical instability of the in vivo lumbar spine: Implications for injury and low back pain. Clin. Biomech. 11, 1–15.

Falla, D., Jull, G., Hodges, P.W., 2004. Feedforward activity of the cervical flexors muscles during voluntary arm movements is delayed in chronic neck pain. Exp. Brain Res. 157 (1), 43–48.

Falla, D., O'Leary, S., Fagan, A., Jull, G., 2007. Recruitment of the deep cervical flexor muscles during a postural correction exercise performed in sitting. Man. Ther. 12, 139–143.

Giles, L.G.F., Singer, K.P., 1998. Clinical Anatomy and Management of Cervical Spine Pain. Butterworth, Oxford.

Hodges, P.W., 1999. Is there a role for transversus abdominis in lumbo-pelvic stability? Man. Ther. 4 (2), 74–86.

Hodges, P.W., Richardson, C.A., 1997a. Feedforward contraction of transversus abdominis is not influenced by the direction of arm movement. Exp. Brain Res. 114, 362–370.

Hodges, P.W., Richardson, C.A., 1997b. Contraction of the abdominal muscles associated with movement of the lower limb. Phys. Ther. 77 (2), 132–142.

Hodges, P.W., Kaigle Holm, A., Holm, S., Ekström, L., Cresswell, T., Hansson, T., Thorstensson, A., 2003. Intervertebral stiffness of the spine is increased by evoked contraction of transversus abdominis and the diaphragm: In vivo porcine study. Spine 28 (23), 2594–2601.

Jacob, H.A.C., Kissling, R.O., 1995. The mobility of the sacroiliac joints in healthy volunteers between 20 and fifty years of age. Clin. Biomech. 10, 352–361.

Johnson, E.F., Chetty, K., Moore, I.M., Stewart, W., Jones, W., 1982. The distribution and arrangement of elastic fibres in the intervertebral disc. J. Anat. 135 (2), 301–309.

Kapandji, I.A., 1974. The Physiology of the Joints, Vol. 3: The Trunk and the Vertebral Column. Churchill Livingstone, Edinburgh.

Kavic, N., Grenier, S., McGill, S.M., 2004. Determining the stabilising role of individual muscles during rehabilitation exercises. Spine 29 (11), 1254–1265.

Mackintosh, J.E., Bogduk, N., 1986. The biomechanics of the lumbar multifidus. Clin. Biomech. 1, 205–213.

Macnab, 1977. Backache. Williams & Wilkins, London.

McGill, S.M., Norman, R.W., 1986. Partitioning of the L4–L5 dynamic movement into disc, ligamentous and muscular components during lifting. Spine 11 (7), 666–678.

McGill, S.M., Juker, D., Kropft, P., 1996. Quantitative intramuscular myoelectric activity of quadratus lumborum during a wide variety of tasks. Clin. Biomech. 1, 170–172.

Mercer, S., Bogduk, N., 2001. Joints of the cervical vertebral column. J. Orthop. Sports Phys. Ther. 31 (4), 174–182.

Middleditch, A., Oliver, J., 2005. Functional Anatomy of the Spine, second ed. Elsevier.

Milne, N., 1991. The role of zygapophysial joint orientation and uncinate processes in controlling motion in the cervical spine. J. Anat. 178, 189–201.

Moore, R.J., 2000. The vertebral end-plate: what do we know? Eur. Spine J. 9, 92–96.

Moseley, G.L., Hodges, P.W., 2006. Reduced variability of postural strategy prevents normalisation of motor changes induced by back pain: A risk factor for chronic trouble? Behav. Neurosci. 120 (2), 474–476.

Moseley, G.L., Hodges, P.W., Gandevia, S.C., 2002. Deep and superficial fibres of the lumbar multifidus muscle are differentially active during voluntary arm movements. Spine 27, E29–E36.

Nachemson, A., 1976. The lumbar spine: an orthopaedic challenge. Spine 1, 59–71.

Oliver, J., Middleditch, A., 1991. Functional Anatomy of the Spine. Butterworth-Heinemann, Oxford.

Palastanga, N., Field, D., Soames, R., 1994. Anatomy and Human Movement: Structure and Function, second ed Butterworth-Heinemann, Oxford.

Panjabi, M.M., Takata, K., Goel, V.K., 1983. Kinematics of lumbar intervertebral foramen. Spine 8, 348–357.

Percy, M.J., Tibrewall, S.B., 1984. Lumbar intervertebral disc and ligament deformation measured in vivo. Clin. Orthop. Relat. Res. 191, 281–286.

Richardson, C.A., Snijders, C.J., Hides, J.A., Damen, L., Pas, M.S., Storm, J., 2002. The relation between the transversus abdominis muscles, sacroiliac joint mechanics, and low back pain. Spine 27 (4), 399–405.

Toshio, M., Murakami, G., Sato, A., Noriyasu, S., 2002. The function of the psoas major muscles: passive kinetics and morphological studies using donated cadavers. J. Orthop. Sci. 7 (2), 199–207.

Vleeming, A., Pool-Goudzwaard, A.L., Stoeckart, R., van Windergarden, J.P., Snidjers, C.J., 1995. The posterior layer of the thoracolumbar fascia: Its function in load transfer from spine to legs. Spine 20 (7), 753–758.

Wilke, H.J., Wolf, S., Claes, L.E., Arand, M., Wiessend, M., 1995. Stability increase of the lumbar spine with different muscle groups. A biomechanical in-vitro study. Spine 20, 192–198.

Williams, P.L., Bannister, L.H., Berry, M.M., Collins, P., Dyson, M., Dussek, M. et al., 1995. In: Gray's Anatomy, thirty eighth ed. The anatomical basis of Medicine and Surgery Churchill Livingstone, New York.

Chapter 13

Human movement across the lifespan

Lyn Horrocks

CHAPTER CONTENTS

LEARNING OUTCOMES

At the end of this chapter, you should be able to:

1. describe the individuality of growth and development as the foundation for human movement
2. explore the intrinsic (physical and personal) and extrinsic (social and environmental) factors that influence human movement throughout life
3. discuss the importance of play and recreation as motivation for movement across the lifespan
4. consider the influence of personal lifestyle choices of work, sport, recreation and social activities on human movement.

INTRODUCTION

This chapter looks at human movement across the lifespan, from the earliest movements before birth, through growth and development, as a child and teenager (adolescent), and continuing acquisition of skills and activities as adults. This chapter will be presented in the stages of human movement across the lifespan:

- before birth
- perinatal (newborn)
- the early years

- preteen years
- teenagers (adolescents)
- adulthood, focusing on the movement of young adults, adults and older adults.

At all stages of life, personal, social and environmental factors influence human movement. Drawing on the themes and model for human movement set out in Chapter 1, this chapter will use the terms as follows.

- *Personal factors*: the individuality of human movement, lifestyle choices and motivation.
- *Social factors*: family, community and cultural influences; social expectations and cultural norms; and political influences.
- *Environmental factors*: the home environment; opportunities for play, education and sport; the work environment; and continual skill acquisition.

This chapter will draw these influences together and explore an individual's movement in the context of movement *motivation, opportunities, challenges* and *consequences*. Recognizing the complexity and interdependence of these factors and others expanded elsewhere in this book, this chapter will not rehearse all the related discussions but rather direct the reader to these further resources.

MOVEMENTS BEFORE BIRTH

WHEN DO HUMANS START TO MOVE?

The movements of the developing fetus can be observed and recorded using ultrasound techniques. By 16 weeks' gestation, isolated leg and arm movements and alternate reciprocal kicking have begun. These early movements are the foundation for normal movement development and by 20 weeks are more vigorous and can be felt by the mother. At this stage, the fetus also has individual finger and toe movements.

HOW DOES THE BABY'S ENVIRONMENT INFLUENCE MOVEMENT?

The baby is in an environment in which the effects of gravity are reduced, surrounded by amniotic fluid and supported by the strong muscles of the uterus. The baby is beginning to move their limbs, bending and kicking legs, flexing and extending arms and moving fingers and toes individually. During the third trimester (24–40 weeks' gestation), the baby grows in size and develops strength in muscles by kicking and pushing against the resistance of the uterine wall (see also Ch. 2). The wide variety and repertoire of gross and fine movements, called *fidgety movements*, are essential prerequisites for the ongoing development of normal human movement (Prechtl, 1997).

In the last few weeks of pregnancy, babies are curled up and supported by the uterus, with flexed spine and limbs (physiological flexion) in preparation for birth and for adaptation to the influence of gravity. Babies born prematurely, before they have developed this physiological flexion, have difficulty moving their arms and legs against gravity. If premature babies are placed supine or prone, their head is turned to the side and their arms and legs are widely abducted and laterally rotated, away from the body, resting on the mattress. This position can lead to poor head control and flattening of the sides of the head, abnormal position of hips and shoulder joints, and difficulty in getting hands together and to the mouth. For this reason, premature babies are usually placed on their side, supported by a specially shaped mattress, with their arms and legs flexed close to their body (Poultney, 2007).

THE PERINATAL PERIOD

BIRTH: THE NEW ENVIRONMENT

The biggest change in environment humans ever experience is during birth: from the warm, fluid-filled supporting uterus in the mother to a separate existence and freedom of movement in air, under the influence of gravity. Each contraction of the uterus during labour interrupts blood flow through the placenta, reducing available oxygen. In response to this stress, adrenaline (epinephrine) and noradrenaline (norepinephrine) (the flight or fight hormones) are released, preparing the lungs for breathing air. The build-up of carbon dioxide stimulates the respiratory centre, causing the respiratory muscles to contract, drawing air into the lungs as the baby is delivered, enabling the first deep breath and cry.

NEWBORN

A newborn baby, born at full term (about 40 weeks' gestation), already shows a unique repertoire of movements: turning the head and sucking for feeding, bringing a hand to mouth, wriggling fingers and toes and strong kicking. The baby looks around and shows a wide variety of facial expressions. As we shall see, these actions are the foundation for the development of normal human movement.

THE EARLY YEARS

There is a wealth of resources available detailing human development (see further reading list), so it is not the intention of this text to rehearse these topics. Rather, the focus of this chapter is the motivation to move at different stages of development, leading to refinement of gross and fine movements, communication and language, and cognitive and social skills.

MOTIVATION FOR MOVEMENT: INTERACTION BETWEEN MOTHER AND BABY

A newborn baby is driven to move by the need to feed and to communicate and interact with their mother or carers. Early reflexes produce the movements needed for feeding: turning the head, rooting and sucking. When held upright with feet on a firm surface and tilted slightly forwards, babies make stepping movements (stepping or walking reflex) that are a precursor for the development of walking (Gallahue and Ozmun, 2002). The individuality of human movement is seen as a baby begins to develop selective movements when responding to the mother, using head control, eye movements and reach and grasp, demonstrating the integration of all body systems: musculoskeletal, sensory, cardiac and respiratory, neurological, integrative and cognitive. Living in the social world of their family, the baby communicates with a variety of sounds (cooing), taking turns vocalizing with another person and smiling readily when handled and played with.

The development of head control in midline, at about 3 months, enables the baby to watch the movements of their hands and fingers. This is the start of the hand–eye coordination needed to explore their environment by touching a person or toy through reach and grasp movements.

STAGES OF DEVELOPMENT OF HUMAN MOVEMENT LEADING TO WALKING

In this section, we will consider the stages of development of human movement leading to independent upright walking, and consider that at the same time, the baby is also developing fine hand function and communication skills. Each of the following developments will be explored in turn:

- patterns of movement and core stability
- balance in sitting to enable fine hand function
- sequences of movement from one position to another
- crawling and exploring
- upright standing and developing balance
- cruising (weight shift and stepping sideways)
- independent walking
- progression of walking.

ACTIVITY 13.1

Readers are referred to the following recommended texts. Please see the further reading list for this chapter for specific details.

- Bellman and Peile (2006) *The Normal Child*
- Cech and Martin (2002a) *Functional Movement Development Across the Life Span*
- Shumway-Cook and Woollacott (2007) *Motor Control: Translating Research into Clinical Practice*
- Chapters in this book: Chapter 5 (*Posture and Balance*) and Chapter 6 (*Motor Learning*)

Developing patterns of movement and core stability

In the development of human movement, muscles and joints need to work together in a combination of actions to produce a functional movement such as reaching for a toy, bringing hands to mouth for feeding, or moving the legs for walking. These are called *patterns of movement* and consist of lots of individual muscle contractions and joint actions, all coordinated to produce a smooth, controlled functional movement (e.g. the movements required at the shoulder, arm, and hand and fingers to reach out and grasp a toy). The hip, leg and foot movements seen when babies are kicking are the patterns of movements that will later develop into walking. These patterns of movements are possible only if the baby's body (trunk) provides a stable but mobile

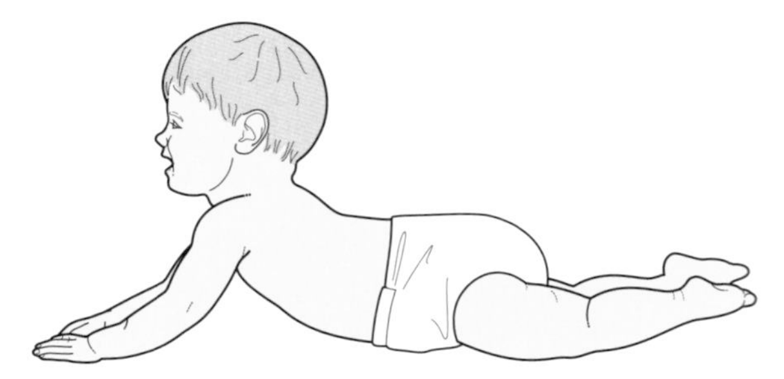

Figure 13.1 A baby developing extension by pushing up on straight arms in prone.

base. This core stability develops as the baby works their abdominal muscles and back extensors against gravity, by lifting legs up and kicking when lying on their back in supine, and extending their back and pushing up on their hands and arms when lying on their front in prone (Fig. 13.1).

Developing sitting balance and fine hand control

Human movement is a sensory–motor experience. The development of sitting balance and fine hand control demonstrates the influence of intrinsic factors (developing core stability and mobility) and extrinsic factors (exploring the environment and toys) on human movement. At about 6 months, the baby sits on the floor with legs out to the side, leaning forwards and propping on hands. Over the next few weeks, gradual development of core stability of abdominal and back muscles enables more upright independent sitting balance. In response to the stimulation of handling toys of different shapes, textures and weights (Fig. 13.2), hand control develops into a range of manipulative skills for play and function, picking up small objects with a fine pincer grip between thumb and tip of forefinger to finger feed, put toys in and out of a box and turn the pages of a small book. These stages in human movement development will prepare the baby for function, such as sitting in a high chair, holding a cup for a drink and finger feeding. But hands are not only used to support feeding. Hand gestures, pointing, copying hand gestures, clapping hands and waving etc. are key elements in human communication in all cultures and languages (see also Ch. 7; Psychosocial influences on human movement).

Figure 13.2 A baby showing balance in sitting and playing with a toy.

Sequences of movement from one position to another

Sequences of movement are the coordinated actions used to change position, such as from sitting into crawling, turning round to look behind you, and changing direction of movement when walking and running. Learning to move from one position to another is an essential stage in the development of human movement. An example of this is rolling over, which requires the coordinated movements of both flexion and extension of the head, body and limbs. Think how you move from lying on your back, turning on to your side and then right over on to your front. Try this in slow motion and feel the rotation (twisting) movements that are needed for this action. Starting with your head turning to one side, the shoulders turn and the body, hips and legs gradually flex (bend) and rotate (twist) to follow the direction of your head. You are now halfway, lying on your side; to get right over on to your front, you need to extend your head and neck, trunk and limbs. So you can see how rotation is a combination of flexion and extension in a sequence of movements.

Babies start learning to roll at about 4–5 months old, and this sequence of movements gradually become smoother, more coordinated and much faster.

Crawling and exploring

Babies are motivated by curiosity and can make choices about the movement needed to play and explore their immediate surroundings (Fig. 13.3).

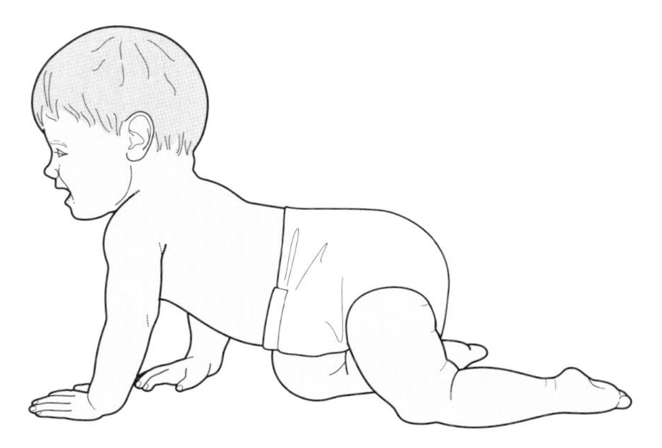

Figure 13.3 A baby crawling and exploring the environment.

The individuality of human movement is seen as babies use sequences of movement to move around their home by rolling or crawling, bottom shuffling in sitting, or bear walking with hands and feet on the floor. This mobile weight bearing through arms strengthens the muscles around the shoulders, which is necessary for the development of hand function. Their fine hand skills develop in response to the experience of sensory stimulation through many different textures – smooth flooring, carpet, sand and grass – and feeling things that are soft, hard, rough, warm, cold or wet. Motor planning develops by moving around the home using gross motor movements and manoeuvring between pieces of furniture, people and toys, which also enhances language development by physically experiencing prepositions such as over, under, round, through and between.

Upright standing and developing balance

At around 9–12 months, babies pull themselves up to stand, experiencing weight bearing through their feet and legs. This gradually strengthens the muscles around the pelvis, hips, knees and feet. Young children have a large-sized head in proportion to their body and legs, causing the centre of gravity to be higher and more anterior than in an adult, so initially they stand with their feet very wide apart and lean forwards on to the furniture to increase the base of support (Fig. 13.4). With increasing core stability and developing strength in muscles around the hips, they gradually need less support and can stand alone.

Squatting (feet flat on the floor with fully flexed hips and knees) is an important position for play

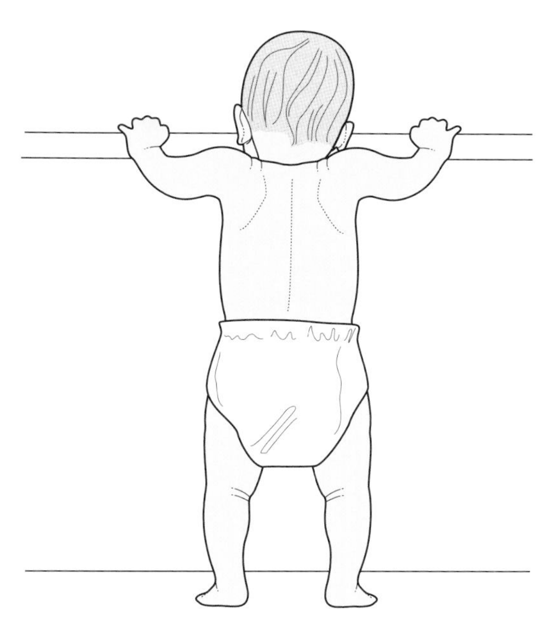

Figure 13.4 A baby in upright standing supported by holding on to the furniture.

Figure 13.5 A child in stable squatting position, which enables concentration on fine hand movement for play.

(Fig. 13.5), and in many cultures this position continues to be used throughout life for activities of daily living such as washing, food preparation and cooking and for skilled work. It is very stable, with a low centre of gravity and wide base, allowing concentration on fine hand function.

Cruising and stepping sideways

The patterns of movement seen earlier in kicking and crawling, combined with weight bearing and increased muscular strength, prepare the baby for walking. The centre of gravity in young children is relatively high because the head is very large and heavy in relation to the trunk, whereas in adults the head and trunk are more in proportion, making the centre of gravity lower. In order to take a step, the child leans over to one side, shifting weight on to one foot to free the opposite leg to move. Initially, this movement is out to the side, so progress, when furniture or other support allows, is often sideways in a pattern of movement called cruising. The ability to balance and move in this upright position against gravity slowly develops with the integration of all systems: musculoskeletal, vestibular, visual and proprioceptive. When learning to walk, the child's initial posture is leaning forwards, because the weight of the head is making the centre of gravity high and anterior, potentially causing a loss of balance. By taking a step forwards to maintain balance, the child gradually moves forwards into independent walking.

Independent walking

The age at which children stand and walk depends on early development and genetic and cultural traits, for example babies from African cultures tend to walk earlier than those from Europe, while children denied opportunities for stimulation and exploration show delay or retardation in all aspects of their development. Walking is a self-directed activity with young children, motivated by curiosity and the desire to explore their surroundings. Children develop body awareness and motor planning skills by walking around the home, manoeuvring between pieces of furniture, people and toys (see also Ch. 6; Motor learning).

Progression of walking in children

Gallahue and Ozmun (2002) have described three stages of development and progression of walking in children: initial (the first year of walking), elementary (the gradual development of the walking pattern) and mature (progression to an established energy-efficient gait). This gradual progression of walking depends on having normal muscles, joints and nervous system. Growth also affects balance and posture in walking, and the head in a small child is very heavy and large in relation to the body, bringing the centre of gravity high and anterior. Growth in the trunk and legs makes the child taller and the head and body are more in proportion. As their legs grow longer, they are able to take longer strides, and refinement of balance mechanisms and motor control mean they can walk with a narrower base and smoother action and can vary direction and speed.

- Initial: difficulty maintaining upright posture, with unpredictable balance; short steps; flat foot contact with flexed knee; hips and feet turned out (lateral rotation).
- Elementary: gradual smoothing of walking pattern, with improved motor control; increase in step length; heel–toe contact; arms at sides, with small swing; base of support hip width; increased pelvic tilt; out-toeing reduced; apparent vertical movement at each step, caused by active pushing off with the toes.
- Mature: reflexive arm swing; narrow base of support; relaxed elongated gait; minimal vertical lift; definite heel–toe contact.

MOVEMENT AS A PRETEEN CHILD: PROGRESSION FROM YOUNGER CHILD

MOTIVATION FOR MOVEMENT: TO FIND OUT ABOUT MYSELF AND OTHERS

Play is a fundamental right for all children that enables them to learn about themselves and their world. The centrality of play to human development is recognized by the United Nations Convention on the Rights of the Child 1989 in the following statement: 'Children have a right to relax, play and join in a wide range of activities' (article 31, UNCRC 1989).

Motivated to move by playing with others, children use play to experience physical challenges as they run, jump and chase each other. Play can be structured (e.g. in an organized game of football) or free moving and creative (e.g. in dance, dressing up, drawing and painting). In addition to physical development, playing with others also provides stimulation for social and communication development. Through play, children learn the vital social skills needed for life in their culture and local communities, i.e. interaction, joining in, taking turns, sharing and being part of a group or team. Children need suitable, safe facilities to be provided locally for them to have the opportunity to enjoy playing with others.

DEVELOPING POSTURAL CONTROL

The intrinsic (internal) factors of improved postural awareness and core stability enable better hand–eye coordination and the development of these skills by interacting with other people, demonstrated by throwing and catching a soft toy, beanbag or big ball. Children continually experience challenges in the physical and social environments of family, home, nursery and kindergarten, providing many opportunities for skill acquisition and refinement. For more information, please see Chapter 5 and Bradley and Westcott (2006).

CASE STUDY 13.1

This case study concerns the importance of play in the development of human movement.

In this case study, consider the intrinsic factors (physical, emotional and personal) and extrinsic factors (the environment and social aspects of playing together) and their influence on the children's movement.

Jenny, who is 9 years old, is playing with her friends in the park. There is a patch of grass, a climbing frame, a slide and a rope swing. What are the human movement skills needed for playing on these pieces of equipment? For example, children need strength in arms and hands to hold on the rope and balance for climbing on the frame. Can you think of any more?

A group of children are playing football on the grass. List the social, communication and physical skills they are demonstrating, such as coordination to kick the ball and team skills. See how many more you can add.

THE ONGOING CHALLENGES OF POSTURAL CONTROL DEVELOPMENT

Growth

A key feature in babies, children and adolescents is the process of continuing growth, a major factor differentiating them from adults. Human movement is essential for stimulation of bone and muscle growth and development of strength and endurance.

The amount of growth in all tissues and systems, including the brain, in the first 2 years of life is remarkable.

There are many factors affecting growth: genetic make-up, nutrition, activity and the environment and hormones, particularly human growth hormone. Bone grows in length at the epiphyseal plates at each end of long bones and in thickness at the periosteum surrounding the bone (appositional growth). Muscle cells are formed in the developing fetus and do not increase in number but grow by increasing in size (hypertrophy) in response to the stretch by bone growth and increase in activity. Organs, soft tissues, nerves and blood vessels also grow at the same time (Tortora and Grabowski, 2003).

Growth can be recorded and monitored from birth through childhood using centile charts. Each culture has developed guidelines for normal growth, based on the average weight and height for age in that population. They can be used to indicate the need for medical intervention if a child is not growing normally because of poor feeding, malnutrition or illness.

Consequences of developmental variations

Variations in body size and shape

Body size and shape are determined by genetics and the child's social and cultural environment: nature and nurture. Society is responsible for changing attitudes to diet, nutrition and activity levels and is influenced by poverty, advertising on the media, shopping on the Internet, fast food, and lower levels of activity together with cultural norms and expectations.

Obesity is a term with very different social and cultural meanings and implications across the globe. In the western world, it can be defined as weight above the 97th centile or 20% above the mean average weight for age in that population (Francis and Jaimeson, 2006). Obesity is valued in some cultures as evidence of a family's wealth (Helman, 1996) and for specific sports, such as sumo wrestling, in which the increased weight and bulk of the competitors are desired as they aim to move their opponent off balance. In the UK, the USA, Australia and Hong Kong, there is concern about the increasing number of children who become obese and the consequences for their long-term health (Gallahue and Ozmun, 2002). Overweight children can become obese adults, with increased risk of heart disease, stoke and diabetes (Francis and Jaimeson, 2006). Osteoarthritis in weight-bearing joints can develop, particularly in hips and knees. Children who are obese cannot move easily, get breathless and have difficulty joining in physical activities and sport. This can affect children's self-esteem, making them vulnerable to bullying and social exclusion from their peer group.

Variations in walking

Walking is a fundamental activity of human movement, and young children show many normal variations depending on their stage of development and skeletal growth. Common variations are walking with feet turned in or out and flat feet. Some children may need assessment to exclude neurological or orthopaedic conditions, but most outgrow these tendencies by adolescence as part of normal growth. By the time children are about 7–9 years old, they have the posture and alignment of an adult but with a greater range of movement at most joints. As they are going through periods of rapid growth, often in 6-monthly bursts, some children experience temporary growing pains in their leg muscles after an active day running around in school. The exact cause is unknown, but the pain is usually relieved by rubbing or massaging their legs (Kennedy, 2006).

Delay or difficulty with human movement in children

Development is a continuum with individuality, but if children are experiencing difficulty or not showing normal progress they may need assessment. The *Movement ABC* (Henderson and Sugden, 2007) is a selection of games and activities for children to enjoy and can be used to assess gross and fine movement function and skills. Difficulty in performing some of these tasks can indicate when specific activities may help the child, which parents, teachers and therapists can incorporate into the school day and at home (see also Chambers and Sugden, 2006).

EDUCATION

All children have a right to education (article 28, United Nations Convention on the Rights of the Child 1989) throughout the world, and education should aim to develop each child's personality and talents to the full (article 29, UNCRC 1989). These guiding principles are interpreted and implemented in different countries in a variety of ways depending on culture, traditions and resources. School is where children spend most of their time and learn the social, cognitive and physical skills that they will continue to develop throughout life.

How does the educational environment influence continuing movement development?

Human movement is an integral part of the school day, giving children opportunities and challenges for learning in a wide range of activities. Structured education is guided by the philosophy and expectations of each culture and community. In the UK, this is directed by the Education Act (Department of Education 1996) and set out in the National Curriculum (Department of Education 1999). The wide range of topics facilitates children's development across all life skills: physical, cognitive, emotional, communication and social.

In the UK, the physical education programme has been designed to facilitate the development of human movement. Activities are appropriate for the different stages of development and gradually introduce new skills (Physical Education Key Stages 1 and 2; Department of Education 1999). Children learn to play games and sports in groups and teams, which can lay the foundations for a lifelong interest in different sports and activities that will influence their fitness and mobility throughout life. By frequently running, jumping and playing games, the young child's musculoskeletal system develops strength and endurance. Just a few minutes in a playground at break time will show how children can run, jump and shout loudly simultaneously, each activity exercising their heart and lungs and improving respiration and circulation.

These activities and gross motor movements are essential for improving the core stability and postural control children need to enable them to sit at a desk and develop fine hand control. They develop the fine hand and finger movements needed for handwriting, drawing and painting, sewing, playing musical instruments and computer keyboard skills.

ADOLESCENCE: A PERIOD OF RAPID GROWTH, SOCIAL PRESSURE AND PERSONAL CHOICES

Adolescence is a time of personal as well as physical growth, when young people start taking responsibility for their lifestyle and priorities. Adolescents develop an increased awareness of their cultural and social identity; image is everything, and they are heavily influenced by their peers to conform. There are many opportunities open to them, and the challenge to succeed in education, sport and social life determines the choices they make, which can have lifelong consequences for their health and well-being (see also Ch. 7).

INTRINSIC PHYSIOLOGICAL AND ANATOMICAL CHANGES (Box 13.1)

Puberty starts at around 10–12 years old with release of hormones that trigger rapid growth and secondary sexual development, all of which affect human movement. Rapid increase in bone growth by length, circumference and density occurs in spurts, with the surrounding soft tissues and muscles stimulated to grow by hypertrophy (muscle fibres increasing in size, not number). These changes affect mobility, as adolescents adapt to their rapidly increasing height and weight. The ratio of height to weight changes during adolescence, from prepubescent children being lightweight for their height to teenagers increasing their weight in relation to height. The effect of gravity on the heavier body increases the effort required to walk, run and jump and may contribute to some teenagers becoming less active.

Rapid growth spurts of the femur and tibia, combined with a lot of sport, can cause some teenagers to experience pain at the insertion of the quadriceps, attached by the patella tendon into the tibial tubercle, just below the knee, and the calf muscles attached by the Achilles tendon into the calcaneum, at the back of the heel. This temporary discomfort can limit their movement, but orthotics, in the form of patella tendon straps and gel heel pads, give support and relieve pain, enabling continued participation in sport.

Young people are also at risk of pain and problems with posture and movement if they are sitting at desks for hours in school and slumped over a computer in the evenings (Gillespie, 2002). Sitting in a flexed posture puts strain on the muscles and ligaments around the spine, as it continues to grow slowly for several years after the long bones in the arms and legs have ceased to grow in length. Human movement, especially up against gravity, such as walking, running and jumping, gradually requires more effort and energy, which may lead some teenagers to adopt a more sedentary lifestyle (see also Chs 5 and 12).

BOX 13.1 Intrinsic and extrinsic factors influencing human movement in adolescents

Intrinsic factors

Physical
- Health status
- Freedom from pain or injury
- Rapid growth and development
- Changing height:weight ratio
- Strength, endurance, coordination and dexterity
- Play and skill acquisition
- Physical challenges and opportunities

Psychological
- Self-motivation and determination
- Success, reward, satisfaction and pleasure
- Self-esteem, image and confidence
- Difficulty, disappointment and failure
- Encouragement, help and support
- Personal lifestyle choices

Extrinsic factors

Social
- Family, extended family and carers
- Peers, friends, colleagues and team members
- Teachers and lecturers at school, college and university
- Local community and volunteers
- Sports coaches, instructors and trainers
- Teachers for dance, yoga, martial arts, music and crafts

Wider social network
- Society expectations and social policy
- Cultural norms and values

Environmental
- Financial stability or relative poverty
- The home: house, flat, bungalow, caravan or shared accommodation
- School and further or higher education
- The workplace

Local environment
- Local policies for housing, traffic control and pollution
- Local facilities for recreation, leisure and sport
- Travel, transport and safety
- Charities, voluntary organizations and community projects

Wider social environment
- Human rights
- United Nations Convention on the Rights of the Child
- Health and social care
- Government campaigns for play, sport and recreation

EXTRINSIC FACTORS: SOCIAL AND ENVIRONMENTAL INFLUENCES (Box 13.1)

During this period of rapid growth, teenagers are particularly vulnerable to physical stress and liable to sustain injuries if they are pushing their bodies to the limit in competitive sport. There is sometimes social pressure from peers, families and sports teams on these young people to compete and succeed. In the wider environment, local facilities for recreation and sport, and policies for housing, traffic control and pollution will influence participation in activities such as walking to school, cycling, play, sport and leisure. Campaigns for safe play and leisure areas and community projects can facilitate adolescents continuing to enjoy physical activities.

CASE STUDY 13.2

Think back to when you were 15 years old. What were your interests, hobbies and activities?

Which factors were influencing you? How did the intrinsic factors (physical, emotional and personal) and extrinsic factors (the environment and social aspects) influence you and affect human movement? For example, maybe you learned to swim because you lived near the swimming pool and your older sister was a lifeguard there, or perhaps you had good balance and sense of rhythm so joined a dance class, or possibly you enjoyed playing a musical instrument.

Write or draw a sample of a typical week including activities you took part in at school and those with your family and friends.

ADULTHOOD

In this section, we will look at human movement in adults, building on the foundations established during childhood and adolescence. Personal lifestyle choices of studies, employment, sport and leisure will influence skill acquisition and refinement, mobility and strength.

In this section, we will:

- consider the effects of intrinsic factors – physiological and physical changes that occur during adult life, the motivation for movement and the influence of personal lifestyle choices
- discuss the social and environmental context for human movement, providing opportunities and challenges for adults in education, work, rest, leisure and sport
- explore the political and social responsibilities to provide facilities, programmes and opportunities for adults to pursue activities and maintain mobility.

YOUNG ADULTS

Challenges for movement

Young adults continue slow growth of bone in circumference and density until their mid twenties, and the spine also continues growing, explaining why some young people grow taller while away at college. This is a crucial time for laying down strong, dense bone with weight-bearing activities, to prevent osteoporosis at a later age (Cech and Martin, 2002b, 2002c). Body shape continues to develop, with broadening of shoulders and chest in young men and of pelvis and hips in women. Muscles hypertrophy, adding bulk and weight, depending on the demands put on them during physical activities such as walking, cycling, swimming and manual work.

Opportunities and choice

As young people leave school and enter further education and employment, they experience great changes in lifestyle and increase in personal responsibilities. There are high expectations of them by their families and society. They need to balance their time between study, work and leisure, and the choices they make affect human movement.

ACTIVITY 13.2

Revisit the chapters on ergonomics and musculoskeletal activities (Ch. 8) and on psychosocial aspects of human movement (Ch. 7).

CASE STUDY 13.3

This case study concerns work, rest and play at university.

Chris (20) is a student at university. Consider the stresses on his body as he sits in lectures and works on assignments at his desk. Think about his position and the effect this has on his neck and back. How could he improve his posture and prevent problems in the muscles and joints of his shoulders and hands?

Chris is keen on sport. How can participation in sport and social life help to counteract these problems? An example would be swimming backstroke to straighten his spine and mobilize his shoulders, or dancing with his friends for mobility and fun. Can you think of any more examples?

You will find more information in Chapter 5 on posture, Chapter 8 on the environment and Chapter 12 on the spine.

ADULTS

What motivates adults to move?

As healthy adults, we tend to take human movement for granted, thinking about it only if injured or trying to acquire a new skill. Just pause a moment and consider the wide range and individual variety of movements you already possess. Human movement is for function, and being personally independent is a strong motivator for human movement, including personal and domestic activities of daily living.

ACTIVITY 13.3

Consider the everyday opportunities for human movement. List the activities you have already done today, starting with getting out of bed. What has motivated you to keep moving?

Think of the basic requirements for those movements, such as core stability of trunk, mobility of joints, strength of muscles and dexterity of fingers. Can you add any more?

Some actions are almost automatic, such as washing your hands; others require skill, coordination and dexterity, such as making a sandwich. Can you add any more examples?

Think back to your childhood and adolescent years and consider how the choices of activities, sport, hobbies and work that you have made in your life have influenced the way you move and the skills you have acquired. For example, learning to ride a bike as a child enables you to keep fit by cycling to work.

The working environment: presenting opportunities and challenges

With increasing time spent in employment, many adults are unable to be as physically active as in their youth. Their work may necessitate long hours in one position, requiring concentration and using repetitive movements, for example typing, telephone switchboard or call centre work, factory assembly work, sewing, writing and driving. Many jobs require a very stable base and strong core stability to allow freedom of movement of shoulders, arms and hands for fine, delicate, precision work. Consider the demands on the body for a dentist, hairdresser, lace maker, computer board assembly worker and laboratory technician.

The consequences can be postural problems, pain in spine and joints, and repetitive strain injuries. Holding a static posture for a long time is tiring and can lead to muscle fatigue, tension and loss of concentration. Adults continue to develop strength, skills and mobility by adapting to the requirements of their chosen work and leisure activities.

A farrier (blacksmith), for example, will develop muscle bulk and strength in trunk and arms after many years shoeing horses and metalwork, while a window cleaner, repeatedly climbing ladders, will have strong legs, mobile shoulders and excellent balance.

The working environment can impose particular physical and psychological stress on people, affecting their movement and well-being. Hazardous environments such as in deep sea fishing, coal mining, the construction industry, farming, the fire service and police work demand fit active people able to react quickly for their own and others' safety.

But of course rest, recovery and relaxation are also essential for continued work, sport and activities of daily living. Local council recreation and leisure facilities such as swimming pools, tennis courts, cycle tracks, parks and gardens offer the opportunity to enjoy a variety of activities. Adults need exercise to keep fit and active, especially if recovering from illness or injury or if obese. An example in the UK is the *Health for Life* scheme that has been introduced to enable doctors to refer people for physical activities at their local leisure centre at concessionary rates (Cardiff Council 2008).

Physiological challenges in adulthood

How do hormonal changes occurring in adults affect human movement?

During pregnancy, the many hormonal changes and the weight of the developing baby affect the mother's posture, ligaments and joints. These changes affect general mobility and activity levels, although many women carry on working until just before their baby is born. After the birth, the mother's body takes many months to regain the normal hormone levels, fitness and mobility.

Hormonal changes experienced by men and women in their fifties and sixties may cause symptoms that deter them from continuing sport or taking up new activities. A reduction in bone mineral density can lead to osteoporosis and vulnerability for fractures. Weight-bearing exercises such as regular walking provide stimulus to bone, helping to maintain density.

Motivation for keeping mobile and active throughout adulthood

People are motivated to keep mobile and active so that they can continue their work and leisure activities for many years. This can have health-related benefits of prevention of heart disease, osteoporosis and stroke (Folsom et al., 1997; British Heart Foundation 2005).

Maintaining joint mobility and muscle strength contributes to older adults remaining independent and enjoying a wide range of activities. Cultures have different attitudes to human movement in older adults, and some society expectations can be limiting, assuming they will not be able to participate in sport. This may be challenged by sports organizations that encourage and welcome older people. An example in the UK is Scotland's national strategy for sport, which emphasizes that age should be no barrier to participation (Nicholson, 2004). This report on older people, sport and physical activities is supported by a comprehensive review of international literature on sport for older people, outlining the health benefits.

OLDER ADULTHOOD

In this section, we consider:

- physiological changes occurring in older adulthood and the challenges to movement that these present
- the benefits of being active
- prevention of falls
- opportunities for physical activity.

What are the physiological changes that affect human movement in older adults?

Ageing at a cellular level occurs gradually throughout adult life and more rapidly during the sixth and seventh decades. There is loss of elasticity in tissues, skin becomes thinner and joints produce less synovial fluid. Muscle mass, strength and endurance are gradually reduced, which may be part of the ageing process or the result of disuse atrophy caused by less physical activity (Woodrow, 2002). An example of this is weakness in hip and knee extensors, so older adults may have difficulty getting up from a low chair or going up and down stairs.

The effect of these physiological changes can be seen in altered posture, balance and postural control of movement. Osteoporosis in the spine causes an increased thoracic kyphosis displacing the centre of gravity forwards, which, combined with decreased range of movement in hip extension, leads to hip and knee flexion and the body leaning forwards. To compensate for this, older people often extend their shoulders to bring their arms back either into their pockets or behind their back. This leaves them vulnerable if they trip and cannot quickly put their hands out for protection. Their ability to adjust to increased body sway using the hip and ankle strategies becomes limited (see also Ch. 5) (Laughton et al., 2003). Step and stride length reduces, and they may have difficulty clearing the ground with their toe in the swing phase of gait, causing slips, trips and falls. They may need to use a stick or walking frame for stability and safety when walking.

CASE STUDY 13.4

Agnes is keeping mobile and independent.

Agnes (85), the grandmother in the family, is fit and well and personally independent. She participates in physical activities in the community, enjoying t'ai chi classes and free swimming sessions. She meets the youngest child after school every day, and they walk home together, shopping for food on the way. She is mobile and active, but physiological changes have led to some loss of range of movement at her hips and strength of hip and knee extensors.

Discuss her lifestyle and how this helps to keep her mobile. List some adaptations she may need in the house to maintain her independence. For example, she uses a bag on wheels to carry the shopping home and needs a handrail on both sides of the stairs. How many other ideas can you add (thinking about your friends and members of your family who are of Agnes's generation may help).

Agnes sees her friend Edith for coffee every Tuesday. They meet at Edith's flat, because she has arthritis and the stiffness in her hips and knees makes walking difficult and painful. The joints of her hands and fingers are stiff, and this affects fine movements such as doing up buttons and preparing vegetables.

Discuss the extra equipment and adaptations Edith may need to help her maintain independence and enjoy activities. For example, she has a special armchair with a seat that tips forwards to help her get up into standing, and as she can no longer knit jumpers, she now enjoys making patchwork cushion covers using her sewing machine.

Challenges to human movement in older adulthood

Compounding factors and threshold of ability

By the time people reach their eighth and ninth decades, they may have compounding factors of visual or hearing difficulties and joint stiffness that affect their ability to move easily and maintain personal independence. A threshold of ability is required for independent living, and a minor incident, such as a muscle strain or illness requiring a few days' rest, can precipitate a rapid decline and loss of personal independence. This may not be inevitable, and carefully graded exercise programmes can be beneficial.

The National Service Framework for Older People in England (Department of Health 2001) and in Wales (Welsh Assembly Government 2006) recommends physical activity and community exercise programmes, supported by evidence of the positive benefits from aerobic and anaerobic exercises (World Health Organization 1996; Munro, 1997; Department of Health 2004). Older adults who have osteoporosis are particularly at risk of fractures, and regular weight-bearing exercises can be beneficial in maintaining bone density. The National Service Frameworks incorporate health and social activities in their health promotion programmes for older people, for example the standards include:

- moving and physical activity
- free swimming and walking initiatives
- green gym movement (gardening and allotments)
- improved nutrition
- emotional well-being
- health protection
- safety promotion
- keep warm, keep safe
- exercise referral schemes.

Prevention of falls in older adults

The prevention of falls is a major public health issue worldwide, involving the health and community services, local councils, social services and the voluntary sector. In the UK, this topic is addressed in the National Service Frameworks for Older People, setting standards for services to maintain the health and mobility of the older population (standard 6, National Service Framework for Older People; Department of Health 2001). Fall prevention programmes target vulnerable people, educating them about the causes of falls, making their environment safe, and the value of physical activity to maintain muscle strength and joint mobility and improve balance (Campbell, 1997; Department of Health 2002; Close and McMurdo, 2003; Howe et al., 2008; McClure et al., 2008). Gardner et al. (2001), working in Dunedin, New Zealand, describe their service for identification and assessment of vulnerable adults and have designed individual programmes that increase in difficulty to promote muscle strength, incorporate walking and can be continued safely at home.

Opportunities for physical activities

Many people remain fit and active throughout their lives, some achieving remarkable feats of physical activity. Others continue a lifelong interest or take up a new sport or hobby when they have more free time. Ongoing personal independence can be helped by maintaining a reasonable level of joint range, muscle strength and cardiovascular fitness by joining group activities or following an individual programme.

Some examples in the UK of these opportunities for continued enjoyment of human movement are as follows.

- The University of the Third Age (University of the Third Age 2008) is a voluntary organization run by and for older adults, offering opportunities for lifelong learning and human movement through a wide variety of topics and group activities such as walking, dancing, gardening, t'ai chi and swimming.
- DVDs and videos of exercises and ch'i kung (Williams, 2008), Pilates (Chandler, 2008), t'ai

chi and yoga are available for people to use in the comfort and privacy of their own home.

- Improving balance and posture and joining in the actions in virtual sport using Nintendo Wii Touch! Generations games (Nintendo, 2008) are very popular and can be used by the whole family.

CONCLUSION

In this chapter, we have discussed human movement across the lifespan, starting with the earliest kicking before birth, through stages of growth and development in children and adolescents, continuing skill acquisition in adults and maintaining mobility in older adults. This provides a foundation for further study into the fascinating topic of human movement.

The influence of personal, social and environmental factors that lead to the individuality of human movement has been considered. Personal lifestyle choices and motivation, at all stages of life, influence the development and enjoyment of human movement. The social world in which children and adults live shapes family, community and cultural norms and expectations. Different environments offer challenges and opportunities for refining skills, increasing strength and endurance, and improving mobility and dexterity.

All these factors enable people to enjoy the benefits of movement at work and recreation across the lifespan.

References

Bellman, M., Peile, E. (Eds.), 2006. The Normal Child. Churchill Livingstone, London.

Bradley, N.S., Westcott, S.L., 2006. Chapter 3 motor control: developmental aspects of motor control in skill acquisition. In: Campbell, S.K., Vander Linden, D.W., Palisano, R.J. (Eds.), Physical Therapy for Children. third ed. Saunders, St Louis.

British Heart Foundation, 2005. http://www.bhf.org.uk/ (Accessed 14/2/09).

Campbell, A.J., 1997. A randomised control trial of a GP home exercise programme to prevent falls in elderly women. Br. Med. J. 315, 1065–1069.

Cardiff Council, 2008. Health for Life. Cardiff Council. http://www.cardiff.gov.uk (Accessed 14/2/09).

Cech, D.J., Martin, S., 2002a. Cech, D. J., Martin, S. (Eds.), Functional Movement Development Across the Life Span. WB Saunders, Philadelphia.

Cech, D.J., Martin, S., 2002b. Skeletal system changes. In: Cech, D.J., Martin, S. (Eds.), Functional Movement Development Across the Life Span. WB Saunders, Philadelphia.

Cech, D.J., Martin, S., 2002c. Theories affecting development. In: Cech, D.J., Martin, S. (Eds.), Functional Movement Development Across the Life Span. WB Saunders, Philadelphia.

Chambers, M., Sugden, D., 2006. Early Movement Skills: Description, Diagnosis and Intervention. Whurr, Chichester.

Chandler, L., 2008. Real People Pilates for over 50s. DVD and VHS video. Real People Pilates, Frome.

Close, JCT, McMurdo, MET on behalf of the British Geriatrics Society Falls and Bone Health Section, 2003. Falls and bone health services for older people. Age Ageing 32, 494–495.

Department of Education, 1996. Education Act. Department of Education, London.

Department of Education, 1999. Key stages 1 and 2. National Curriculum. Department of Education, London. http://curriculum.qca.org.uk/key-stages-1-and-2/index.aspx (Accessed 12/2/09).

Department of Health, 2001. National Service Framework for Older People February 2001. http://www.rcgp.org.uk/pdf/ISS_SUMM01_02.pdf (Accessed 14/2/09).

Department of Health, 2002. Information strategy for older people. Department of Health, London.

Department of Health, 2004. 5 a week: evidence on the impact of physical activity and its relationship to health, Report to the Chief Medical Officer. Department of Health, London.

Folsom, A.R., Arnett, D.K., Hutchinson, R.G., Liao, F., Clegg, L.X., Cooper, L.S., 1997. Physical activity and incidence of coronary heart disease in middle aged women and men. Med. Sci. Sports Exerc. 29, (7), 901–909.

Francis, M., Jaimeson, J., 2006. Feeding. In: Bellman, M., Peile, E. (Eds.), The Normal Child. Churchill Livingstone, London.

Gallahue, D.C., Ozmun, J.C., 2002. Selected factors affecting motor development. In: Gallahue, D.C., Ozmun, J.C (Eds.), Understanding Motor Development: infants, children, adolescents, adults. McGraw-Hill, Boston.

Gardner, M.M., Buchner, D., Robertson, M.C., Campbell, A.J., 2001. Practical implementation of an exercise-based fall prevention

programme. Age and Ageing 30, 77–83.
Gillespie, 2002. The physical impact of computer and electronic games use on children and adolescents: a review of the current literature. Work 18, 249–259.
Helman, C.G., 1996. Culture. Health and Illness. Butterworth-Heinemann, Oxford.
Henderson, S.E., Sugden, D.A., 2007. Movement Assessment Battery for Children, second ed (Movement ABC-2)http://www.psychcorp.co.uk/?gclid=CMWxwOyL35gCFQsyQgodCHu0dA (Accessed 14/2.09).
Howe, T.E., Rochester, L., Jackson, P.M.H., Banks, P.M.H., Blair, V.A., 2008. Exercise for improving balance in older people. Cochrane Database Syst. Rev. 2.
Kennedy, N., 2006. Boundaries of normal in health. In: Bellman, M., Peile, E. (Eds.), The Normal Child. Churchill Livingstone, London.
Laughton, C.A., Slavin, M., Katdare, L., Nolan, L., Bean, J.F., Kerrigan, D.C., et al., 2003. Aging, muscle activity, and balance control: physiologic changes associated with balance impairment. Gait Posture 18 (2), 101–108.
McClure, R., Turner, C., Peel, N., Spinks, E., Eakin, E., Hughes, K., 2008. Population-based interventions for the prevention of falls-related injuries in older people. Cochrane Database Syst. Rev. 2.
Munro, J., 1997. Physical activity for the over 65's. J. Public Health Med. 19 (4), 397–402.
Nicholson, L., 2004. Physical activity and older people in Scotland. A research review for Sport Scotland, Glasgow.
Nintendo, 2008. Nintendo Wii Touch! Generations games and activities. http://www.nintendo.co.uk/NOE/en_GB/touch_generations_5027.html (Accessed 14/2/09).
Poultney, T., 2007. Physiotherapy for children. Butterworth-Heinemann, London.
Prechtl, H.F., 1997. State of the art of a new functional assessment of the young nervous system. An early predictor of cerebral palsy. Early Hum. Dev. 50 (1), 1–11.
Shumway-Cook, A., Woollacott, M.H., 2007. Motor Control, Translating Research into Clinical Practice. Lippincott Williams & Wilkins, Philadelphia.
Tortora, G.J., Grabowski, S.R., 2003. Principles of Anatomy and Physiology, tenth ed John Wiley, Hoboken.
United Nations, 1989. United Nations Convention on the Rights of the Child. http://www.everychildmatters.gov.uk/uncrc/ (Accessed 14/2/09).
University of the Third Age, 2008. http://www.u3a.org.uk/ (Accessed 14/2/09).
Welsh Assembly Government, 2006. National Service Framework for Older People in Wales. http://www.wales.nhs.uk/sites3/home.cfm?orgid=439&redirect=yes (Accessed 14/2/09).
Williams, G., 2008. Exercises sitting down: Chi Kung seated exercises for energy and wellbeing. http://www.exercisesittingdown.com/ (Accessed 14/2/09).
Woodrow, P., 2002. Ageing: issues for physical, psychological and social health. In: Woodrow, P. (Ed.) Whurr, London.
World Health Organization, 1996. Heidelberg Guidelines for promoting physical activities among older persons. World Health Organization, Geneva.

Further reading

Age Concern, 2000. http://www.ageconcern.org.uk/ (Accessed 14/2/09).
Bellman, M., Peile, E., 2006. Tables of normal neurodevelopmental data. In: Bellman, M., Peile, E. (Eds.), The Normal Child. Churchill Livingstone, London.
Bernard, M., 2000. Promoting health in old age. Open University Press, Buckingham.
Campbell, S.K., 2006. The child's development in functional movement. In: Campbell, S.K., Vander Linden, D.W., Palisano, R. J. (Eds.), Physical Therapy for Children. third ed. Saunders, St Louis.
Chartered Society of Physiotherapy, 1999. Guidelines for the management of osteoporosis. http://www.csp.org.uk/director/libraryandpublications/publications.cfm?item_id=74C877EFBBC09C2035BE2BD690FAB375 (Accessed14/2/09).
Close, J., 1999. A Randomised Control Trial: Prevention of Falls in the Elderly Trial (PROFET). The Lancet 353, 93–97.
Department for Children, Schools and Families Every Child Matters, http://www.everychildmatters.gov.uk/ (Accessed 14/2/09).
Donnellan, C., 2005. Ageing Issues, Issues Series, vol. 105. Independence, Cambridge.
Hardman, A.E., Stensel, D.J., 2003. Physical activity and health – the evidence explained. Routledge, Oxford.
Help the Aged, 2003. http://www.helptheaged.org.uk/en-gb (Accessed 14/2/09).
Morris, M., Schoo, A., 2004. Optimising exercise and physical activity in older people. Butterworth-Heinemann, Oxford.

Pre School Playgroups Association (Gwent, South Wales), 2000. http://www.touchnewport.com/business/list/bid/3801743 (Accessed 14/2/09).
Squire, A., 2002. Health and wellbeing for older people. Foundations for practice. Baillière Tindall, London.
Trew, M., 2005. Human movement through the life span. In: Trew, M., Everett, M. (Eds.), Human Movement, fifth ed. Churchill Livingstone, Oxford.
Welsh Assembly Government, 2005. National Service Framework for Children in Wales. http://www.wales.nhs.uk/sites3/home.cfm?OrgID=441 (Accessed 14/2/09).
Woodrow, P., 2000. Ageing: Issues for physical, psychological and social health. In: Woodrow, P. (Ed.), Whurr, London.

Chapter 14

Measuring and analysing human movement

Tony Everett

CHAPTER CONTENTS

LEARNING OUTCOMES

When you have completed this chapter, you should be able to:
1. discuss the concepts of measuring human movement
2. discuss a range of measurement techniques available
3. choose the appropriate tool for analysing and measuring movement
4. evaluate the different measurement techniques.

INTRODUCTION

When considering human movement, it is immediately evident that each movement performed is complex. This complexity involves not just the degree and sequence of joint motion or the amount and type of muscle work but a host of other parameters as well. Some of these include initiation, control and stopping; voluntary or involuntary components; intentional or non-intentional movement; speed; direction; balance and equilibrium; and patterned or isolated movements. There are also other dimensions to movement, such as the social context, the environment and the health status of the mover. Some of these are considered within this book and some are not.

It is relatively straightforward to measure some of the physical aspects of movement such as joint range or muscle strength in a non-functional context, but difficulty arises when quality of movement must also be considered. Some aspects of quality and health status are considered in Chapter 15 (*Scales of Measurement*), but on the whole there is no consensus as to what constitutes quality of movement. The task of measuring and analysing movement therefore seems very daunting. Until there is consensus on what constitutes movement and what its components are, absolute measurement and analysis will be difficult.

Throughout this book, there have been attempts to define aspects and components of human movement, and the increasing sophistication of technology means that progress in analysis is being made. This chapter will explore methods of measuring some of the components of movement and suggest ways in which movement can be analysed.

ANALYSIS OF MOVEMENT

The method of analysing movement will depend on the purpose for which the analysis is taking place. A judgement will have to be made as to the required accuracy, validity and reliability of the result obtained. Will the results be used to determine what normal movement is, to establish deviations caused by pathologies, to measure the outcome of interventions, as a research tool or to inform the patient? Each of these will require subtly different uses of the results as well as possibly different methods to obtain the outcomes.

ACTIVITY 14.1

Working in small groups, discuss some of the activities that the family have performed in the previous chapters of this book. Using these as a base, describe any other activities that the family may undertake within their home, work or leisure time. Following your discussion, try to simulate one or two of the activities. Can you list the parameters you need to identify in order to fully describe the movement?

In your discussion in Activity 14.1 of what we mean by movement, you may have included topics such as mobility, strength, coordination, endurance, balance, posture, function, occupation, communication, leisure and goals. I am sure that you have come up with a lot more. This shows the complexity of human movement, and you can begin to appreciate the difficulty in analysing it. How you discussed the quality of movement would be interesting!

Despite all the difficulties, we still need to try to analyse movement with the most accurate results. Therefore movement analysis may be defined as the subjective and objective measurement of:

- the activity
- its components
- goals obtained.

METHODS OF ANALYSIS

There are a variety of methods that can be used to analyse movement or to measure the components of the movement. Measuring these components will be considered later in this chapter, although it is difficult to separate the analysis of movement from the measurement of its components.

The methods of analysis can be split into two broad categories:

1. observational
2. mechanical or instrumental.

OBSERVATIONAL ANALYSIS

Observational analysis is what most therapists, ergonomists and coaches have at their disposal.

As you will have experienced in the above activity, describing an observed movement is very complex. After the movement has been performed several times, then you can begin to see trends occurring. The problem is that the more subjects repeat the movement, the more tired they will become so the initial movement may change. A method of limiting this problem could be the use of video recording. Video recording reproduces the movement, but it becomes two-dimensional, and this in itself will have complications. Video recording will be discussed later.

To enable us to optimize the use of visual analysis, it is important to develop a framework on which to build the analysis. A suggested framework could be:

- the starting position
- the movement
- the finishing position.

The starting position

The starting position can be defined as a position of readiness from which the movement can take place. It can be used as:

- a foundation for the activity
- a point of fixation for one part of the body
- a training for posture and balance.

There are four functional fundamental starting positions. These are:

1. lying
2. kneeling
3. sitting
4. standing.

All other starting positions are termed *derived starting positions*. Analysis of the starting position can take place by considering the following criteria.

- Joint position: position in degrees (visual estimation).
- Muscle work: name the muscle, type of work (static usually) and range (inner, outer or middle; see Ch. 3, *Joint Mobility*).

The movement

This is analysed sequentially in time and order.

1. Segment movement
   a. type of movement
   b. plane in which the movement takes place
   c. axis around which the movement takes place
2. Joint action
   a. type (flexion, extension, etc.)
   b. approximate range (in degrees)
   c. sequence
3. Muscle work
   a. function
   b. range (inner, middle or outer)
   c. type (concentric, eccentric or static)
   d. sequence
4. The finishing position
   a. return to the original position
   b. beginning of a new phase
   c. starting position for a new movement
   d. a position of rest

Using something like the above list allows us to get some order into our movement analysis, and we can optimize our success by repetition, breaking down the components and being systematic.

> **ACTIVITY 14.2**
>
> Think about an activity that you have talked about in Activity 14.1, walking up the cinema steps as discussed in Chapter 5 (*Posture and Balance*), for example. Use the above grid to visually analyse this movement. Try analysing other activities in this way – it will be a good test of your anatomy.

MECHANICAL ANALYSIS

Use of film for analysis of movement

While experienced observers can obtain a substantial amount of subjective information about human movement, they do not have the ability to observe and remember all the complex multijoint movement patterns that occur in even the simplest functional activities. The unassisted eye functions at the equivalent of 1/30th of a second exposure time and can only see details of slow motion; the brain too, despite its amazing ability, has a limit on the amount of information it can absorb and remember. As a consequence, when observing complex movement only a limited amount of the detail is actually seen (Terauds, 1984). The other major drawback to unaided visual observation is that only subjective information can be obtained, and without baseline data, reliable measurement of change is impossible.

Film has been used to enhance understanding of human movement for more than 100 years. Cine, video and still photography have all been valuable, as they all enabled movement to be observed in much more detail than is possible on unaided visual analysis; they also permitted measurement and provided a permanent record. While cine is still very accurate, it is expensive and difficult to use, and the process of developing films is time-consuming. Still photography is limited by the fact that it captures only one instant in time and the totality of movement cannot be seen. Video, on the other hand, is cheap, easy to use and very portable, and the results are immediately available. The advent of digital recording has made the quality much better, and digital recording on to compact discs has become the norm.

Compact discs can be stored over many years and replayed repeatedly. This is particularly valuable, as it allows the analysis of movement after the patient or subject has left, when there is time for uninterrupted observation and analysis. Visual analysis of film is enhanced by the use of freeze-frame, slow-motion facilities and computer-aided analysis software, which will be discussed later.

The qualitative and quantitative use of film

A vast amount of qualitative information can be obtained from film. Human movement as a total pattern can be observed and reobserved. The relationship of all body parts to each other can be seen, as can the quality of the movement – whether it is fast or slow, uncoordinated or smooth. The client, patient or athlete can be shown the film as part of the rehabilitation process or skills coaching, and this greatly facilitates understanding of movement difficulties. The film can be kept and used for subjective comparison with films taken at a later date, enabling judgements to be made about progression or deterioration of the activity or skill.

Digital recording makes the image much more compliant to computer-aided analysis. In the recent past, digitization was a manual and laborious process, but the current digitized image can be in several commercially available processing packages. Computer-aided analysis of images can give a wide range of information, and most systems now allow the analysis of movement in more than one plane. The computer analysis software makes it possible to plot body coordinates (centre of gravity etc.). Knowledge of the position of the centre of gravity is important when considering the efficiency of movement. For example, smooth displacements of the centre of gravity tend to indicate a more efficient movement than those in which the centre of gravity is subjected to extensive vertical displacement. Figure 11.4 illustrates the displacement of the centre of gravity; the original information that enabled this figure to be drawn was taken from computer analysis of videotape. The computer can also generate stick diagrams, and these are valuable as an initial qualitative analysis of the sequence of movement. A stick diagram of a jump is shown in Figure 14.1.

Data on joint angles in one or more planes of movement can be collected, and the pattern of movement at a joint can be graphically represented and related to other joints or the whole body. Joint angle data are available for any instance in the movement sequence.

The velocity and acceleration of limb segments can be measured, and the data give useful information about patterns of movement, for example when comparing the acceleration of the tibia in the swing phase of normal gait with the acceleration of the shank of an artificial limb in amputee gait.

Using video, it is also possible to calculate cadence, stride length and velocity in gait, but to do this it is necessary to provide some form of scaling in the filming area (Whittle, 1991). Filming with video is a relatively simple and cheap technique that can be undertaken almost anywhere, and because it does not require measuring equipment to be attached to the subject, it does not disrupt the movement being analysed. (Whether there is a psychological effect on movement patterns brought about by the self-consciousness of being filmed is not known, but this should not be discounted.)

For meaningful data collection, great care must be taken in setting up the filming site and arranging the camera. For accurate spatial and temporal measurements, the camera must be positioned carefully in relation to the subject and timing and scale devices must be included in the field of view.

CASE STUDY 14.1

John's sister Pat is 59 and had had rheumatoid arthritis for approximately 20 years when she suddenly noticed an increased clumsiness when undertaking functional activities using her hands. She had learned to cope with limited movement and pain at both shoulder and elbow joints and severe ulnar deviation of her metacarpophalangeal joints. For many years, she had managed most activities of daily living despite her substantial problems, but now she found she was

Figure 14.1 A stick diagram of a jump. This was generated from data obtained by digitizing a video film and gives an overall subjective impression of the movement.

knocking over items as she went to pick them up, and her accuracy at putting down objects such as cups or a vase of flowers was seriously compromised. She reported that she had noted no change or deterioration in her physical condition, so it was decided to evaluate the total upper limb joint movement patterns by use of video. When viewed in slow motion, the tape revealed an inability to fully extend the joints of her index finger beyond the resting position. As she performed grip activities, the index finger became caught on the object she wished to pick up, and because of the lack of extension, she was unable to let go of objects at the end of a task. A physical examination subsequently showed total rupture of the extensor indices tendon.

Computerized kinetic analysis system

These systems are able to collect, display and analyse three-dimensional movement data from many joints and body segments simultaneously. The systems use high-resolution cameras that capture infrared light reflected from markers placed at predefined points on the subject. Strobe lights are situated around the cameras, and as the subject moves, the reflected light is captured by the cameras. The data are then digitized and processed through specifically designed software and displayed as movement diagrams and as calibrated data. Such systems are very sophisticated but therefore expensive and usually available only in specialized centres.

The difficulties in obtaining quantitative data when analysing movement have been discussed. Collecting data on the individual parameters is easier and therefore more technically advanced. It must be remembered, however, that many of these measurements are taken when the parameter is isolated and not within the functional activity.

CASE STUDY 14.2

Kevin, a good friend of Liz and keen motorcyclist, was struck from the side while on the motorbike by a car that joined a main road without stopping. Apart from the injuries received as a consequence of the impact, Kevin was then run over by an oncoming bus. He suffered multiple fractures and avulsion of soft tissue but despite his injuries made a remarkable recovery.

During his rehabilitation, force traces taken from a Kistler force plate showed considerable variation from the ground reaction forces that would normally be expected in walking (a normal trace taken from a force plate is shown later in the chapter; see Fig. 14.5). His trace indicated a marked reduction in initial foot to floor contact force and also in the force that should have been generated on push-off. Kevin was reluctant to approach heel strike at normal velocity, because the forces generated caused him considerable pain at the tibial fracture sites. He also experienced difficulty in generating force at push-off because of tissue damage to the plantar flexor muscles.

MEASURING JOINT RANGE

Handheld goniometers

Traditionally, joint motion has been investigated by measuring the maximum range of movement available at individual joints. This is a static measurement of the end of range position, and a handheld goniometer is used for the purpose. These simple goniometers have to be aligned over the joint axis, and this introduces a potential source of error if the instantaneous centre of rotation changes throughout the movement or if the goniometer becomes misaligned. Nevertheless, the reliability and validity of handheld goniometers have been shown to be fairly good (Gajdosik & Bohannon, 1987), although their usefulness is limited by the fact that they can record only static position and therefore have little value in the description of continuous or functional movement. Despite being used as objective measures for testing the efficacy of therapeutic intervention, they give no indication of the functional range of joint movement.

Normally, static goniometric measurements are taken with the joint in a non–weight-bearing position that allows full range of active or passive movement to be measured. However, very few functional activities of the lower limb are performed in a non–weight-bearing position, and few upper limb functions are performed without the limb holding a weight, so the results obtained are not an accurate reflection of the subject's capabilities in a functional activity.

Electrogoniometers

Electrogoniometers have the ability to measure joint movement during a functional activity. The electrogoniometer, which was introduced by

Karpovitch in the 1950s, can take a number of forms (Rothstein, 1985). At its most simple, it can consist of two end blocks joined by an electronic potentiometer that is encased within a protective spring. More sophisticated devices may use up to three potentiometers for each joint, thus allowing simultaneous measurement of movement in three planes.

Two different types of goniometer are shown in Figure 14.2. In both cases, the way in which they are designed allows measurement to take place regardless of whether the centre of rotation of the

Figure 14.2 Two different types of electrogoniometer.

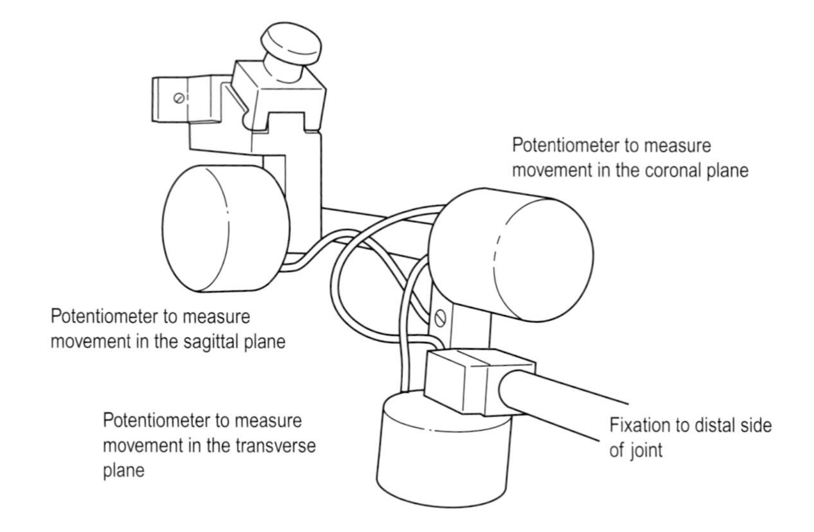

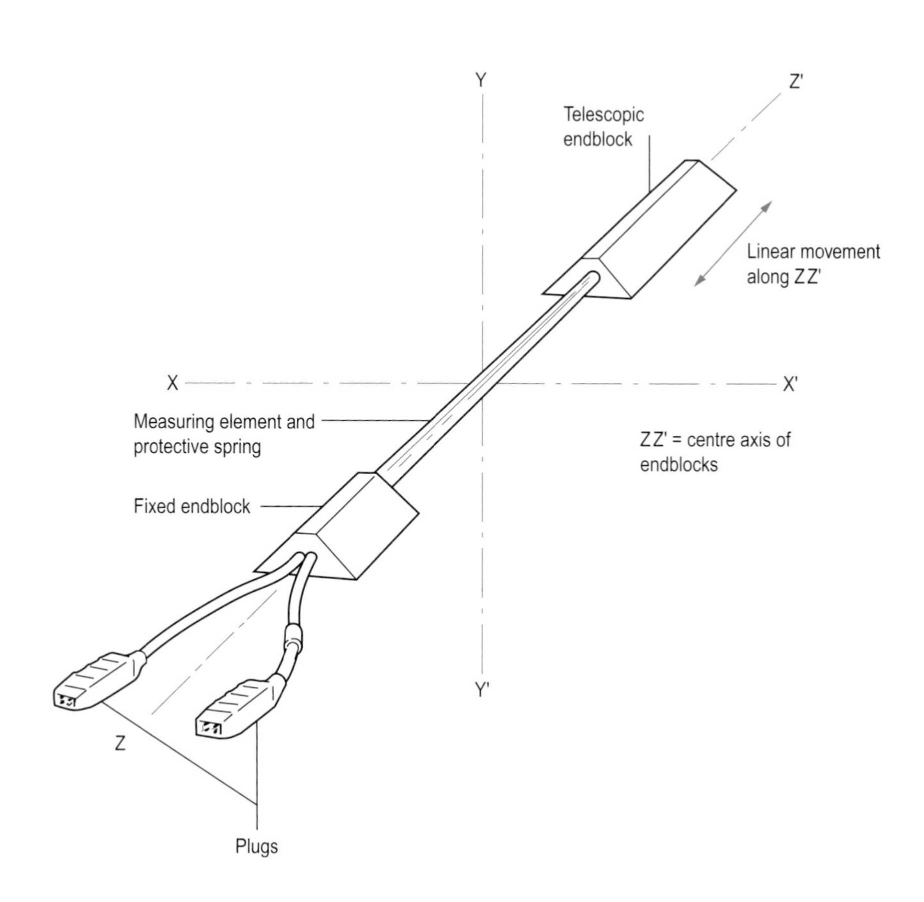

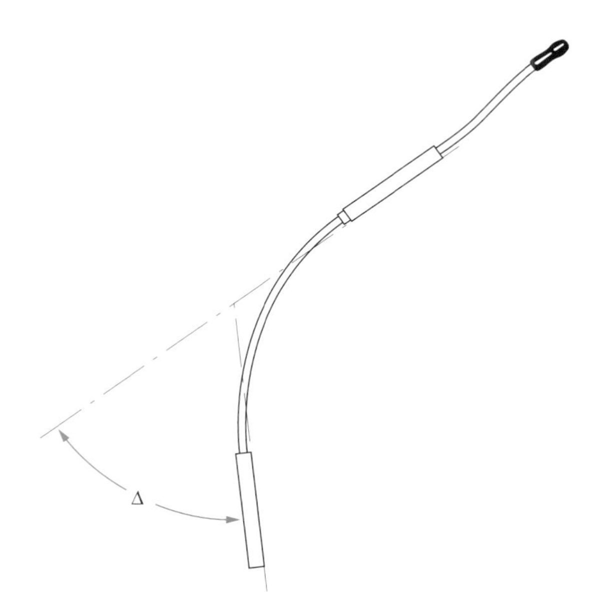

Figure 14.3 This electrogoniometer does not need to be aligned over the axis of the joint. Movement of one end block in relation to the other enables calculation of the position of the joint.

goniometer coincides with the centre of rotation of the joint. Figure 14.3 illustrates how this is possible. With these types of electrogoniometer, movement of a joint will result in movement of the potentiometer and the resultant strain on the potentiometer generates electrical signals, i.e. the resistance in the potentiometer is changed. These signals, in the form of voltage and, less commonly, current, are plotted and, after calibration, represent the angular displacement of the joint. Only angular displacements are measured. Linear movements that result in telescoping of the potentiometer do not produce strain and consequently no voltage is recorded. The joint displacement curves for the hip, knee and ankle joint that are shown in Chapter 11 (*Function of the Lower Limb*) were taken from data obtained from an electrogoniometer (see Fig. 11.6).

Electrogoniometers are lightweight and do not interfere to any great extent with the activity being tested. Only a small force is needed to distort the potentiometer, making the instrument very sensitive. Error associated with the use of electrogoniometers comes mainly from the means by which they are fixed to the subject and the method by which data are relayed to the computer for analysis. Fixation has proved to be a problem over many years, as human limbs are normally conical in shape, and unless the goniometer is stuck directly to the skin, it is in danger of sliding down or round the limb during movement. When straps or bands are used to hold the goniometer on to the limb, they normally need to be so tightly fastened that they are uncomfortable and restrict movement. Using adhesive tape to affix the electrogoniometer directly to the skin is a more satisfactory means, although skin movement over the joint may present a problem. Reliability of results using this method of fixation is reasonably good (Troke et al., 1998).

Many researchers have demonstrated the reliability of electrogoniometers (Rowe et al., 1989; Troke et al., 1998). Rome and Cowieson, (1996) investigated the reliability of electrogoniometers when measuring the full range of movement at the ankle joint and found that there was no significant difference in the results obtained on separate days, thus suggesting that the goniometer was highly reliable. Myles et al. (1995) found the hysteresis effect to be 3.6°, with a residual error of 2.9° for repeated measurements of large ranges. Smaller joint ranges, however, showed discrepancy in the order of only 1° for hip and knee flexion during walking. Hazelwood et al. (1995) tested the construct validity of the electrogoniometer and found the measures to be highly repeatable, with little variation. All these errors can be kept to a minimum if the operational definition is implemented with care. Electrogoniometers have been found to be valid, reliable and easy to use. They help to measure joint ranges during activity and therefore represent a good picture of the functional capabilities of that joint. Although electrogoniometers are mostly used within the field of research, it is hoped that the costs will decrease, bringing them within the price range of clinical therapists and sport scientists.

Optoelectronic devices and polarized light goniometers

These devices use the radiation of light in the measurement of movement. Light is either reflected from or transmitted from markers placed on the subject's skin. These devices come in several forms and tend to be costly but can produce detailed and highly accurate information relating to the movement of segments of the human body. Optoelectronic devices are based on the same principle as cine and video in that they require markers to be placed on the body; the coordinates of the markers are tracked throughout the movement and calculations can then be made. Unlike cine and video, the systems do not give a visual image of the subject but simply a frame by frame representation of the position and change in position of each marker. From this, it is normally possible to produce computer-generated stick figures or graphs of the position of a joint showing range plotted against time. This gives good quantitative information but does not address the issue of quality of movement, as there is no visual representation of the actual subject. The markers can either be active or passive.

Systems using passive markers rely on reflective markers placed on the subject's skin. Some form of light, often infrared, is transmitted towards the subject and the rays are reflected back off the markers to a series of 'cameras' that record the marker position. Sufficient cameras need to be placed around the subject so that each marker is visible to a minimum of two cameras. Sampling frequency is normally 50 Hz, which enables the system to track the change in position of the markers and produce a reasonable record of the gross pattern of movement. The markers have no identity, leaving the system vulnerable if crossover of markers occurs. This is commonly seen when, for example, the marker on the wrist crosses the marker on the greater trochanter during the stance phase of gait. The accuracy of the system relies on human input to ensure that the computer accurately identifies which is the wrist marker and which is the marker on the greater trochanter, or it is dependent on a good-quality computer program that is able to correctly process the incoming data.

The more expensive systems use active skin markers, each of which has its own small power pack that enables it to actively transmit infrared rays to a receiving system of several cameras. Because each marker has its own transmitting signal, the receiver not only picks up the position and displacement of the marker but can identify which marker it has picked up. This gives the advantage of differentiating between markers and removes the potential source of error that can occur when two markers cross over each other.

For both the active and passive systems, it is necessary to identify a ground reference point before measurement takes place. This enables the computer to calculate the absolute and relative positions of the markers in three dimensions.

If only one marker is placed on a segment of the body, the system can measure and record displacement of that segment in three dimensions, giving the segment's absolute and relative position. The application of two markers enables the system to calculate the distance between them relative to time, and this enables calculations of velocity and acceleration. If the two markers are placed to represent the two ends of a long bone, then the computer can calculate the displacement of that bone relative to the floor or another reference point. With three or more markers, the angles at joints can be measured as well as the accelerations and velocities of limb segments.

Digital inclinometer

This is usually a battery-operated device with an angle display in degrees and/or an inclination display in percentage. It is based on the principle of a builder's spirit level. The device is placed along the segment to be moved and can be fixed during the movement or replaced post movement and the angle or inclination calculated. It is reported to be a very accurate recording of the movement.

MEASURING THE FORCE GENERATED BY MUSCLES

At this point, it would be worth referring back to Chapter 2 (*Skeletal Muscle, Muscle Work, Strength, Power and Endurance*) and reading the discussion on muscle strength, force and torque.

Manual testing

Many therapists and coaches test muscle strength manually. This will give a crude and subjective description of the force that can be generated by the muscle. Manual muscle testing can be carried out on an individual muscle or a group of muscles. The therapist or coach has to resist the action of the muscle being tested throughout its range (isotonic or concentric strength) or in a fixed position (isometric strength). The result obtained is purely subjective on behalf of the tester, and this places obvious limitations on the usefulness of this measure. If the muscle or muscle group is bilateral, then the tester can test it against the other side. This is more useful, as there is a baseline to test against, although the result is still fairly subjective and not so useful if there is a bilateral problem!

Mechanical testing

Use of free weights

As was discussed in Chapter 2, finding the 1 repetition maximum by calculating the maximum weight that can be lifted for one repetition only is a useful quantitative measure. It gives an objective measurement and involves the use of equipment that is readily available within most departments or sporting areas. The drawback is that it takes a long time to elicit the measurement, as it has to be done by trial and error. This has implications for inducing fatigue in the patient, safety issues and time constraints. Once obtained, it gives an objective baseline of muscle strength that can be used in rehabilitation or in a sporting context.

Isokinetic dynamometry

There are no direct methods of measuring the work undertaken or force generated by individual muscles during functional movements, although there are mathematical approaches that can be employed to provide data on the net muscle moment at a joint (Winter, 1990). Isokinetic dynamometers can record the variation in muscle torque throughout a range of joint movement, and this is a major advance on manual assessment of force or handheld dynamometry. Unfortunately, isokinetic dynamometers are large machines, and in order for measurements to take place, the subject has to be attached to the machine; clearly, this is not going to permit the measurement of muscle torque in functional activity. The isokinetic dynamometer requires the subject to be seated, with the axis of the joint to be tested aligned with the axis of the machine. The limb must be strapped tightly to the dynamometer chair to ensure that misalignment of these two axes does not occur and movement can occur only through the range permitted by the fixation.

Modern isokinetic dynamometers are able to measure muscular torque, work, power, the rate of torque production (explosiveness) and endurance in movements that involve the muscle concentrically and eccentrically. The force generated in isometric muscle activity can also be measured. Information gained from the static testing position is then used to extrapolate to the moving human being.

Peak torque is measured in newton.metres (N.m) and represents the highest torque output achieved by a muscle as it moves its joint through range of motion. It would appear to be an accurate and reproducible measure and the most commonly collected data using the isokinetic dynamometer (Kannus, 1994). The peak torque generated by a muscle varies according to the velocity of the movement, and this is known as the torque–velocity relationship. It is greatest at the lower velocities and declines as the velocity increases. Kannus (1994) suggests that with increasing angular velocity, the point at which peak torque is achieved occurs later in the range. In normal muscle, there is an optimum part of the range when muscles are able to generate maximum force (midrange). It is speculated that on the faster angular velocities, the muscle may not have recruited all possible fibres by midrange and the angle to peak torque may be changed. If peak torque is to be considered at different velocities, then angle to peak torque should also be taken into consideration.

Torque can be measured anywhere within the range of movement, and this is called angle-specific torque. When angle-specific torque is measured in inner or outer range, the accuracy of the measurement decreases and may fall below an acceptable level, making results inconsistent (Kannus, 1994).

Figure 14.4 shows the traces taken from a normal subject who was measured during concentric contractions of the shoulder joint abductors and adductors. The traces clearly indicate that peak torque for both abduction and adduction was achieved towards the early part of the measured range. As the torque recordings do not start until the preset velocity has been achieved, the actual joint and muscle range is greater than that shown on the traces.

Work and power measurements can be obtained from the isokinetic dynamometer. In isokinetics, work is defined as the area under the torque curve where the torque curve is torque against angular displacement (work is torque ∞ angular displacement). Work is measured in joules. Power is the rate of muscular work and increases with angular velocity. Average power is total work for a given contraction divided by the time taken and is measured in joules per second or watts.

It is also possible to measure peak torque acceleration energy, which is the greatest amount of work performed in the first 125 ms of a contraction. It is measured in joules, and it is suggested that this measurement is indicative of explosive ability, as it gives the rate of torque production. There is some doubt about the reliability of these data and their repeatability, especially at low speeds (Kannus, 1994).

Endurance indices can be defined as the ability of muscle to perform repeated contractions against

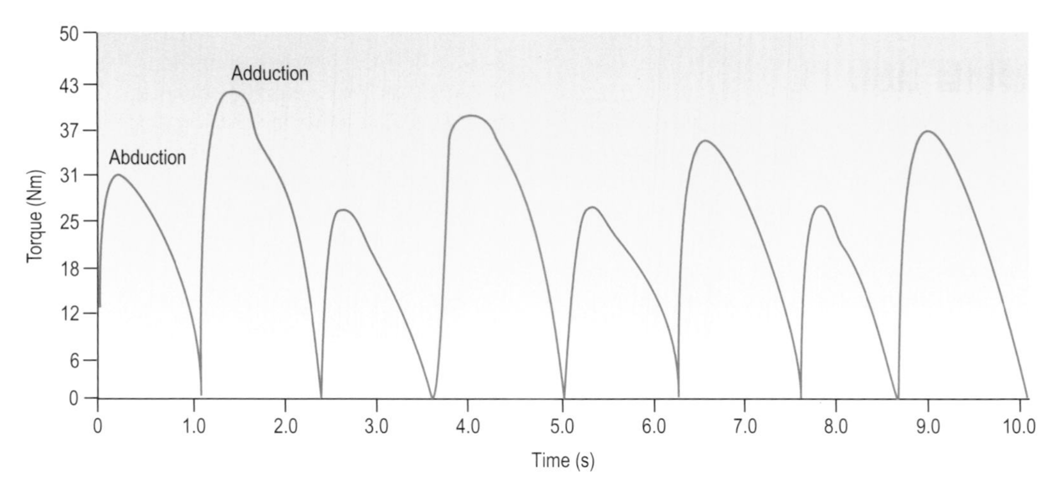

Figure 14.4 Torque curves of repeated concentric activity of the shoulder abductor and adductor muscles. Data were obtained using an isokinetic dynamometer.

a load. An endurance index is supposed to indicate the rate of fatigue. There appears to be no agreement as to the best way to test for endurance, although most tests work on a reduction from peak torque over a specified period of time. Isokinetic dynamometers can provide this information.

The isokinetic dynamometer is a popular tool because it provides information about muscle groups that may be functioning isometrically, isotonically and isokinetically. It is the only device that makes concentric and eccentric measurements possible. Unfortunately, the data obtained do not relate to human function because they have been collected from a single joint or single muscle group activity. Extrapolating from isokinetic dynamometry data to function must therefore be viewed with caution.

Handheld dynamometers

Handheld dynamometers are used to measure the reactive force of a segment when it is prevented from moving by the dynamometer held by the operative. It measures the isometric muscle strength (torque) in a specific range of the movement. The dynamometer is portable and usually a digital read-out. It uses an inbuilt force transducer and relays the value to the display, usually via a memory. The handheld dynamometer has been shown to have high test–retest reliability (Dunn and Iversen, 2003; Taylor et al., 2004).

FORCE AND PRESSURE MEASUREMENTS

GROUND REACTION FORCES

It is possible to measure reaction forces between the human being and the supporting surface in a number of functional activities. Chapter 9 (*Biomechanics of Human Movement*) introduced the mechanics that underpin this method of measurement. This technique is most commonly seen in the evaluation of activities such as locomotion, getting out of a chair and postural sway in standing.

There are a number of different ways in which ground reaction forces can be measured; the most complex methods measure vertical forces and shear forces in the horizontal plane. In the horizontal plane, they measure forces both in an anterior–posterior direction and in a medial–lateral direction. From these three forces, it is possible to calculate a single point about which these forces are said to act and also to represent these forces by a single ground reaction vector of given magnitude and angle.

Typically, force plate data are plotted against time, showing the patterns of change of force during the period of time that the foot is in contact with the supporting surface. Figure 14.5 shows the normal vertical reaction forces seen in walking plotted in two different ways. Major movement abnormalities are apparent both on visual observation and

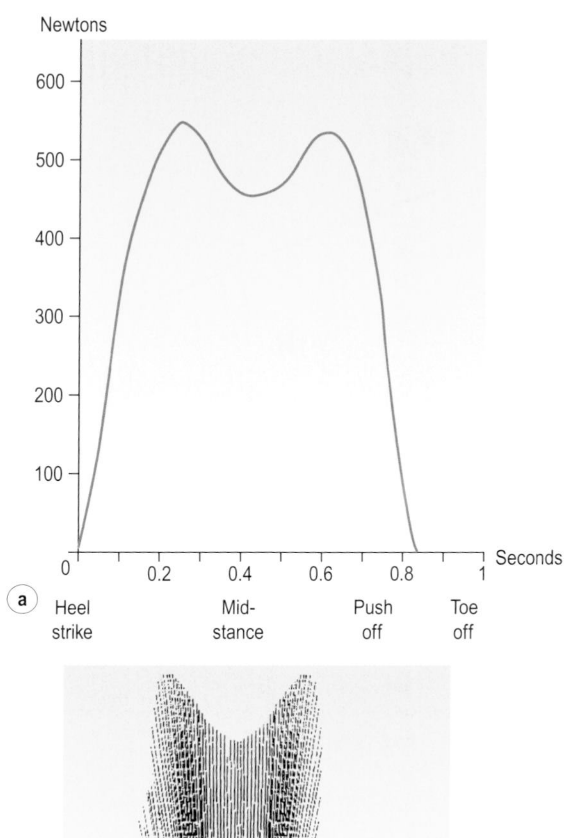

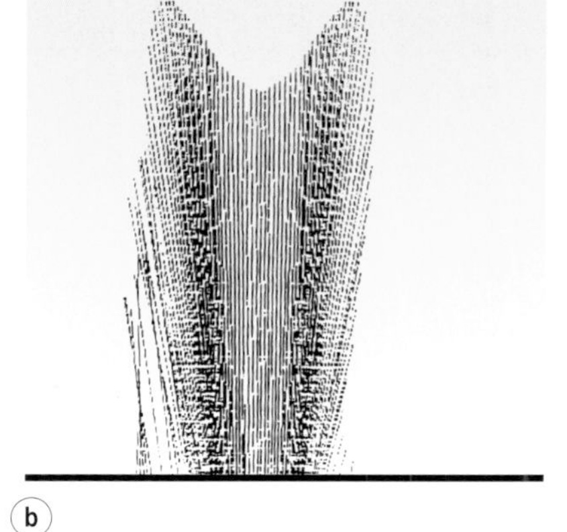

Figure 14.5 Normal force traces taken from the stance phase of walking: (a) a force curve, and (b) a vector diagram.

on detailed examination of a force trace (Fig. 14.6). One advantage of using a force plate is that the data obtained enable quantification of change over time. Force plates are, however, very expensive, and their use as a clinical tool in most outpatient departments is likely to remain limited.

Other objective information can be collected using force plates. Centre of pressure measurements are useful in determining time to stability following a functional activity or other perturbation to balance. This measure can be used in helping to diagnose anatomical, physiological and biomechanical anomalies in the planning of therapeutic interventions or to enhance coaching skills.

ACTIVITY 14.3

In Figure 14.6, the effects of abnormal gait are illustrated. Work out how the normal gait pattern has been changed in order to produce the two force traces. What might be the possible causes of these two changes in gait pattern?

PURE PRESSURE MEASUREMENTS

These can be made via pressure plates or in-shoe devices. In both cases, pressure sensors are distributed across the whole load-bearing surface of the measuring device. This enables measurements of pressure to be made over the whole of the area that is in contact with the device. Figure 14.7 shows a printout from a Musgrave footprint pressure plate. The printout is colour-coded to indicate different levels of pressure, and the load during any part of the stance phase can be obtained.

Pressure plates take a variety of forms, and different models are available to measure foot pressures and the pressures in sitting and lying. Floor-mounted pressure plates provide information about standing and the different forms of locomotion, but they are restrictive in that they may have to be set into the floor or into a walkway and are normally directly linked to a computer. Pressure distribution in sitting and lying has proved useful in the design and evaluation of beds and chairs and has also provided an insight into how pathologies can alter the distribution of pressure.

If a device is placed inside a shoe, the foot–shoe interface pressures can be measured, and these provide information on pressure distribution across the foot in all functional situations (see Fig. 14.8). These in-shoe devices enable data to be collected both on pressure distribution across the whole of the plantar surface of the foot and also across time. Most of these devices are connected by a short lead to a data logger that is normally worn on the waist belt, and the data can be downloaded on to a computer for analysis at a later date.

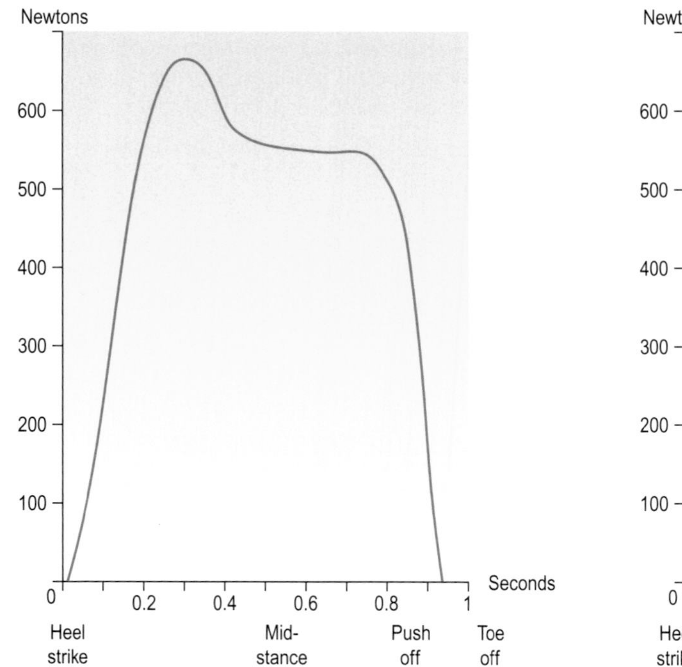

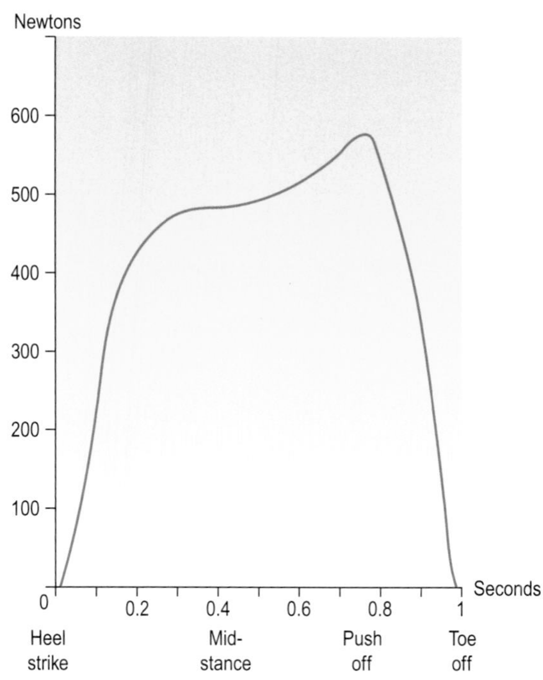

Figure 14.6 Force traces taken from patients with abnormal gait patterns.

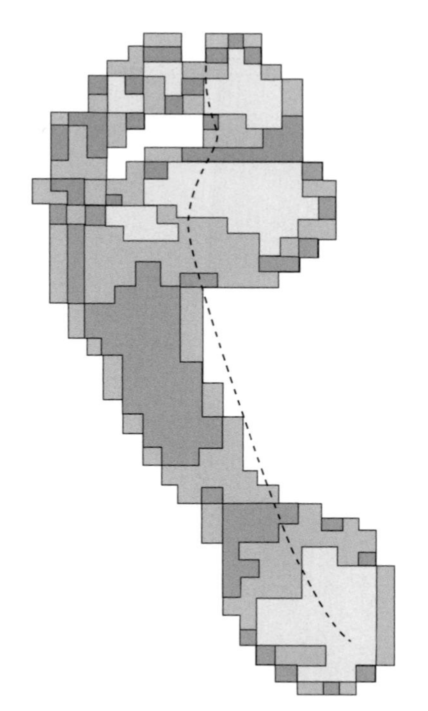

Figure 14.7 Distribution of pressure across the sole of the foot during walking. The light areas represent the greatest pressure.

Figure 14.8 An in-shoe pressure device.

ELECTROMYOGRAPHY

Electromyography (*EMG*) is the term used to describe not only the electrical signals produced as a result of the contraction of a muscle but also the method of collecting these signals and the data that are produced.

When a muscle is quiescent, there is little electrical activity. However, during muscular activity, electrical signals are produced and can be recorded. Electromyography will show if a muscle is active or not, the duration of that activity, and as the EMG increases in magnitude with tension, the signals will also give an indication of how much torque is being generated.

THE BASIS OF ELECTROMYOGRAPHY

At a cellular level, the muscle fibre or cell is the unit of contraction. During muscle activity, there is an electrical potential change and depolarization and repolarization of the surface membrane of the cell. There is transmission of impulse across the sarcolemma to the interior of the muscle cell via a complex system of tubules. When a neural impulse reaches the motor endplate, a wave of depolarization spreads across the cell, resulting in a twitch followed by relaxation. This twitch can last from a few milliseconds to 0.25 s. The depolarization is followed by a wave of repolarization.

The muscle electrical potentials, called muscle action potentials, will result in a small amount of the electrical current spreading away from the muscle in the direction of the skin, where electrodes can be used to record the electrical activity. The nearer the electrode is to the muscle, the larger the recorded signal. If muscle fibres some distance from the electrodes are conducting electrical current, the muscle action potentials recorded will be smaller than they would be for similar-sized motor units closer to the electrodes.

Values actually obtained from muscle vary between 100 μV (microvolts) and 5 mV (millivolts). The signal can be very small, and a problem may exist because electrical activity from sources other than the muscle can overwhelm the desired signal. This unwanted electrical activity is called *noise*, and a number of strategies have to be adopted to eliminate unwanted noise.

TYPES OF ELECTRODE

Either surface or indwelling needle electrodes can be used, although surface electrode EMG is most common in the analysis of human movement. Both types of electrode not only pick up electrical activity that passes over their conducting surface but can also register electrical currents nearby. Surface electrodes are normally small metal discs of about 1-cm diameter, although they can be smaller if tiny muscles are being tested. The electrodes are usually made of silver or silver chloride and are sensitive to electrical signals from superficial muscles. They give a reading corresponding to the average electrical activity.

Needle electrodes are normally fine hypodermic needles containing a conductor that is insulated except for its protruding end. As two electrodes are needed, the outer part of the hypodermic forms one and the conductor inside the hypodermic forms the other. Otherwise, fine wires can be used and these are less intrusive. There is some evidence that the surface electrodes are more likely to give reliable data, and as a non-invasive technique, they are preferable (Arokoski et al., 1999; Winter, 2004).

ELECTROMYOGRAPHY PROCESSING

The EMG signal can be viewed as a raw signal or processed to enable it to be compared or correlated with other physiological and biomechanical signals (Winter, 2004). Computers are used for this purpose, and it is important to be aware that the original signals will have been subjected to a number of manipulations before the final data are produced. This should not normally be a problem, but when the output is not what was expected then the signal processing should be checked.

The traces in Figure 14.9 were taken from a normal shoulder during the movement of flexion and extension and were part of a study to identify the sequence of activation of the various shoulder muscles. The trace illustrates the difficulty often experienced when trying to identify the instant of muscle activation.

ELECTROMYOGRAPHY AND THE PHASIC ACTIVITY OF MUSCLES

Electromyography can provide information on whether or not a muscle is active and for how long the period of activity or inactivity continues. There is always a small lag between the onset of electrical activity in a muscle and perceived movement of the limb; this is in the region of 30 ms and is probably not significant in terms of analysis of the phasic activity of muscles. The lag is due partially to the chemical changes that must take place before the muscle can contract and partially to the need for the muscle to take up slack before joint movement can occur.

A similar lag occurs at the end of muscle activity. With the cessation of electrical activity, the muscle continues to contract for a short period while the chemical changes stabilize and the muscle is able to relax. It is likely that the period between the end of electrical activity and the cessation of contraction will vary between muscle groups and will also be dependent on the type of muscle contraction. In the normal human being, the quadriceps has been studied most often, and it shows a lag period of between 250 and 300 ms (Inman et al., 1981).

It is valuable to know the duration of involvement of various muscle groups during movement, but there can be practical problems. It is not always clear from the EMG traces when a muscle starts to contract and when contraction ends. Figure 14.9 illustrates this well. All current information on the phasing or sequencing of muscles in functional activity is taken from EMG studies. These studies show when the muscles are activated and when they cease activity, but EMG is not able to provide information on whether the activity is isometric, concentric or eccentric. EMG studies of muscles involved in normal walking have shown that the input of individual muscles may last for only a very short time, often in the region of 0.2 s. As this would not give the muscle sufficient time to produce a movement at a joint, it leads to speculation that these muscles are doing no more than producing an isometric action (Inman et al., 1981). With such minor involvement of muscles in normal gait, it is no wonder that people are able to walk for many hours before experiencing fatigue.

ELECTROMYOGRAPHY AND FORCE PRODUCTION

While it is commonly agreed that EMG can distinguish between a working muscle and a quiescent muscle, there is disagreement about whether the relationship between EMG activity and torque is consistent and linear. If the relationship is linear, EMG can be used to calculate the forces generated by muscles during functional activities. If, however, the EMG signal has an inconsistent, non-linear relationship with muscle torque, then it has little value as a measurement tool.

ELECTROMYOGRAPHY AND ISOMETRIC TENSION

Early experiments by Lippold (1952) found that for the gastrocnemius muscle, a linear relationship could be shown between the average amplitude of EMG and the tension developed in muscle. This appeared to indicate that EMG could be used to measure the force generated, but unfortunately, in subsequent decades, other workers found different results. Although the EMG increased, it did not do

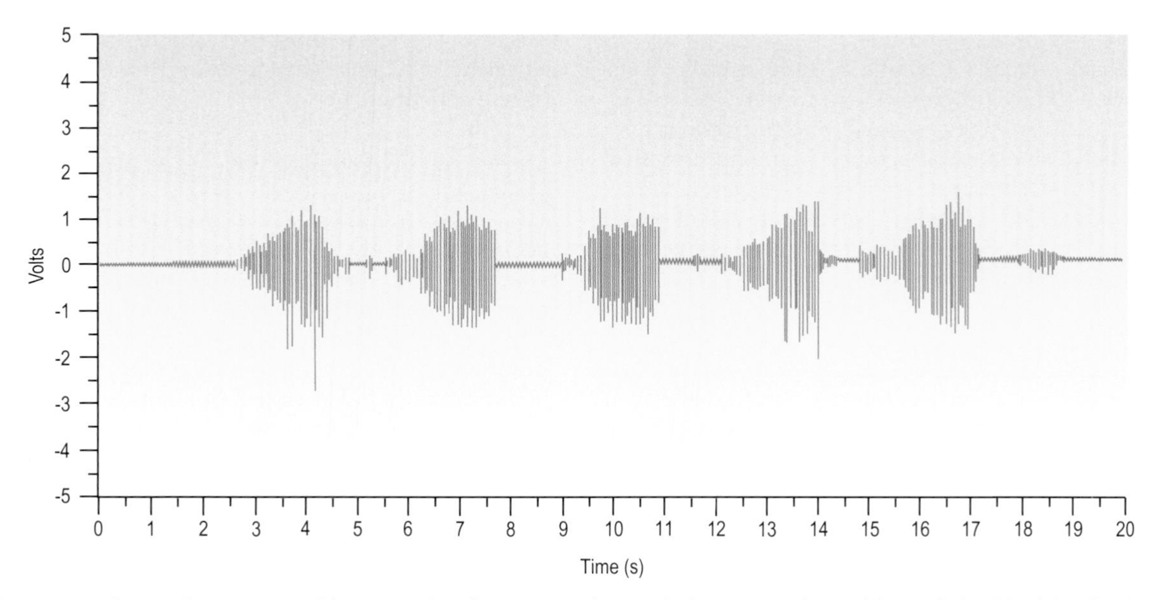

Figure 14.9 A raw electromyographic trace taken from supraspinatus during repeated repetitions of shoulder joint flexion at 608/s.

so in direct relationship to the actual amount of force generated by the muscle (Zuniga and Simons, 1969; Rau and Vredenbregt, 1973; Lawrence and De Luca, 1983). Winter (2004) suggests that there is no more than a 'reasonable relationship' between isometric force generation and EMG activity. If this is true, the results from EMG can be used to give only a general prediction of muscle tension.

ELECTROMYOGRAPHY AND ISOTONIC TENSION

Much less experimental activity has been undertaken in the field of EMG–isotonic force relationships than that of EMG–isometric relationships. On low-velocity contractions, there is an identifiable but inconsistent relationship between EMG and muscle torque, and the faster the contraction the more difficult it becomes to see any relationship (Komi, 1973).

Mathematical calculations of muscle moment, based on force plate and joint angle data, can be made. Information gained from these calculations and from simultaneous EMG correlates very closely, suggesting that EMG is an accurate method of collecting information about whether a muscle is contracting or not during a functional activity (Olney and Winter, 1985). EMG will also give some indication of the magnitude of the force being generated by that muscle, although no detail.

ENERGY EXPENDITURE ANALYSIS

There are a number of approaches that can be used to calculate energy expenditure during movement; the details are beyond the scope of this book but can be found in any advanced physiology text. Direct calorimetry produces highly accurate and repeatable results but is largely impractical, as it requires the use of an airtight insulated chamber in which the activity is performed. Indirect calorimetry most commonly relies on the analysis of the oxygen and carbon dioxide content of expired air and is valuable and reasonably practical but still requires the subject to be attached to some form of device in which expired air can be collected. The need to wear a nose clip and to breathe through a fairly large mouthpiece can be rather daunting and needs quite lengthy acclimatization. In addition, upper limb activities may be restricted by the tube leaving the mouthpiece (Whittle, 2002; McArdle et al., 2007).

Consideration of energy expenditure is important in patients who are frail or disabled. When individuals are already functioning at the limit of their ability, it is essential that they are encouraged to undertake activities in the most energy-efficient way possible. Analysis of human movement in terms of energy expenditure provides essential information that will then inform treatment approaches.

CONCLUSION

This chapter has shown that technological advances have made it possible to get good objective data for many of the individual parameters involved in human movement. These data have a variety of uses in many aspects of movement analysis. The fact that many of these take place in isolation from the functional activity is a drawback, and caution must be used when the results are extrapolated to the movement itself. The increasing use of video and computerized kinematic systems is making an impact on the understanding of human movement and leading to some exciting areas of research into human movement.

References

Arokoski, J.P.A., Kankaanpaa, M., Valta, T., et al., 1999. Back and hip extensor muscle function during therapeutic exercise. Arch. Phys. Med. Rehabil. 80 (7), 842–850.

Dunn, N.F., Iversen, M.D., 2003. Inter-rater reliability of knee muscle forces obtained by hand held dynamometer for elderly subjects with degenerative back pain. J. Geriatr. Phys. Ther. 26 (3), 23–29.

Gajdosik, R.L., Bohannon, R.W., 1987. Clinical measurement of range of motion: review of goniometry emphasising reliability and validity. Phys. Ther. 67, 1867–1872.

Hazelwood, M.E., Rowe, P.J., Salter, P.M., 1995. The use of electrogoniometers as a measurement tool for passive movement and gait analysis. Physiotherapy 81 (10), 639.

Inman, V.T., Ralston, H.J., Todd, F., 1981. Human Walking. Williams & Wilkins, Baltimore.

Kannus, P., 1994. Isokinetic evaluation of muscular performance: implications for muscle testing and rehabilitation. Int. J. Sports Med. 15, S11–S18.

Komi, P.V., 1973. Relationship between muscle tension, EMG and velocity of contraction under concentric and eccentric work. In: Desmedt, J.E. (Ed.), New developments in electromyography and clinical neurophysiology, vol. 1. Karger, Basel.

Lawrence, J.H., De Luca, C.J., 1983. Myoelectric signal versus force relationship in different human muscles. J. Appl. Physiol. 54 (6), 1653–1659.

Lippold, O.C.J., 1952. The relationship between integrated action potentials in a human muscle and its isometric tension. J. Physiol. 117, 492–499.

McArdle, W.D., Katch, F.I., Katch, V.L., 2007. Exercise Physiology, Energy, Nutrition and Human Performance, sixth ed. Lea & Febiger, Philadelphia.

Myles, C., Rowe, P.J., Salter, P., Nicol, A., 1995. An electrogoniometry system used to investigate the ability of the elderly to ascend and descend stairs. Physiotherapy 81 (10), 640.

Olney, S.J., Winter, D.A., 1985. Predictions of knee and ankle moments of force in walking from EMG and kinematic data. J. Biomech. 18 (1), 9–20.

Rau, G., Vredenbregt, J., 1973. EMG force relationship during voluntary static contractions (M. biceps). Med. Sport Biomech. 3 (8), 270–274.

Rome, K., Cowieson, F., 1996. A reliability study of the universal goniometer, fluid goniometer and electrogoniometer for the measurement of ankle dorsiflexion. Foot Ankle 17, 28–32.

Rothstein, J.M. (Ed.), 1985. Measurement in Physical Therapy. Churchill Livingstone, Edinburgh.

Rowe, P.J., Nicol, A.C., Kelly, I.G., 1989. Flexible goniometer computer system for the assessment of hip function. Clin. Biomech. 4, 68–72.

Taylor, N.F., Dodd, K.J., Graham, H. K., 2004. Test retest reliability of hand held dynamometer strength testing in young people with cerebral palsy. Arch. Phys. Med. 85 (1), 77–80.

Terauds, J., 1984. Sports biomechanics. Proceedings of the International Symposium of Biomechanics in Sport. Academic Publishers, Del Mar.

Troke, M., Moore, A.P., Cheek, E., 1998. Reliability of the OSI CA 6000 spine motion analyzer with a new skin fixation system when used on the thoracic spine. Man. Ther. 3 (1), 27–33.

Whittle, M., 1991. Gait Analysis, an Introduction. Butterworth-Heinemann, Oxford.

Whittle, M., 2002. Gait Analysis, an Introduction, third ed. Butterworth-Heinemann, Oxford.

Winter, D.A., 1990. Biomechanics and Motor Control of Human Movement, second ed. Wiley Interscience, New York.

Winter, D.A., 2004. Biomechanics and Motor Control of Human Movement, third ed. John Wiley, New York.

Zuniga, E.N., Simons, D.G., 1969. Non-linear relationship between averaged electromyogram potential and muscle tension in normal subjects. Arch. Phys. Med. Rehabil. 50, 613–620.

Chapter 15

Scales of measurement

Susan Corr and Roshni Khatri

CHAPTER CONTENTS

LEARNING OUTCOMES

When you have completed this chapter, you should be able to:

1. understand that loss of movement has an impact on individuals' ability to carry out everyday activities
2. recognize that loss of movement can affect functional status and quality of life
3. understand that measuring the effect movement loss has on mobility, function and quality of life is difficult
4. identify appropriate measurement scales needed to measure the impact of movement loss.

INTRODUCTION

Previous chapters have outlined common methods of measuring human movement. When considering human movement, it is important to be aware that a loss of movement, minimal or severe, potentially has an impact on the wider aspects of an individual's life. This chapter will outline two areas that could be affected when looking at movement in a more global context. As a consequence, it is necessary to consider movement as a multifactorial, multidimensional, complex concept. It is important to know how the loss of movement affects an individual's ability to carry out everyday activities and the effect loss of movement has on how individuals perceive their quality of life. In essence, this type of measurement is moving away from measuring impairment to measuring disability or handicap. Identifying symptoms is not enough to establish how an illness, disorder or injury actually affects a person's life (Üstün et al., 2003). The World Health Organization's *International Classification of Functioning, Disability and Health* reflects a change towards considering functioning and disability (World Health Organization, 2001).

There is a broad range of choices when looking at measurement scales, and it is important to consider these when selecting a way of measuring the impact of loss of movement. It is critical that therapists choose one that is designed for the purpose for which it is required (Fisher, 1992). A lot of measurement scales can also be used to assess the outcome of treatment and therefore are suitable to measure the impact of loss on more than one occasion. Gompertz et al. (1993) suggest that in stroke rehabilitation, for example, broader issues such as mood and perceived health should be measured and not just movement loss or functional status.

GENERAL PRINCIPLES OF MEASUREMENT SCALES

VALIDITY AND RELIABILITY

No matter what measurement scale is being selected, it is essential to consider the reliability and validity of that scale. Validity of a measurement scale relates to whether it measures what it is intended to measure (Bowling, 2005). A scale needs to appear relevant and clear (face validity) and to examine comprehensively the concept that it intends to measure (content validity) (Bowling, 2005). Scales also need to be reliable, i.e. consistent at producing the same results whether at repeated intervals (test–retest), by the same rater at different times (intra-rater) or by different raters (inter-rater) (Bowling and Ebrahim, 2005). If a measurement scale has high inter-rater reliability, different raters when measuring the same individual will produce the same results (McDowell, 2006). It is also important to question the sensitivity of the measurement scale. This question addresses whether the measure is able to identify changes that may occur over time (Bowling, 2005). A further issue to consider is whether the scale is relevant to every person being measured (McDowell, 2006). Measurement scales that have been developed for specific client groups include the Glasgow assessment schedule for head injuries (Livingstone and Livingstone, 1985), Parkinson's disease disability index (McDowell et al., 1970) and Robinson Bashall functional assessment for arthritis patients (McCloy and Jongbloed, 1987).

If a measurement scale is shown to be reliable and valid, then it is considered to be standardized. This standardization process allows scales to be used to compare individuals (McDowell, 2006). Once a measurement scale is standardized, the process of conducting it will be formalized, meaning that there will be procedural instructions outlining what environment the scale should be conducted in, what materials are required and the sequencing of the scale. These procedural arrangements should be clear and concise, and if followed strictly, ensure that the scale remains reliable and valid (Burton, 1989; Law and Letts, 1989).

The King's Fund (1988) recommends that measurement, using standard scales, should be undertaken on all patients. Although this is desirable, it is important to consider Barer's (1989) suggestion that formal measurement scales by a therapist may show what patients can do under test conditions but informal measurements made by carers are more likely to indicate what the patients do in real life. He goes on to suggest that postal surveys or interviews of patients at home may reveal what they think they can do.

METHODS OF SCORING

There are several different ways in which a measurement scale may be scored. The types of scale being described in this chapter are not interval or ratio measurements, in which the distance between two numbers on the scale are of a known size, such as exact degrees of movement or length in centimetres (Streiner and Normand, 2003). On the whole, a verbal description will be used to identify how able an individual is to carry out an activity, or a continuum of agreement is used, based on verbal expressions such as 'disagree', 'unsure' or 'agree'. Scales using verbal definitions such as this are ordinal scales, i.e. the options for responses are in some kind of order and have a relationship to each other (Streiner and Normand, 2003). A third type of scale exists but is not commonly used in the type of measure being discussed in this chapter. These are nominal scales, in which numbers are used just for labels, such as male or female, and the numbers have no value in relation to each other. It is necessary to be able to identify how a scale is measured in order to understand the appropriateness of the measurement to the question being asked. Regardless of which scoring system is used, it should be straightforward and quick. Also, it needs to be easy to interpret so that the results can be used for treatment planning or measuring the outcome of treatment (Law and Letts, 1989).

ACTIVITY 15.1

Try to identify some examples of each of the three different types of scale. You may find some examples in everyday situations such as in supermarkets or questionnaires in magazines. For interval scales, there needs to be an equal gap between two scores; for nominal scales, numbers are used to label the items in the scales but there is no relationship between the numbers, and for ordinal scales there is a grading system although the intervals are not equal.

DATA COLLECTION METHODS

Further consideration needs to be given to how information is gathered when using measurement scales. There are several options, including observation, interview or self-rating. Each method has strengths and limitations.

Observation

When observing, it is possible to see exactly how able an individual is to carry out an activity. However, a limitation may be the effect being watched has on individuals as they demonstrate their ability to perform a task. This may be positive or negative, i.e. they may make greater efforts to do something if being watched. Alternatively, they may feel more anxious and lack confidence if being observed. A key to using observation as an assessment method is being clear about what is to be observed and how to match what is seen with possible scores. For example, it is important to know exactly what needs to be observed in order to rate an individual as able to manage most of the task.

Interview

Interviews are another mechanism for gathering information. If this is the method used for a measuring scale, often the interview is structured. This means there are particular questions to ask, and even how these should be phrased may have been identified. One limitation of interviews is the reliance on individuals' perception of their ability rather than actually observing their ability. Alternatively, if it is very difficult for individuals to carry out an activity it may be easier for them to inform you of this rather than struggle to demonstrate their difficulties.

Self-rating

Similarly, self-rating methods of gathering information are usually very structured. The measurement scale is most likely to be in questionnaire format, when the individual will answer a number of questions. It may be valid and reliable for use by post, which makes it a useful tool for research as well as clinical practice. To be used in this way, it will need to be easy to complete, will need to have clear instructions and should not take long to complete.

Some measurement scales are reliable for use by observation, interview and self-rating. The Barthel Activities of Daily Living Index (BI), which is an example of this, is outlined in the next section.

MEASUREMENT SCALES FOR MOBILITY

ACTIVITY 15.2

Draw the following table on a sheet of paper and use it to consider the strengths and limitations of using interviews, observation and self-rating scales to measure an individual's ability to ride a bicycle.

Method	Strengths	Limitations
Interviews		
Observation		
Self-rating		

DEFINING MOBILITY

The ability to move around with ease, i.e. to mobilize, enables individuals to carry out everyday activities. However, this requires the consideration of a range of concepts including lying to sitting, sitting to standing, turning, stopping, balance, gait, walking speed, balance and functional reach (Smith, 1994; Wall, 2000). Independent mobility incorporates all of these critical tasks.

MOBILITY SCALES

A wide range of mobility scales are available for use in clinical settings. Some have been developed to measure a particular aspect of mobility such as balance. The Berg Balance Scale (Berg et al., 1992) is an example of this. Others have been developed to measure the impact of diseases such as cardiovascular disease and lung disease on the exercise capacity of older people. An example of this is the 6-min walk test (Enright et al., 2003).

More commonly, mobility scales have been developed to measure it as a global concept. Two such scales that will be outlined further in this section are the Get Up and Go Test (Mathias et al., 1986) and the Elderly Mobility Scale (EMS) (Smith, 1994).

Get Up and Go Test

The Get Up and Go Test was developed in the UK by Mathias et al. (1986). It was designed to challenge an elderly person's sense of balance, and it requires the individual to stand up from a chair, walk a short distance, turn around, return and sit down again. It is considered a simple test to administer and uses a five-point ordinal scale in which 1 equals normal (patient is not at risk of falling), 2 equals very slightly abnormal, 3 equals mildly abnormal, 4 equals moderately abnormal and 5 equals severely abnormal. Face, content and concurrent validity have been established, as have inter-rater and test–retest reliability for this measure (Nakamura et al., 1998). It is considered as a good tool for use with the elderly (Nakamura et al., 1998). More recently, Wall (2000) has revised this measure and developed the Expanded Timed Get Up and Go (ETGUG). The ETGUG uses a multimemory stopwatch and an extended walkway, resulting in a more sensitive and clinically useful measure of mobility. It requires minimal equipment, professional expertise or training to administer.

Elderly Mobility Scale

The EMS was developed by Smith (1994) to assess mobility in elderly patients. It tests lying to sitting, sitting to lying, sitting to standing, standing, gait, walking speed and functional reach (see Table 15.1). Therefore it covers locomotion, balance and key position changes, which are components of more complex activities of daily living (ADL). The maximum score possible, which represents independent mobility, is 20, and the minimum score is 0.

The concurrent validity and inter-rater reliability of the scale have been established, and it is easy to use in practice (Smith, 1994; Prosser and Canby, 1997). Smith (1994) suggests that individuals scoring under 10 on the EMS are likely to need help with mobility and ADL, while those scoring 14 or more are likely to be independent in mobility.

It provides an effective and ready objective measure of the change in an individual's mobility (Prosser and Canby, 1997; Spilg et al., 2001). A modified version of the Elderly Mobility Scale has also been found to be valid and reliable and

Table 15.1 Elderly Mobility Scale showing interpretation of the scores

CONCEPT	0	1	2	3	4
Lying to sitting	Needs help of two or more people	Needs help of one person	Independent	NA	NA
Sit to stand	Needs help of two or more people	Needs help of one person (verbal or physical)	Independent in over 3 s	Independent in under 3 s	NA
Stand	Stands only with physical support[a] (i.e. help of another person)	Stands but needs support[a]	Stands without support[a] but needs support to reach	Stands without support[a] and able to reach	NA
Gait	Needs physical help to walk or constant supervision	Mobile with walking aid but erratic or unsafe turning	Independent with frame	Independent (including use of sticks)	NA
Timed walk	Unable to cover 6 m	Over 30 s	16–30 s	Under 15 s	NA
Functional research	Under 8 cm or unable	NA	8–16 cm	NA	Over 16 cm

NA, not applicable.
[a]Support means needs to use upper limbs to steady self.
(After Smith 1994, with permission.)

quick to use (Kuys and Brauer, 2006). Its use in determining the effectiveness of differing approaches to rehabilitation can be considered, although it must always be remembered for those requiring rehabilitation that mobility does not stand alone in determining need and outcome. It can also predict the probability of a patient having two or more falls (Spilg et al., 2003). A modified version of the Elderly Mobility Scale has also been found to be valid and reliable and quick to use (Kuys and Brauer 2006).

ACTIVITY 15.3

Using your local workplace or academic library, conduct a search for mobility scales that may be of use to you. Before you start, establish what criteria you are looking for in the scale, for example suitable for use with a particular population, standardized, or developed in the past 5 years. Use these criteria to help you draw up relevant keywords for your search.

MEASUREMENT SCALES FOR FUNCTIONAL STATUS

DEFINING FUNCTION

There are several different interpretations of functional status. According to the Collins dictionary (Harper Collins, 1987), *function* means a natural action. This can be interpreted in humans to be the natural everyday actions that are carried out by individuals, such as walking, toileting and eating. The perspective taken on function can also be influenced by the different philosophies of the professions within multidisciplinary teams. Fisher (1994) outlines them in relation to the goal of intervention (see Table 15.2). There are, of course, many overlaps among the professions, but the table highlights the most unique aspects of each profession. All the disciplines share concerns related to function and functional independence; how they differ is on how they frame the patient's problems and the goal of treatment. These differences may have an influence when choosing a measurement scale. Also, it is important to have an awareness of the interpretation of function that may have been adapted by those who have developed scales for measuring function. By developing scales to measure an individual's functional status, the expectation is that it will be possible to get a current picture of the individual's abilities to carry out the actions that are natural to all. According to McDowell (2006), an individual is healthy if he or she is physically and mentally able to do the things he or she wishes and needs to do.

ACTIVITIES OF DAILY LIVING SCALES

A large number of scales have been developed over the past few decades, mainly in rehabilitation medicine. This section describes two in detail: the BI, which measures ADL, and the Nottingham Extended ADL Scale, which measures extended activities, i.e. more community-based activities that also require movement in order for them to be carried out. The scales chosen are selected to illustrate the types available and are by no means the only ones available or that should be considered when identifying a measurement scale to assess function. Healthcare professionals need to carefully consider the options available when they are selecting a measurement scale. The choice will depend on why it is being used. It may be needed to measure ability, to evaluate the amount of help patients may need on discharge, to identify treatment goals or as an outcome measure.

Barthel Activities of Daily Living Index

One of the most widely known and used measures of ADL is the BI. It was developed in 1965 by Mahoney and Barthel as a simple index of independence to score the ability of patients to care for themselves and, by repeating the test, to assess

Table 15.2 Unique perspectives of medicine, nursing, occupational therapy, physiotherapy and social work

PROFESSION	FRAME	METHODS	GOAL
Medicine	Illness	Pharmacology, surgery	Symptom elimination
Nursing	Health	Helping and caring, interprofessional interactions	Healthy function, wellness
Occupational therapy	Occupation	Therapeutic activity, adaptation	Occupational function
Physiotherapy	Physical capacity	Therapeutic exercise and agents	Physical function, mobility
Social work	Systems	System change, helping process	Social function

(After Fisher 1994, with permission.)

their improvement (Mahoney and Barthel, 1965). It contains 10 items and has both a self-care and a mobility component. This scale does not measure movement in isolation but the output of movement, that is, the ability to carry out activities. The mobility component contains items related to transfers and ambulation. The original scoring system was from 0 to 100; this has been modified so that scores range from 0 to 20, with higher scores signifying better functioning (Collin et al., 1988; McDowell, 2006). This modified version is most commonly used in practice. Table 15.3 shows the 10 items in the scales and the scoring system used. Two items, transfer and mobility, can be scored 0, 1, 2 or 3, while the remainder just 0, 1 and 2. The table indicates the interpretation of each score for all the items. A total score is then calculated.

The BI is not hierarchical, meaning that it does not result in a total score that gives a clear indication of the level of disability (Gibbon, 1991). This means that two people may score the same, for example 10, but it would not be possible to assume that both are independent in the same activities. The items are not in an order that reflects the complexity and difficulty of carrying out particular activities. As can be seen from Table 15.3, it also has different interpretations for the score values. For example, for bathing, being independent scores only 1, whereas being independent in transfers scores 3. This does not necessarily reflect the complex movements and skills required to carry out each of these activities. Wade (1992) provides working definitions for the scores so that therapists are clear about what the individual needs to be able to do in order to be scored as independent. Using these will ensure a standardized use and understanding of the scale.

When the scores are totalled, there is no single way of interpreting this total. Granger and Hamilton (1990) found that no specific total score could be regarded as adequately specific or sensitive to be used as a criterion for admission for rehabilitation services or as an indication of readiness for discharge. McDowell (2006) suggest that the BI is most useful in assessing patients who are moderately or severely disabled. There is a ceiling and flooring effect, meaning that those scoring the maximum (20) may still be significantly handicapped but with the potential of improvement beyond the limits of the scale. As a result, the BI lacks sensitivity to change in its upper range (Rodgers et al., 1993). The same occurs at the lower end. To overcome this, patients with high BI scores should be subsequently assessed with an ADL scale, i.e. a scale that assesses a broader range of activities.

The BI is most frequently used in clinical practice by therapists and other members of the multidisciplinary team. It has been shown to be reliable for use in this format and also for use in formal research, by post and over the telephone (Wade, 1992). Law and Letts (1989) indicated that more specifically it has been shown to have adequate observer and test–retest reliability. It is a valid measurement scale for function in ADL and is sensitive to measure changes in ADL after treatment in controlled research settings.

Eakin (1993) suggests that the appeal of the BI lies in the fact that it is simple and quick to use, its results can be easily understood and communicated between different professions, and its content

Table 15.3 Barthel Activities of Daily Living Index showing the interpretation of the scores

CONCEPT	0	1	2	3
Bowels	Incontinent	Occasional accident	Continent	
Bladder	Incontinent	Occasional accident	Continent	
Grooming	Dependent	Independent		
Toilet use	Dependent	Needs some help	Independent	
Feeding	Unable	Needs help	Independent	
Transfer	Unable	Major help	Minor help	Independent
Mobility	Dependent	Wheelchair-independent	Walks with one person	Independent
Dressing	Dependent	Needs help but can do half	Independent	
Stairs	Unable	Needs help	Independent	
Bathing	Dependent	Independent		

(After Wade 1992, with permission.)

is perceived as relevant to both clinicians and patients. However, it also needs to be remembered that it has been thought not to have enough items to account for the impact of rehabilitation and that the grading system is not sufficiently sensitive to reflect change, particularly in the short term.

ACTIVITY 15.4

Read below about Carol and Ian, friends of John and Liz. Using the information given, calculate their scores using the BI. Once you have completed this, consider the usefulness of the index to establish levels of function.

Carol is a 45-year-old teacher who recently had a right cerebrovascular accident (stroke). She has been discharged back to her two-storey house and is beginning to settle into some routine at home. She manages to wash and dress independently. With the aids provided by the social services occupational therapists, she is independent going up and down stairs and getting in and out of the bath. Although she can manage to make a hot drink, she is unable to cook a meal or cut up her food. At the moment, her husband does the cooking and shopping. Carol manages to walk around the house but is unable to walk on uneven ground. She has not been outside and does not feel confident to go to the shops. She is glad that they have a downstairs toilet, which she can manage to get to and use independently.

Ian is a 55-year-old car mechanic who runs his own small business. He has rheumatoid arthritis. He is currently unable to work or tend his garden because of pain and stiffness mainly in his hands. He walks to his local shop daily to get the paper. He has always been a casual dresser (T-shirts and deck shoes rather than shirt and tie or suits) and can get dressed if wearing his normal clothes. He is unable to drive at the moment but uses public transport. He has difficulty turning on the taps but can make a cup of tea if the kettle is filled for him. He is also independent showering; he has a walk-in shower rather than a bath. He is independent going up and down the stairs. Ian recently bought an electric shaver and manages to shave independently with this.

Extended activities of daily living scale

The Extended ADL Scale was developed in Nottingham by Nouri and Lincoln (1987) to assess the activities that may be important to stroke patients who are living in the community. It includes activities that relate to carrying out domestic tasks and other activities that take place outside the home environment. It has been validated for administering by interview and post (Nouri and Lincoln, 1987; Lincoln and Gladman, 1992; Wade, 1992). More recently it has been tested and found to be reliable and valid with individuals with arthritis of the hip (Harwood and Ebrahim, 2002). It consists of a questionnaire of 22 activities divided into four groups: mobility, kitchen, domestic and leisure.

The Extended ADL Scale is a ranked scale, meaning that all patients with the same scores are independent in the same items. Lincoln and Gladman (1992) found that an overall total score could provide an indication of overall independence in the activities if comparing groups of people. They recommend that with individual patients, section scores, for example mobility or leisure scores, rather than overall totals should be used when identifying a patient's progress or change over time. They recommend this as they found discrepancies when using the total score. The scale is appropriate to use in research into the evaluation of rehabilitation.

Table 15.4 outlines the items in each of the four sections of the scale. The response options are the same for each item. The score 1 is given if activities are performed by patients on their own or on their own with difficulty. For activities that patients are unable to perform or for which they require help, the score is 0.

ACTIVITY 15.5

Here is some additional information about Carol and Ian. Use this information, along with what you already know about them, to calculate their scores using the Nottingham Extended ADL Scale.

Carol loves reading but finds she takes a long time to read the newspaper. She feels her concentration is too poor to read books. She has yet to go out anywhere, mainly because she is unsteady walking and lacks confidence. She is anxious about whether she may ever return to work. She has been overwhelmed with the good wishes from her pupils and colleagues and has written several thank you notes.

Ian is anxious about his business, which his son is currently running. He has been struggling of late to hold a pen to write and has done most of his communication over the telephone. His wife has traditionally done all the household tasks, with the garden being his domain. He has needed help with the garden lately. At the moment, his wife drives when they go out shopping or visiting friends.

Table 15.4 Extended Activities of Daily Living Scale

CONCEPT	NO (0)	WITH HELP (0)	ON MY OWN WITH DIFFICULTY (1)	ON MY OWN (1)
Mobility				
Do you:				
• walk around outside?				
• climb stairs?				
• get in and out of the car?				
• walk over uneven ground?				
• cross roads?				
• travel on public transport?				
In the kitchen				
Do you:				
• manage to feed yourself?				
• manage to make yourself a hot drink?				
• take hot drinks from one room to another?				
• do the washing up?				
• make yourself a hot snack?				
Domestic tasks				
Do you:				
• manage your own money when you are out?				
• wash small items of clothing?				
• do your own housework?				
• do your own shopping?				
• do a full clothes wash?				
Leisure activities				
Do you:				
• read newspapers or books?				
• use the telephone?				
• write letters?				
• go out socially?				
• manage your own garden?				
• drive a car?				

(After Nouri & Lincoln 1987, with permission.)

MEASUREMENT SCALES FOR QUALITY OF LIFE

DEFINING QUALITY OF LIFE

Like functional status, quality of life is not easy to define. There are a number of different views on the scope of what should be included in the broad consideration of quality of life. Fallowfield (1990) and De Haan et al. (1993) suggest that four domains make up quality of life. These are physical, functional, psychological and social health. The physical health dimension refers to disease-related and treatment-related symptoms as well as pain and sleep. The functional dimension comprises self-care, mobility and physical activity level, as well as the capacity to carry out various roles in relation to family and work. The psychological dimension includes issues such as depression, anxiety and adjustment to illness, cognitive functioning, well-being, life satisfaction and happiness. Finally, the social health dimension includes qualitative and quantitative aspects of social contacts and interactions such as relationships and participation in leisure and social activities. The inclusion of these dimensions reflects the need to ensure that when thinking of quality of life as a concept, it is seen as a broad spectrum of consequences of disease including elements of impairment, disabilities and handicaps, as well as patient's perceived health status and well-being. Quality of life relating to health is distinct from quality of life as a whole, which would also include adequacy of housing and income and perceptions of the local environment (Bowling and Normand, 1998).

What is not clear is the required balance of the four dimensions that needs to be present in order to ensure quality of life. Diener (1984) suggests that part of the influence of health on quality of life is not simply the direct effect on how people feel physically but also on what their health allows them to do. For example, in two separate studies of stroke patients, Niemi et al. (1988) and Wyller et al. (1997) found that even when patients had a good recovery in terms of physical movement they reported a poor quality of life. This shows that the impact of a disease on health-related quality of life is important but can be difficult to understand and to measure. Conversely McDowell (2006) suggests that a patient's ability to adapt to their illness may result in scores for quality of life remaining stable even though functional measures record decreases in abilities.

ACTIVITY 15.6

Ask a number of friends and family of different ages to sum up what issues contribute to their definition of quality of life. If they include health as one issue, ask them to clarify what they mean by this. Also find out what level of ill health they need to experience in order for their quality of life to be affected. Are there differences in views between people of different age groups?

QUALITY OF LIFE MEASURES

Quality of life measures range from being broad general health profiles (generic) to disease-specific scales. The broad health profiles are those that have not been developed for specific target populations or patient groups. A strength of these is that comparisons of quality of life results across patient populations can be made. A limitation is that they do not always focus on the specific problems of a given patient group. Two examples of these are the Short Form 36 (SF-36) Health Survey and the Nottingham Health Profile (NHP). Both of these will be discussed later in this section. Disease-specific quality of life scales exist for a range of specific patient groups including those who have had strokes (Holbrook and Skilbeck, 1983) or who have rheumatic disorders (Liang et al., 1990), cardiovascular diseases (Wenger and Furberg, 1990) or cancer (Van Knippenberg and de Haes, 1985). Disease-specific scales do not allow cross-disease comparisons but are often more sensitive to the quality of life issues particularly relevant to specific populations of patients.

Most of the available quality of life scales depend on patients to rate themselves. Quality of life is a very personal issue, and therefore getting individuals to rate themselves is the preferred method of administration. It is also possible to use structured interviews or written questionnaires. It can be difficult for patients with serious cognitive, speech and language disorders to complete these (De Haan et al., 1993).

Quality of life measures can be used for a number of reasons. They may be useful for patients with chronic conditions when recovery is not expected and when success of treatments may best be measured in terms of maintaining an acceptable quality of life for the patient as the disease progresses (Talamo et al., 1997). They can also be used to facilitate the process of identifying which patients will probably benefit from which type of rehabilitative procedure (Mathias et al., 1997). Treatment from a multidisciplinary team may include a range of interventions that individually are difficult to measure directly or demonstrate outcome, or for which there is no sensitive measure. In these situations, a quality of life measure may be the most appropriate option to demonstrate outcome from the treatment. Quality of life measures can also be used to evaluate treatment programmes. Baker and Intagliata (1982) suggest that if patients' life situations are not improved in some way and they are not happier or more satisfied after participating in treatment, then it is difficult to ultimately justify the treatment.

The Short Form 36

The Short Form 36 Health Survey is an example of a measurement scale for quality of life that includes physical functioning as a component of the measurement. It has been developed from a longer medical outcomes study questionnaire, which has had the number of items reduced to 36, hence the title Short Form 36. The aim of reducing the items was to develop a scale that could be conducted in a short period of time and therefore lend itself to be used in a broad range of settings (Ware and Sherbourne, 1992).

The SF-36 was developed to be an indicator of quality of life for population studies as well as an outcome measure in clinical practice and research (McDowell, 2006). It is suitable for use with all patients, as the measure addresses aspects of health that are important to all patients rather than just those with a particular condition. It is a questionnaire that can be completed by anyone over 14 years of age in a clinical setting or at home. It can also be administered in an interview. It is easy to use and takes between 5 and 10 min to complete. This has made it popular (Larson, 1997).

The SF-36 categorizes the 36 items into eight areas relating to health concepts. These are physical functioning (10 items), role limitations caused by physical problems (four items), social functioning (two items), bodily pain (two items), general mental health (five items), role limitation caused by emotional problems (three items), vitality (four items) and general health perceptions (five items) (Bowling, 2005). The inclusion of bodily pain and vitality as concepts of health is unique to the SF-36 scale (Ware and Sherbourne, 1992). The final item asks about health change over the past year. The items relating to movement (see Tables 15.5–15.7) ask about levels and types of limitation when lifting and carrying groceries, climbing stairs, bending, kneeling and walking moderate distances.

Each of the eight different sections produces an individual section score ranging from 0 to 54. For each item, there is a choice of responses on a Likert scale, ranging from 'limited a lot' to 'not limited at all' or 'all of the time' to 'none of the time' (Brazier, 1995). These are not combined to form an overall score, and therefore it is hard to make comparisons (Bowling and Normand, 1998). It is possible to identify the scores relating to movement separately from those relating to the other concepts relating to health. Like all scales that contain sections, it is important to carry out the whole scale and not just select the aspects relating to movement. These sections are not validated to be used in isolation. Higher scores indicate a perception of good quality of life (Talamo et al., 1997).

Table 15.5 The 10 items of the Short Form 36 Health Survey that relate to physical functioning

The following questions are about activities you might do during a typical day. Does your health now limit you in these activities? If so, how much?

ACTIVITIES	YES LIMITED A LOT	YES, LIMITED A LITTLE	NO, NOT LIMITED AT ALL
Vigorous activities such as running, lifting heavy objects or participating in strenuous sports	1	2	3
Moderate activities such as moving a table, pushing a vacuum cleaner, bowling or playing golf	1	2	3
Lifting or carrying groceries	1	2	3
Climbing several flights of stairs	1	2	3
Climbing one flight of stairs	1	2	3
Bending, kneeling or stooping	1	2	3
Walking more than a mile	1	2	3
Walking half a mile	1	2	3
Walking 100 yards	1	2	3
Bathing and dressing yourself	1	2	3

(From Ware 2000, with permission.)

Table 15.6 The four items of the Short Form 36 Health Survey that relate to role limitations caused by physical problems and pain

During the past 4 weeks, have you had any of the following problems with your work or regular daily activities as a result of your physical health?

	YES	NO
Cut down on the amount of time you spent on work or other activities	1	2
Accomplished less than you would like	1	2
Were limited in the kind of work or other activities	1	2
Had difficulty performing the work or other activities (e.g. it took extra effort)	1	2

(From Ware 2000, with permission.)

Table 15.7 The two items of the Short Form 36 Health Survey that relate to pain

How much bodily pain during the past 4 weeks?	
	SCORE
None	1
Very mild	2
Mild	3
Moderately	4
Severe	5
Very severe	6
During the past 4 weeks, how much did pain interfere with your normal work (including work both outside the home and housework)?	
	SCORE
Not at all	1
A little bit	2
Moderately	3
Quite a bit	4
Extremely	5

(From Ware 2000, with permission.)

There is considerable evidence for the validity and reliability of the SF-36 and its ability to measure changes in health status over time (Brazier et al., 1992; Ware and Sherbourne, 1992; Garratt et al., 1994; Jenkinson et al., 1994). This supports its use as a routine scale for monitoring and assessing quality of life in both clinical practice and research. Anderson et al. (1996) found it to be valid for use in stroke rehabilitation, and Talamo et al. (1997) with patients with rheumatoid arthritis. However, as with other measurement scales, ceiling and flooring effects can occur.

Nottingham Health Profile

Hunt, McEwen and McKenna developed the NHP in 1980 as a scale to measure health status (Hunt et al., 1985). To develop it, 768 patients with a variety of health problems generated over 2000 statements. These statements were then reduced to 38. The profile consists of two parts, the first consisting of 38 statements addressing the following areas: energy, pain, emotional reactions, sleep, social isolation and physical mobility. The second part has seven statements concerning paid employment, jobs around the house, social life, personal relationships, sex life, hobbies and interests, and holidays. Box 15.1 includes examples of statements that address issues relating to movement used in the profile and again, like the SF-36, these examples should not be used out of the context of the whole measure.

BOX 15.1 Examples of statements

- I can only walk about indoors.
- I find it hard to bend.
- I'm unable to walk at all.
- I have trouble getting up and down stairs or steps.
- I find it hard to reach for things.
- I find it hard to dress myself.
- I find it hard to stand for long (e.g. at the kitchen sink or waiting for a bus).
- I need help to walk about outside (e.g. a walking aid or someone to support me).

(With permission of SP McKenna, 2000.)

The NHP is short and simple and can be self-administrated or carried out by interview. It takes about 5 min to complete. It is sensitive to change and has been tested extensively for reliability and validity (Bowling, 2005). Scores ranging from 0, indicating no problem, to 100, when problems in all areas have been identified, are used to score it. As a result, a higher score reflects severe problems.

There are different views on whether the NHP is actually measuring quality of life. Wade (1992) suggests that it may be recording mood rather than global quality of life, and Ebrahim et al. (1986) suggest it is an indicator of depressed mood, while Bowling (2005) suggests that it is identifying how people feel when they are experiencing various states of ill health.

CONCLUSION

Movement cannot be considered as an isolated component of life. If there is a loss of movement for any reason, there will be implications on how everyday activities are carried out and for how quality of life is perceived. The difficulty is in quantifying or measuring these implications. Scales have been developed to try to measure these issues. Careful consideration needs to be given when selecting a measurement scale to ensure it fulfils the purpose for which it is needed.

References

Anderson, C., Laubscher, S., Burns, R., 1996. Validation of the Short Form 36 (SF36) Health Survey Questionnaire among stroke patients. Stroke 27, 1812–1816.

Baker, F., Intagliata, J., 1982. Quality of life in the evaluation of community support systems. Eval. Program Plann. 5, 69–79.

Barer, D.H., 1989. Use of the Nottingham ADL scale in stroke: relationships between functional recovery and length of stay in hospital. J. R. Coll. Physicians Lond. 23 (4), 242–247.

Berg, K.O., Maki, B.E., Williams, J.I., et al., 1992. Clinical and laboratory measures of postural balance in an elderly population. Arch. Phys. Med. Rehabil. 73, 1073–1080.

Bowling, A., Ebrahim, S., 2005. Handbook of Health Research Methods. Investigation measurement and analysis. Milton Keynes, Open University Press.

Bowling, A., 2005. Measuring health. A review of quality of life measurement scales 3rd Edition, Milton Keynes, Open University Press.

Brazier, J., 1995. The Short-Form 36 (SF-36) Health Survey and its use in pharmacoeconomic evaluation. Pharmacoeconomics 7, 403–415.

Brazier, J., Harper, R., Jones, N., et al., 1992. Validating the SF-36 health survey questionnaire: new outcome measure for primary care. Br. Med. J. 305, 160–164.

Burton, J., 1989. The model of human occupation and occupational therapy practice with elderly patients. Part 2: application. Br. J. Occup. Ther. 52, 219–221.

Collin, C., Wade, D.T., Davies, S., et al., 1988. The Barthel ADL Index: a reliability study. Int. Disabil. Stud. 10, 61–63.

De Haan, R., Aaronson, N., Limburg, M., et al., 1993. Measuring quality of life in stroke. Stroke 24, 320–327.

Diener, E., 1984. Subjective well-being. Psychol. Bull. 95, 542–575.

Eakin, P., 1993. The Barthel Index: confidence limits. Br. J. Occup. Ther. 56 (5), 184–185.

Ebrahim, S., Barer, D., Nouri, F., 1986. Use of the Nottingham Health Profile with patients after a stroke. J. Epidemiol. Community Health 40, 166–169.

Enright, P., McBurnie, M., Bittner, V., et al., 2003. The 6-min walk test, a quick measure of functional status in elderly adults. Chest 123, 387–398.

Fallowfield, L., 1990. The quality of life. The Missing Measurement of Health Care. Souvenir, London.

Fisher, A.G., 1992. Functional measures, part 2: selecting the right test, minimizing the limitations. Am. J. Occup. Ther. 46 (3), 278–281.

Fisher, A.G., 1994. Functional assessment and occupation: critical issues for occupational therapy. Keynote Address at the Annual Conference of the New Zealand Association of Occupational Therapists.

Garratt, A., Ruta, D., Abdalla, M., et al., 1994. SF-36 Health Survey Questionnaire: II. Responsiveness to changes in health status in four common clinical conditions. Qual. Health Care 3, 186–192.

Gibbon, B., 1991. Measuring stroke recovery. Nurs. Times 87 (44), 32–34.

Gompertz, P., Pound, P., Ebrahim, S., 1993. The reliability of stroke outcome measures. Clin. Rehabil. 7, 290–296.

Granger, C.V., Hamiliton, B.B., 1990. Measurement of stroke rehabilitation outcome in 1980's. Stroke 21 (9 Suppl.), II–46–47.

Harper, C., 1987. The New Collins Dictionary and Thesaurus in One Volume. Harper Collins, Glasgow.

Harwood, R.H., Ebrahim, S., 2002. The Validity, Reliability and Responsiveness of the Nottingham Extended Activities of Daily Living Scale in Patients undergoing Hip Replacement. Disability and Rehabilitation 24 (7), 371–377.

Holbrook, M., Skilbeck, C.E., 1993. An activities index for use with stroke patients. Age Ageing 12, 166–170.

Hunt, S.M., McEwen, J., McKenna, S.P., 1985. Measuring health status: a new tool for clinicians and epidemiologists. J. R. Coll. Gen. Pract. 35, 185–188.

Jenkinson, C., Wright, L., Coulter, A., 1994. Criterion validity and reliability of the SF-36 in a population sample. Qual. Life Res. 3, 7–12.

King's Fund 1988. Consensus conference. Treatment of stroke. BMJ 297, 126–128.

Kuys, S., Brauer, S., 2006. Validation and Reliability of the modified Elderly Mobility Scale Australasian Journal of Ageing 25 (3), 140–144.

Larson, J., 1997. The MOS 36-Item Short Form Health Survey. A conceptual analysis. Eval. Health Prof. 20, 14–27.

Law, M., Letts, L., 1989. A critical review of scales of activities of daily living. Am. J. Occup. Ther. 43 (8), 522–528.

Liang, M.H., Katz, J.N., Ginsburg, K.S., 1990. Chronic rheumatic disease. In Spiller, B. (Ed.), Quality of life assessments in clinical trials. Raven, New York, pp. 441–458.

Lincoln, N., Gladman, J., 1992. The extended activities of daily living scale: a further validation. Disabil. Rehabil. 14, 41–43.

Livingstone, M.G., Livingstone, H.M., 1985. The Glasgow assessment schedule: clinical and research assessment of head injury outcome. Int. Rehabil. Med. 7, 146–149.

Mahoney, F.I., Barthel, D.W., 1965. Functional evaluation: the

Barthel Index. Md. State Med. J. 14, 61–65.

Mathias, S., Nayak, U.S.L., Isaacs, B., 1986. Balance in elderly patients: the 'Get-up and Go' test. Arch. Phys. Med. Rehabil. 67, 383–389.

Mathias, S., Bates, M., Pasta, D., et al., 1997. Use of Health Utilities Index with stroke patients and their caregivers. Stroke 28, 1888–1894.

McCloy, L., Jongbloed, L., 1987. Robinson Bashall Functional Assessment for arthritis patients: reliability and validity. Arch. Phys. Med. Rehabil. 68, 486–489.

McDowell, I., 2006. Measuring health. A guide to rating scales and questionnaires, 3rd ed. Oxford, Oxford University Press.

McDowell, F., Lee, J.E., Swift, T., 1970. Treatment of Parkinson's syndrome with L-dihydroxyphenylalanine (Levodopa). Ann. Intern. Med. 72, 29–35.

Nakamura, D., Holm, M., Wilson, A., 1998. Measures at balance and fear of falling in the elderly: a review. Phys. Occup. Ther. Geriatr. 15, 17–32.

Niemi, M., Laaksonen, R., Kotila, M., et al., 1988. Quality of life four years after stroke. Stroke 19 (9), 1101–1106.

Nouri, F., Lincoln, N., 1987. An extended activities of daily living scale for stroke patients. Clin. Rehabil. 1, 301–305.

Prosser, L., Canby, A., 1997. Further validation of the Elderly Mobility Scale for measurement of mobility of hospitalised elderly people. Clin. Rehabil. 11, 338–343.

Smith, R., 1994. Validation and reliability of the Elderly Mobility Scale. Physiotherapy 80, 744–747.

Spilg, E., Martin, B., Mitchell, S., et al., 2001. A comparison of mobility assessments in a geriatric day hospital. Clin. Rehabil. 15, 296–300.

Spilg, E., Martin, B., Mitchell, S., et al., 2003. Falls risk following discharge from a geriatric day hospital. Clin. Rehabil. 17, 334–340.

Streiner, D.L., Norman, G.R., 2003. Health measurement scales: a practical guide to their development and use, 3rd edition. Oxford: Oxford University.

Talamo, J., Frater, A., Gallivan, S., et al., 1997. Use of the Short Form 36 (SF36) for health status measurement in rheumatoid arthritis. Br. J. Rheumatol. 36, 463–469.

Üstün, T., Chatterji, S., Bickenbach, J., et al., 2003. The international classification of functioning, disability and health: a new tool for understanding disability and health. Disabil. Rehabil. 25, 565–571.

Van Knippenberg, F.C.E., de Haes, J. C.J.M., 1985. The quality of life of cancer patients: a review of the literature. Soc. Sci. Med. 20, 809–817.

Wade, D.T., 1992. Measurement in Neurological Rehabilitation. Oxford University Press, Oxford.

Wall, J., 2000. The timed get up and go test revisited: measurement of the component tasks. J. Rehabil. Res. Dev. 37 (1), 109–114.

Ware, J., 2000. QualityMetric. Online. Available: http://www.qmetric.com.

Ware, J., Sherbourne, C., 1992. The MOS 36-Item Short Form Health Survey (SF36). 1. Conceptual framework and item selection. Med. Care 30, 473–481.

Wenger, N.K., Furberg, C.D., 1990. Cardiovascular disorders. In: Spiller, B. (Ed.), Quality of life assessments in clinical trials. Raven, New York, pp. 335–345.

World Health Organization, 2001. International Classification of Functioning, Disability and Health. World Health Organization, Geneva.

Wyller, T., Sveen, U., Sodring, K., et al., 1997. Subjective well-being one year after stroke. Clin. Rehabil. 11, 139–145.

Chapter 16

Conclusion

Tony Everett and Clare Kell

Now we have completed our study of human movement and have an understanding of its multifaceted complexity, it is a good time to return to the family scene that was introduced in Chapter 1.

The family have just finished their evening meal and are about to go off to do their own activities when John asks everyone to help clear the table. It has been a large meal, and there are plenty of dishes to be cleared away. Dan stands up in an aggressive manner and his chair is pushed backwards quite a long way. He picks up his plate, which then knocks against his chest, spilling the remains of his meal over him and the floor. Jenny, at the same time, picks up her plate and runs to the kitchen. In her haste, she also pushes her chair with too much force, which results in it completely toppling over. Agnes, who is sitting between Dan and Jenny with her back to the kitchen door, rises slowly, collects a couple of dishes and turns to head for the kitchen. She makes very slow progress. Liz, dressed in her smart tight-fitting skirt and high heels in readiness for her theatre visit, has taken a couple of serving dishes to the kitchen and is about to load them into the dishwasher. As she bends forwards, she loses her balance and has to steady herself with her hand. John, seeing the food drip from Dan's plate on to the new carpet, has rushed to the kitchen, is back very quickly with a cloth, and is squatting beside Dan's chair, mopping up the spillage. Chris, who is sitting on the opposite side of the table, facing the kitchen, surveys the scene of chaos with much amusement. After a little chuckle, he collects up the remaining plates and serving dishes and carries the large pile deftly to the kitchen.

The scene may resemble a slapstick movie, but it is one that could quite easily happen. There is a lot of activity happening here, and a lot of interaction between the family and their environment. Analysing the scene may be difficult, but if we think of all the chapters we have read a series of explanations may emerge.

Let us begin with Dan. Dan has got up from the table thousands of times, and very rarely has he pushed his chair away in such an uncontrolled manner. There must be other factors impinging on this relatively simple learned experience that Dan can normally perform very easily. Physiologically, Dan has the control of the concentric contraction of his quadriceps from the middle to their inner range. There have been no changes to the physical environment to cause a change in the biomechanics of his joints – a change in the friction acting on his base of support, for example. We must therefore look at the non-physical environment for anything that can be affecting his movement. Dan is normally an organized teenager, but during the meal he has realized that he has a large piece of homework to complete for handing in first thing in the morning. He is very eager to get going and has planned in his head that he will go straight to his room and complete his homework task. Logic dictates that helping to clear the table will take only a few minutes and not really have an impact on his homework schedule. A teenager does not always work on logic, however, and Dan has become irrationally and unusually irritated. The fine neuromuscular interplay that normally operates in controlling movement has been disrupted by the signals produced in the limbic system, the area of the brain that controls his emotion, and there has possibly been a release of adrenaline (epinephrine). This adrenaline rush and its effects on the autonomic nervous system increase his normal muscular tone and disrupt the fine motor control usually expected with this learned skill. There may also have been a deliberate demonstration of displeasure, which manifested itself in the aggressive pushing back of the chair!

Similar processes were probably happening when he was lifting up his plate. The action should have been straightforward, but the hormonal and psychological influences overestimated the amount of force needed to lift the plate. As we have seen, the moment around his elbows is calculated using the formula moment = force × perpendicular distance, and as the distance the plate is from the pivot (elbow) remains constant and the force (supplied by the muscle) increases, there will inevitably be an increase in the resulting moment, the consequence of which is Dan's leftovers spilling over himself and the floor.

Jenny has very positive motivation for quickly completing her task. Unlike Dan, she wants to finish clearing her crockery so that she can see the beginning of her favourite television programme. She is sitting cross-legged on the dining chair and is therefore rotated at the trunk to enable her to face the table. As she rises, she has to produce a manoeuvre that only her great joint and muscle mobility coupled with her youth could perform. Consequently, she pushes over the chair. Jenny appears to have boundless energy, and in her haste has not exercised sufficient control over her activity. She is young and has yet to acquire the innate skills that will ensure her own safety and that of those surrounding her. Consequently, her movement is fast and erratic, which results in her chair being knocked over.

Agnes, although the oldest in the family, still has the ability to carry the crockery out to the kitchen without a problem. Why then is her progress so slow on this occasion? Agnes is experienced in her movements and therefore knows her limitations. She knows that her age, with its attendant physical and physiological deterioration, has limited her ability to make sudden changes to her movement as a result of perturbations to her equilibrium. When she gets up from the table, she bends forwards to pick up a few plates. This action immediately throws her centre of gravity outside her base. She corrects for this by leaning backwards and increasing her tone in her trunk. Making these changes limits her ability to quickly add any other changes, so the movement of the chairs on either side and the body of Jenny zooming past create disturbances and obstacles that she has to plan for to ensure her own stability. Consequently, her progress is deliberate and slow.

Liz has reached the dishwasher and has performed the next activity safely many times. The difference this time is that she is wearing clothes that she wears only occasionally and that are totally unsuitable for the task she is carrying out. The height of her heels has altered her posture so that her pelvis has been tilted anteriorly to such an extent that she has had to increase her lumbar curve to compensate and bring her centre of gravity back within her base. When she begins to bend forwards, her back muscles perform eccentric work to allow flexion. This immediately lets her centre of gravity fall outside her base. This problem is increased, as her skirt does not allow for sufficient flexion of the hips and knees to compensate, so much of the movement is flexion of the trunk. She is not able to compensate for this by posterior tilt of the pelvis, because of the height of her heels changing her posture. Consequently, she falls forwards and has to put her hand out to support herself, thereby increasing her base and stability. A prime case of psychosocial influence!

John is probably wishing he had cleared the table himself. Cloth in hand, he is hurrying from the kitchen. His gait still has the recognizable cycle of heel strike, flat foot, mid stance, heel off and toe off within the stance phase. Also recognizable are the acceleration, mid and deceleration phases of the swing phase. As John has increased his gait velocity, the relationship of his stance and swing phase has changed. There is an increase in his swing phase while the time his feet spend on the floor, the stance phase, is decreasing. If we look closely, we will see that there is still a double stance phase, albeit very short, which definitely puts John in the walking category. Both temporal and spatial parameters have altered if compared with John's comfortable walking speed. Stride and step length have increased while the deceleration and push off phases have decreased. Although these two latter temporal parameters have decreased, the force production in both phases is greater, thus allowing greater forward momentum.

Chris has sat back and surveyed the scene. He has noted the obstacles, both inanimate and animate, and has programmed them into his plan of action. Being a good sportsman, he is experienced in adding physical changes to his game plan. He has trained his coordination along with his muscle strength and is thus well able to cope with the changes to his centre of gravity and the demands of muscular strength placed on his body by the increased load of the dishes. He easily collects the dishes and walks effortlessly to the kitchen to deposit them on the worktop before making a quick exit to his badminton match!

This example has shown us that human movement has many complex components and many influences on it. As shown above, movement can be analysed and interpreted many ways depending on the viewpoint of the analyser, the information that needs to be extracted, and the mechanisms available to analyse it with. It is hoped, however, that whatever viewpoint your study of human movement is, this book has given you an appreciation of its many and varied manifestations, and that you feel better equipped to rise to the challenges presented by studying human movement.

Index

NB: Page numbers in **bold** refer to boxes, figures and tables

M

N

O

P

Q

R

S

T